Materials Challenges in Alternative and Renewable Energy II

Materials Challenges in Alternative and Renewable Energy II

Ceramic Transactions, Volume 239

A Collection of Papers Presented at the Materials Challenges in Alternative and Renewable Energy Conference, February 26–March 1, 2012, Clearwater, Florida

Edited by
George Wicks
Jack Simon
Ragaiy Zidan
Robin Brigmon
Gary Fischman
Sivaram Arepalli
Ann Norris
Megan McCluer

The American Ceramic Society

WILEY

A John Wiley & Sons, Inc., Publication

Published by John Wiley & Sons, Inc., Hoboken, New Jersey.
Published simultaneously in Canada.

For general information on our other products and services or for technical support, please contact our
Customer Care Department within the United States at (800) 762-2974, outside the United States at
(317) 572-3993 or fax (317) 572-4002.

Wiley also publishes its books in a variety of electronic formats. Some content that appears in print may
not be available in electronic formats. For more information about Wiley products, visit our web site at
www.wiley.com.

Library of Congress Cataloging-in-Publication Data is available.

ISBN: 978-1-118-58098-1
ISSN: 1042-1122

Printed in the United States of America.

10 9 8 7 6 5 4 3 2 1

Contents

WIND

Preface

Materials Challenges in Alternative & Renewable Energy (Energy 2012) was an important meeting and technical forum held in Clearwater, Florida, on February 26–March 1, 2012. This meeting, organized by The American Ceramic Society (ACerS), represented the third conference in a new series of inter-society meetings and exchanges, with the first of these meetings held in 2008, on "Materials Innovations in an Emerging Hydrogen Economy." The current Energy Conference- 2012 was larger in scope and content, and included 238 participants from 19 countries and included more than 200 presentations, tutorials and posters. The purpose of this meeting was to bring together leaders in materials science and energy, to facilitate information sharing on the latest developments and challenges involving materials for alternative and renewable energy sources and systems.

Three of the premier materials organizations in the U.S. combined forces with ACerS to co-sponsor this conference including ASM International, The Minerals, Metals & Materials Society (TMS), and the Society of Plastics Engineers (SPE). Between these four societies each of the materials disciplines, ceramics, metals and polymers, were represented. In addition, we were also very pleased to have the support and endorsement the Materials Research Society (MRS) and the Society for the Advancement of Material and Process Engineering (SAMPE).

Energy 2012 was highlighted by eight plenary presentations on leading energy alternatives. In addition, the conference included technical sessions addressing state-of-the art materials challenges involved with Solar, Wind, Hydropower, Geothermal, Biomass, Nuclear, Hydrogen, the Electric Grid, Materials Availability for Alternative Energy, Nanocomposites and Nanomaterials, and Batteries and Energy Storage. This meeting was designed for both scientists and engineers active in energy and materials science as well as those who were new to the field.

We are very pleased that ACerS is committed to running this materials-oriented conference in energy, every two years with other materials organizations. We believe the conference will continue to grow in importance, size, and effectiveness and provide a significant resource for the entire materials community and energy sector.

GEORGE WICKS
Savannah River National Laboratory
Energy Conference- 2012 Co-Organizer/President ACerS

JACK SIMON
Technology Access
Energy Conference- 2012 Co-Organizer/Past President ASM International

Acknowledgements

CONFERENCE CO-CHAIRS

George Wicks, *Savannah River National Laboratory, Aiken, SC*
Jack Simon, *Technology Access, Aiken, SC*

TECHNICAL ADVISORY BOARD AND TOPIC CHAMPIONS

Thad Adams, *SRNL*
Jim Ahlgrimm, *U.S. Department of Energy*
Sivaram Arepalli, *Sungkyunkwan University*
Ming Au, *SRNL*
Steve Bossart, *U.S. Department of Energy*
Geoffrey Brennecka, *Sandia National Labs*
Robin Brigmon, *Savannah River National Lab*
Subodh Das, *Phinix*
David Dorheim, *DWD Advisors*
Gary Fischman, *Future Strategy Solutions*
Bernadette Hernandez-Sanchez, *Sandia National Lab*
Natraj Iyer, *Savannah River National Lab*
Bruce King, *Sandia National Lab*
Dan Laird, *Sandia National Lab*
Edgar Lara-Curzio, *Oak Ridge National Lab*
Megan McCluer, *U.S. Department of Energy*
Rana Mohtadi, *Toyota Research Inst. of N.A.*
Gary Mushock, *ASM International*
Ralph Nichols, *Savannah River National Lab*
Ann Norris, *Dow Corning*

Gary Norton, *U.S. Department of Energy*
Ravi Ravindra, *N.J. Institute of Technology*
Steven Sherman, *Savannah River National Lab*
Bob Sindelar, *Savannah River National Lab*
Jay Singh, *NASA*
Rick Sisson, *Worcester Polytechnic Institute*
Ned Stetson, *U.S. Department of Energy*
Greg Stillman, *U.S. Department of Energy*
Hidda Thorsteinsson, *U.S. Department of Energy*
Mike Tupper, *Composite Technology Development, Inc.*
Eric Wachsman, *Univ. of Maryland*
Jy-An (John) Wang, *Oak Ridge National Lab*
Ragaiy Zidan, *Savannah River National Lab*
Mark Mecklenborg, *The American Ceramic Society*
Greg Geiger, *The American Ceramic Society*

CONFERENCE SPONSORS

Harper International
Savannah River National Laboratory
Dow Corning
Sandia National Laboratories
Lanzhou University of Technology
Savannah River National Laboratory
Solar Solutions

Nuclear

STATE OF NUCLEAR ENERGY IN THE WORLD

Thomas L. Sanders
Past President, American Nuclear Society
La Grange Park, IL, 60526, USA
E-mail: Thomas.Sanders@srnl.doe.gov

ABSTRACT

U.S. President Dwight D. Eisenhower's 1953 "Atoms for Peace" speech at the United Nations laid the foundation for the present global nuclear enterprise. In his speech, he recommended the creation of the International Atomic Energy Agency. He offered nuclear technology developed in the United States to other nations as part of a broad nuclear arms control initiative. Since 1953, the world has produced over 400 nuclear power reactors and all but three nations have signed the nuclear nonproliferation treaty (NPT). Significant nuclear arms control treaties have been signed. Important international organizations related to nuclear matters have been established. Perhaps the most important is that neither World War III nor nuclear conflict has occurred.

The end of the Cold War, the events of September 11, 2001, and almost global support for the resurgence of nuclear energy have created a new opportunity to reinvigorate our commitment to peace and prosperity built around a new "Global Nuclear Future." For the U.S. to return to its former position as a visionary leader in the beneficial use of nuclear technology and materials on a global scale, it is imperative that steps be taken to reverse the conditions and decisions that led to the present situation—for the most part, the U.S. nuclear supply industry has moved offshore. This will require an integrated or holistic view of the global nuclear enterprise, from the cradle-to-the grave. Some of the realities of the global nuclear state are outlined in the paper.

INTRODUCTION

The global nuclear picture is complex and changing almost daily, as illustrated by Figure 1. We first performed a global assessment in 1997 working with the U.S. Center for Strategic and International Studies.[1] While the influence of these factors has changed over the past fifteen years, we still need to use civilian nuclear energy as an arms reduction vehicle and consume significant quantities of excess nuclear materials. Reapplication of excess defense nuclear assets in several countries to open and transparent civilian nuclear services is also needed to separate defense and civilian infrastructures. Since the mid 1980's, several emerging nuclear suppliers and users have become capable of competing on the global marketplace, including China. World-wide pressures have changed the energy cost/risk picture. The price of oil in 1997 was one-tenth of what it is today. Our concerns about clandestine nuclear trade in 1997 centered mainly on the need to prevent global trade in "loose nukes" from the Former Soviet Union. Today we have a similar concern with North Korea and Iran. The end of the Cold War and growth of the EU to include former Soviet Bloc countries has resulted in new energy stresses. We still argue over whether excess nuclear weapon materials are an asset to be productively used or a liability requiring large investments in safeguards and security. In 1997, Iraq and Libya were considered the primary proliferators. Today we are equally concerned about Iran, North Korea, and terrorism and three countries remain that have not signed the nonproliferation treaty.

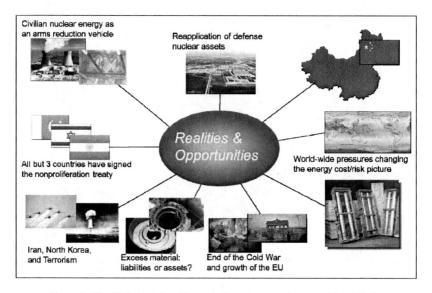

Figure 1. The Global Nuclear Picture is Complex and Changing Almost Daily

DISCUSSION

As illustrated in Figure 2, addressing our collective energy future is on the critical path to global peace and prosperity.[2,22] Energy supply and use have many ties. The prosperity of any nation depends on using energy to produce exportable goods and services. Protecting energy supplies and deliveries drives the national security strategy of many countries, including the U.S. In the "globalization trend," no market is more globalized than energy markets. However, free energy markets are disappearing as more governments decide to control the supply side.

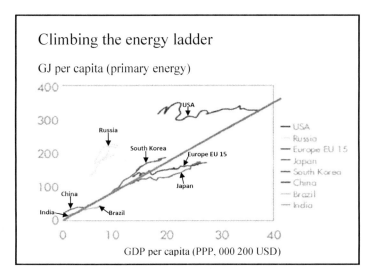

Figure 2. Climbing the Energy Ladder

From a U.S. perspective, there are two sides of the energy conundrum. Energy availability is directly tied to any nation's economic health. Developed nations like the U.S. must change their energy posture to continue to sustain and grow their own prosperity. At the same time, other nations must climb the energy ladder to achieve prosperity and reduce the stresses that lead to poverty, despair, and susceptibility to radical movements. This trend will be a hallmark of the 21st century—today's "have-nots" will demand and must have access to adequate and secure energy supplies. However, an order of magnitude increase in today's energy consumption would be needed to achieve a global minimum standard of living near that of Malaysia's by 2050. While doing so could be key to achieving global peace and prosperity, there is a huge potential for conflict over access to conventional, finite energy resources. Fifty-four percent of the world's natural gas is located in Iran and Russia and almost two-thirds of the world's oil supplies are located in the Mideast.[3,23,25] This conflict potential is directly responsible for a significant fraction of the world's defense posture and expenditures. And, as abundantly illustrated by today's economic stresses, energy scarcity slows world GDP growth (with major impacts on global economies).

Since this curve was developed, China has been "climbing the ladder" at almost a 9% growth rate. According to an article in the Washington Post in 2004,[4] China consumed ½ of the world's cement, 1/3 of the world's steel, 1/5 of the world's aluminum, and approximately 1/4 of the world's copper. China became the second largest importer of oil.[24] As China becomes much more dependent on foreign resources, it too will have to develop a naval capability to protect the pipeline to those resources.[5]

As illustrated by the impact of China's growth above, and recognizing that another two billion people have little access to energy, all forms of energy will be required to meet the global needs of this century. Fossil resources must be used and demand for access to oil and natural gas supply sources in many unstable regions of the world will continue to increase.

Another issue in the U.S. and other countries is the impact of our domestic energy mix on growing trade deficits in U.S. manufactured products.[6] The U.S. trade balance in chemical products is driven by natural gas prices. During the 1½ to 15 years between 1990 and 2005, this trade balance went from a $15B trade surplus to a $10B trade deficit.[4] "Why did this happen"— hundreds of electric-power plants (~300 GWe) built in recent years are fired by natural gas and this increase in demand has made other goods that are dependent on gas non-competitive on the global marketplace.

A single pound of low enriched uranium contains energy equivalent to 33 million cubic feet of natural gas, 250,000 gallons of gasoline, or 4 to 6 million pounds of coal.[7,8] There is no doubt that this resource will be exploited. Expanded use of nuclear energy will provide many options (and also many challenges) and we must simultaneously address nuclear proliferation concerns as we address these future energy needs. The global inventories of fissile materials has grown substantially over the last 50 years. Recent estimates of the current inventories of HEU and Pu are 64 tonnes civilian HEU, 1,475 tonnes military HEU, 200 tonnes military Pu, and 1708 tonnes of civilian Pu (19% separated).[9,21]

This amount of material could supply U.S. reactors for many years. On the other hand, this material could be used for hundreds of thousands of nuclear weapons. Therein lies the paradox and we need to solve it now—either promote and enable the peaceful use of these assets or worry about their existence forever.

Most nations using or desiring nuclear energy resources have renounced nuclear weapons and entered the Non-Proliferation Treaty. Many "so-called" threshold States of the 1980's have signed the NPT. These include South Africa—the first nation to actually disassemble a nuclear weapons stockpile—and Argentina, Brazil, Algeria, South Korea, and others. Several of these countries could become very competitive global nuclear suppliers.[10] For example Argentina has bilateral nuclear cooperation agreements with Algeria, Brazil, Peru, Romania, Turkey, Yugoslavia (Serbia), India, Italy, Iran, Israel, Pakistan, Libya, the Czech Republic, and Germany. Argentina is also developing a small standardized reactor for export to developing nations and has developed indigenous capabilities in uranium enrichment, reprocessing, reactor design, fuel design, and waste management. Other emerging supplier nations with indigenously developed capabilities include China, South Korea, Japan, Kazakhstan, Ukraine, 'Russia,' South Africa, India, and Brazil.

Nuclear fuel cycle technology and enrichment and reprocessing capabilities are widespread. If one considers the former Soviet Union nations in the "developing" category, more than one half of the world's uranium resources are in the "developing world." In fact, more than 30% of the world's uranium deposits are located in Africa.

Today, over 400 power reactors supply 17% of the world's electricity and most are in developed countries. Table 1 lists the currently operating power reactors across the globe and country-by-country plans for expansion.[11,26]

Table 1. World list of Nuclear Power Plants

Argentina – 2 operating, 1 forthcoming	Mexico – 2 operating
Armenia – 1 operating	Netherlands – 1 operating
Belgium – 7 operating	Pakistan – 3 operating, 1 forthcoming
Brazil – 2 operating, 1 forthcoming	Romania – 2 operating, 6 forthcoming
Bulgaria – 2 operating, 2 forthcoming	Russia – 33 operating, 10 forthcoming
Canada – 18 operating,	Slovakia – 4 operating, 2 forthcoming
China – 16 operating, 26 forthcoming	Slovenia – 2 operating
China (Taiwan) – 6 operating, 2 forthcoming	South Africa – 2 operating
Czech Republic – 6 operating	South Korea – 21 operating, 5 forthcoming
Finland – 4 operating, 1 forthcoming	Spain – 8 operating
France – 58 operating, 1 forthcoming	Sweden – 10 operating
Germany – 9 operating	Switzerland – 5 operating
Hungary – 4 operating	Ukraine – 15 operating, 1 forthcoming
India – 20 operating, 6 forthcoming	United Kingdom – 19 operating
Iran – 1 operating	USA – 104 operating, 1 forthcoming
Japan – 50 operating, 2 forthcoming	

As noted above, global nuclear energy expansion can create proliferation concerns. Today's (and tomorrow's) "hot spots" may see significant nuclear energy implementation throughout the 21st century. Table 2 lists the countries participating in the December 2006 IAEA meeting on global nuclear expansion.[12]

Table 2. Countries participating in December 2006 IAEA meeting on Global Nuclear Expansion

Algeria	Argentina	Australia	Bahrain
Belarus	Cameroon	Canada	Chile
China	Croatia	Czech Republic	Egypt
Finland	France	Germany	Georgia
Ghana	Greece	India	Indonesia
Islamic Republic of Iran	Japan	Jordan	Kenya
Republic of Korea	Lithuania	Malaysia	Mexico
Morocco	Namibia	Nigeria	Poland
Russian Federation	South Africa	Sudan	Syrian Arab Republic
Tanzania	Tunisia	USA	Uruguay
United Arab Emirates	Venezuela	Vietnam	Yemen

In 1953, U.S. President Dwight D. Eisenhower started the Atoms for Peace Program to address a number of U.S. national security problems:[13]

1. Increasing global competition over energy resources and a need to fuel rebuilding Europe and Japan after WWII;
2. The need to divert Soviet materials, technology, people, and infrastructure into peaceful purposes; and
3. The need to "manage" the likely spread of nuclear know-how and technology.

Nuclear energy in the U.S. has performed very well over the past few decades. However, the U.S. must meet major 21st century challenges regarding the future global nuclear enterprise; in particular, American competitiveness in the global nuclear marketplace and global nuclear weapon proliferation.

In 2004, U.S. President George Bush announced support for new measures to counter the threat of weapons of mass destruction, including a global nuclear fuel cycle model for the 21st century based on "cradle-to-grave" materials and technology partnerships.[14] Fuel suppliers would operate reactors and fuel cycle facilities, including fast reactors to transmute the actinides from spent fuel into less toxic materials. Fuel users would operate reactors, lease and return fuel, and not have to worry about disposal of radioactive materials. The IAEA would provide safeguards and fuel assurances, backed up with a reserve of nuclear fuel for states that do not pursue enrichment and reprocessing

The "supply and return" concept addresses a major potential proliferation concern with expanded use of nuclear power. Developing such a comprehensive fuel cycle service capability would provide market advantages superior to the current approach, virtually defining how nuclear trade in the 21st century will evolve and enable the nuclear powers to help the developing world acquire the energy resources necessary for achieving a prosperous future and for globally controlling environmental impacts.[15] From a global security perspective it would eliminate the need for customers of exportable nuclear systems to have enrichment and reprocessing capabilities.

Most of the emerging market opportunity across the world is for smaller reactors.[16] According to the IAEA, a small reactor is 0-300 MW(e), while a medium sized reactor generates 300-700 MW(e). Fundamentally, most countries cannot really absorb large thousand Mega watt nuclear systems. Of 435 nuclear power plants around the world last year, 138 were small and medium sized reactors (SMRs). Table 3 lists the world's operating SMRs. These reactors generated 60.3 GW(e) or 16.7% of the world nuclear electricity production. Of 31 recently constructed NPPs, 11 were SMRs.

Table 3. World's operating Small and Medium Reactors[17,27]

Developing/Transitioning Countries		
Argentina – 2	Armenia – 1	Hungary – 4
Czech Republic – 4	Slovakia – 4	Romania – 2
Slovenia – 1	Russia – 11	China – 7
Ukraine – 2	Pakistan – 2	
India – 18	Brazil – 1	
Developed Countries		
Britain – 18	Belgium – 2	Canada – 8
Finland – 2	Japan – 18	Netherlands – 1
South Korea – 6	Spain – 1	Sweden – 2
Switzerland – 3	Taiwan – 2	US – 16

According to one evaluation of the emerging world market, almost 80% of 226 countries are limited by infrastructure to small to medium sized nuclear systems. In fact, only 16% of Mexico's generating systems are greater than 250 MWe in capacity. One could argue that smaller nuclear systems also make sense in large markets such as the U.S. Since 1993, almost 300 GWe of small-to-medium sized natural gas fuel generating systems have been added to the U.S. generating capacity.[18] If smaller long-lived nuclear systems could compete with the rising and unpredictable cost of natural gas, U.S. utilities would have a "modular" capability for meeting increasing demand and for distributed non-electric applications such as oil shale development.

Large-scale development of advanced, right-sized reactors for the emerging world market is the key to enabling nuclear energy to grow as needed and exploit nuclear energy's million fold advantage in energy intensity. More than 50 concepts and designs of innovative SMRs are being developed by *Argentina, Brazil, Canada, China, Croatia, France, India, Indonesia, Italy, Japan, the Republic of Korea, Lithuania, Morocco, Russian Federation, South Africa, Turkey, USA, and Vietnam.* Most of these innovative SMRs also can provide for *non-electric applications* such as water desalination.[19]

Right-sized reactors could be sized to developing country energy grids, factory produced, fueled, sealed, and transported to the site. Some designs could result in secure, highly proliferation-resistant exportable reactors that require no refueling for up to 30 years and would enable the expansion of nuclear-based energy services to most developing countries and produce hydrogen and drinking water, in addition to electricity. These long-lived reactor concepts employ multiple approaches to coolants (H_2O, Na, Pb, Pb-Bi); spectrum characteristics (thermal, epithermal, fast); fuels (metal, particle, nitride); power outputs (1 to 300 MW_E); and specific applications (electricity, hydrogen production, district heating, desalination).

SUMMARY

The global nuclear enterprise will rapidly change over the next quarter century. The existing nuclear states must focus on the future to be able to influence the coming global challenges. The developed countries must enable the emerging world to access clean, reliable energy supplies to fuel their economies. A global nuclear services supply and return system must be created that provides the benefits of nuclear energy to all nations while eliminating any need for production of materials of nuclear proliferation concern. Partnerships among nuclear power states could establish a new paradigm for incorporating advanced manufacturing and information technologies to improve safety, reliability, security, and transparency of fuel cycle systems. Today's research will provide a longer term foundation for creating right-sized nuclear systems that are much more efficient, create 90% less waste, and enable the cradle-to-grave export of long-lived reactors to developing markets in the world. More importantly, such systems could eliminate the need for every user of nuclear technology to develop a waste repository by pursuing multi-national enterprise concepts that provide significant safety, security, economic, and nonproliferation advantages.

Nuclear energy is already an important contributor to power generation in many countries. However, its expansion is somewhat limited by continuing focus on very large generating systems. Worldwide energy demand will grow and could grow with substantial downsides without a robust global nuclear enterprise. Oil and gas will continue to dominate, will cost more, and likely cause additional conflict. Coal can contribute more, but clean coal technology will require nuclear heat to be successful. Renewable energy sources (wind, solar) must be developed but will continue to be a niche contribution. Nuclear energy must grow to fill the needs of the 21st century and additional suppliers to the global marketplace must be developed. In fact, in the absence of near-term action, the U.S. itself will become primarily a major consumer of imported nuclear goods and services with little opportunity for nuclear exports.[20] A half-century after President Eisenhower posed his vision of "Atoms for Peace," we may at last be in a position to help launch a new, "Atoms for Peace and Prosperity" program in partnership with other nations around the world. It is a vision of the future that could lead to realistic, inexpensive, long-lived energy supplies to eradicate the underlying seeds of terrorism, convert "swords into plowshares," and provide a basis for lasting peace and prosperity.

REFERENCES

[1] Conference Chairman: Senator Sam Nunn, *Global Nuclear Materials Management*, A CSIS Conference Report, Energy and National Security Program, December 4, 1998.

[2] Project Co-chairs: Sam Nunn, James r. Schlesinger, Project Director: Robert E. Ebel, Project Executive Director: Guy Caruso, Congressional Co-chairs: Senator Joseph L. Lieberman, Senator Frank H. Murkowski, Representative Benjamin Gilman, Representative Ellen O. Tauscher, *The Geopolitics of Energy into the 21st Century, Volume 1: An Overview and Policy Considerations*, A Report of the CSIS Strategic Energy Initiative, November 2000.

[3] Project Co-chairs: Sam Nunn, James r. Schlesinger, Project Director: Robert E. Ebel, Project Executive Director: Guy Caruso, Congressional Co-chairs: Senator Joseph L. Lieberman, Senator Frank H. Murkowski, Representative Benjamin Gilman, Representative Ellen O. Tauscher, *The Geopolitics of energy into the 21st Century, Volume 2: The Supply-Demand Outlook*, 2000-2020, A Report of the CSIS Strategic Energy Initiative, November 2000.

[4] Peter S. Goodman, Booming China Devouring Raw Materials, *Washington Post*, May 21, 2004.

[5] Bill Gertz, China Builds Up Strategic Sea Lanes, *The Washington Times*, January 18, 2005.

[6] Thaddeus Herrick, Natural Gas Cooks Chemical Sector, *Wall Street Journal*, June 2003.

[7] J. F. Kotek et al., *Nuclear Energy Power for the Twenty-First Century*, A Six-Lab Report, ANL-03/12, SAND2003-1545P, May 2003.

[8] Gamini Seneviratne, Research Chiefs Forge a "Realistic" Nuclear Plan, *Nuclear News*, October 2004.

[9] David Albright et al., *Plutonium and Highly Enriched Uranium 1996 World Inventories, Capabilities, and Policies*, Oxford University Press, 1997.

[10] William C. Potter, *International Nuclear Trade and Nonproliferation, The Challenge of the Emerging Suppliers*, Lexington Books, 1990.

[11] World List of Nuclear Power Plants, *Nuclear News 10th Annual Reference Issue*, March 2008.

[12] Vladimir Kuznetsov, *Opportunities, Challenges, and Strategies for Small and Medium Sized Reactors*, presented at ANS 2006 Winter Meeting, Albuquerque, NM, November 2006.

[13] R.G. Hewlett and J.M. Holl, *Atoms for Peace and War, 1953-1961: Eisenhower and the Atomic Energy Commission*, University of California Press, Berkeley, CA, 1989.

[14] George W. Bush, *President's Description of GNEP*, Radio Address, February 18, 2006.

[15] A joint document of Directors of Russian and U.S. National Laboratories-State Research Centers, Concerning Sustainable Nuclear Energy for the 21st Century, *Toward a Global Nuclear Future*, Vienna, Austria, July 2004.

[16] Vladimir Kuznetsov, Opportunities, Challenges, and Strategies for Innovative SMRs Incorporating Non-electrical Applications, *International Conference on Non-electric Applications of Nuclear Power*, Oarai, Japan, April 2007.

[17] World's Operating SMRs, *Nuclear News General Tables*, March 2007.

[18] Richard Myers, *The Nuclear Energy Renaissance: Realistic Expectations*, NEI, November 15, 2007.

[19] The Russian Research Center, Kurchatov Institute, *Power Provision of Mankind's Sustainable Development, Cardinal Solution of the Nuclear Weapons Non-Proliferation Problem and the Problem of the Environmental Recovery of the Earth Planet*, Moscow, 2001.

[20] The Global Nuclear Future, The Next Era of Nuclear Power, *Sandia Technology*, Volume 4, No. 1, Spring 2002.

[21] The Global Fissile Material Report 2012, Balancing the Books: Production and Stocks, © 2010 International Panel on Fissile Materials, 5th Annual Report of the International Panel on Fissile Materials

[22] Energy Balances of OECD and Non-OECD Countries © OECD/IEA 2006.

[23] U.S. Energy Information Administration Country Analysis Briefs, IRAN, November 2011.

[24] Energy Information Administration, Short-Term Energy Outlook, April 2011.

[25] U. S Energy Information Administration, Oil and Gas Journal Reserves, January 2005.

[26] European Nuclear Society, Nuclear Power Plants Worldwide, February 2012.

[27] The Guardian News and Media, Nuclear Power Stations and Reactors Operational Around the World: Listed and Mapped, 2012.

Batteries and
Energy Storage

CORRELATION BETWEEN MICROSTRUCTURE AND OXYGEN REMOVAL IN SOLID-OXIDE-FUEL-CELL-MODEL ELECTRODES Pt(O₂)/YSZ AND Pd(O₂)/YSZ

G. Beck
Research Institute Precious Metals & Metals Chemistry, Katharinenstrasse 17, 73525 Schwaebisch Gmuend, Germany, correspondence author, phone: +49 7171 1006402, fax: +49 7171 1006900, email: beck@fem-online.de

ABSTRACT
    Platinum and Palladium films were prepared on (111) and (100) orientated yttrium-stabilized zirconia (YSZ) by pulsed laser deposition (PLD) and then subsequently annealed. These metal films are all (111) orientated, but the detailed microstructures depend on the microstructure of the YSZ. On single crystalline (111) orientated YSZ the films are single crystalline. On twin-rich YSZ(111) twin grains (accordingly with 60° grain boundaries) can be found in the films. The films on (100) orientated YSZ are polycrystalline, mostly with 30° and 60° grain boundaries. The palladium films show stronger de-wetting during annealing, but behave similar. The Pt/YSZ and Pd/YSZ systems were electrically polarised in the manner that oxygen was built-out at the metal film (= anodic polarisation). The oxygen removal during the anodic polarisation occurs mainly at the triple phase boundary, but also at the grain boundaries. In the case of metal films with twin grains, bubbles are formed within the grains. The bubbles crack and holes are built instead. In the case of other grain boundaries, these are widened due to oxygen removal. Palladium films are also oxidised during polarisation even at the surface. Accordingly, the metal films are aging during polarisation and the electrochemical behaviour is changing.

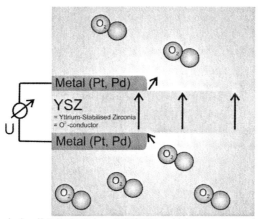

Figure 1. Electrochemical cell Pd/YSZ/Pd or Pt/YSZ/Pt, which separates a chamber into two sub-chambers of different oxygen activity.

1. INTRODUCTION
    In figure 1 an electrochemical cell is shown, built by a solid system of Pt/YSZ/Pt or Pd/YSZ/Pd (YSZ = yttrium-stabilized zirconia) which separates a chamber into two sub-chambers with different oxygen activities. The sub-chamber with the higher oxygen activity works as cathode (=built-in of

oxygen) and the other sub-chamber as anode (=oxygen removal). At the anode side the oxygen, which leaves the solid system, can also react with hydrogen or other gaseous reactants. Therefore, these systems are model electrodes for solid-oxide-fuel-cells or the electro catalysis. The oxygen exchange at the anode and cathode takes place mainly at the triple phase boundary between metal, YSZ and the gas phase [1-11], but Ryll et al. [3] showed, the grain boundaries within platinum are also oxygen permeable. Furthermore, Opitz and Fleig [4] found that oxygen is also stored at the Pt/YSZ interface as chemisorbed oxygen or in oxygen-filled voids. Foti et al. proposed that Pt-O type species were stored at the Pt/YSZ interface – even as a platinum oxide layer - and at the triple phase boundary [5]. In the case of palladium films such oxygen storage is more probable, since the oxygen affinity of palladium is much higher than that of platinum [6]. In figure 2 possible oxygen removal at a grain boundary and at the metal/YSZ phase boundary are illustrated.

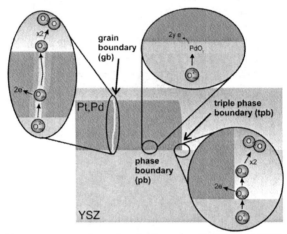

Figure 2. Supposable oxygen removal reactions in the system Pt/YSZ or Pd/YSZ at the triple-phase boundary (tpb), a grain boundary (gb) and at the interface.

Mutoro et al. [7] and Pöpke et al. [8] investigated the oxygen removal during anodic polarisation in situ by light and scanning electron microscopy, respectively. They found formation and cracking of oxygen bubbles within dense platinum films, but not for differently prepared porous platinum films [7]. These dense platinum films were prepared by pulsed laser deposition (PLD) of platinum on (111) orientated YSZ substrate and then subsequently annealed. Such platinum films are described to be (111) orientated and single crystalline or – in the case of a non-perfect (111) YSZ single crystal as substrate - nearly single crystalline [9,10]. Due to hole formation during anodic oxidation, the length of the triple phase boundary is extended – since within such holes the triple phase contact between the platinum, YSZ and the gas phase is given - and, accordingly, the electrode resistance is decreased. Mutoro et al. and Pöpke et al. found that favoured positions for the bubble and hole formation are close to scratches within the platinum film. Other defects within the platinum films have not been identified for preferred bubble formation. Since at defects in a crystal lattice – i.e. point defects, dislocations, grain boundaries and pores or inclusions - the crystal lattice is widened and, therefore, the energy for removing an ion or atom from a defect position is less than for removing it from a lattice position, the bubble and hole formation should take place preferred at such defects. Therefore, in this work the oxygen removal at different types of grain boundaries within platinum and palladium films on YSZ are studied. In addition, the influence of the higher oxygen affinity of palladium compared to platinum is also investigated.

2. EXPERIMENTAL

2.1 PREPARATION OF THE PLATINUM AND PALLADIUM FILMS

The platinum and palladium films were prepared by pulsed laser deposition (PLD) with a KrF laser ($\lambda = 248$ nm) and Argon as background gas (2 Pa, purity 99.95%). The laser had a repetition rate of 6 Hz and pulse energy of 450 mJ in the case of platinum deposition and 5 Hz and 200 mJ in the case of palladium deposition. The targets were cylindrical and polished and had a diameter of 1 cm. They were placed at a distance of 4 cm from the substrate in the PLD chamber. Both the platinum (Ögussa Austria) and palladium (Chempur, Germany) had a purity of 99.95 %. The substrates were commercially polished (111) orientated YSZ single crystals with different fractions of twins as defects and (100) orientated YSZ single crystals. In the case of platinum the substrate temperature was about 400 °C. Palladium could oxidise at this temperature and, therefore, during palladium deposition the substrate temperature was only 200 °C. The growth rate of the films was about 1 μm/h and we prepared films with a thickness of about 500 nm.

2.2 ANNEALING

The as-deposited platinum films were annealed in air at a temperature of 750 °C for 48 hours with a heating and cooling rate of approximately 2 °C/min. The palladium films show stronger de-wetting due to annealing. Here the twin grains always completely disappear after annealing. Therefore, as-deposited twin-rich palladium films were also used in the polarisation experiments.

2.3 POLARISATION

The polarisations were performed in a cyclovoltammetry experiment with the electrochemical cells Pt/YSZ/Pt-paste and Pd/YSZ/Pt-paste, respectively. Accordingly, the Pt/YSZ or Pd/YSZ side was alternatingly polarised as anode or as cathode and analogous oxygen was built-out and built-in there. The experiments were conducted within a heating chamber (Anton Paar, Austria) for the X-ray diffractometer at 400 °C in air ($p(O_2)$=0.21 bar). The chamber was modified for electrical measurements with three copper wire contacts. The electrochemical measurements were performed with a VersaSTAR 3 measurement device (Princeton Applied Research). We measured 10 cycles in the voltage range between 500 mV (= anodic case) and -1000 mV (= cathodic case) with a rate of 10 mV/s.

2.4 CHARACTERISATION

The morphology of the metal films was investigated by scanning electron microscopy (SEM; field emission microscope: Zeiss, Supra 55 VP). The phase composition was investigated by X-ray $\theta/2\theta$-scans within the heating chamber in a D5000 diffractometer (Siemens, Germany). The diffractometer is equipped with an Eulerian cradle and an X-ray lens. The texture was investigated in the same diffractometer by pole figure measurements.

3. RESULTS AND DICUSSION

3.1 PREPARED PLATINUM AND PALLADIIUM FILMS

The platinum and palladium films on all used substrates are (111) orientated, but - as also shown elsewhere [10,11] - the detailed microstructure depends strongly on the microstructure of the substrate. On single crystalline (111) orientated YSZ the metal films are also single crystalline. On twin-rich (111) orientated YSZ, twin grains (accordingly with 60° grain boundaries) with sizes between 100 nm and 300 nm can be found in the metal films. The films on (100) orientated YSZ are polycrystalline with grain sizes between 500 nm and 10,000 nm and mainly 30° and 60° grain boundaries. The palladium films show stronger de-wetting during annealing. Accordingly, the twins in twin-rich as-deposited films on (111) orientated YSZ completely disappear after annealing.

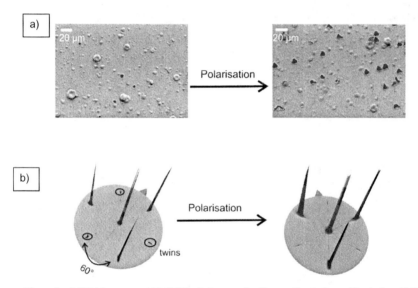

Figure 3. a) SEM images and b) {111} platinum poles figure of a platinum film before (left) and after (right) polarisation. Before polarisation the film had a small fraction of twins. Due to polarisation a small number of large holes were formed and the twins disappeared.

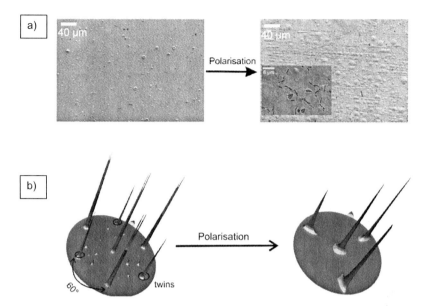

Figure 4. a) SEM images and b) {111} platinum pole figures of a platinum film before (left) and after (right) polarisation. Before polarisation the film had a large fraction of twins. Due to polarisation a large number of small holes were formed and the twins disappeared.

3.2 OXYGEN REMOVAL AT TWIN GRAINS

In the first polarisation experiments the twin behaviour during oxygen removal was studied. Therefore, annealed platinum films with different fractions of twins and an as-deposited (and therefore twin-rich) palladium film were polarised. Figures 3 and 4 show the SEM images and the corresponding pole figures of a platinum film with only a few twins and of a platinum film with several twins, respectively, before and after polarisation. The pole figure measurements show that in both cases the twins have disappeared completely after polarisation and in the SEM images holes can be seen after polarisation. In the case of a small fraction of twins within the platinum film (fig. 3) only a few - but large - holes are formed. In the case of a large fraction of twins (fig. 4) a large number of small holes are formed. Figure 5 illustrates the same result for the as-deposited palladium film: the twins within the film have disappeared and holes are formed due to polarisation. Furthermore, here the twins can be seen before polarisation as a weak visible substructure within the metal film in the SEM image. After polarisation holes of a similar shape are formed.

These results lead to the assumption that the formation of holes, due to cracking of oxygen bubbles, takes place within twin grains.

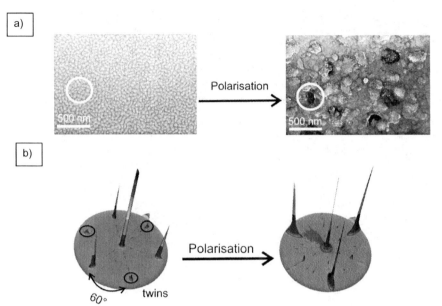

Figure 5. a) SEM images and b) {111} palladium pole figures of a palladium film before (left) and after (right) polarisation. Within the palladium film before polarisation twins are visible in the SEM image as a weak substructure. These twins can also be seen in the pole figure. After polarisation in the SEM image holes of comparable shape to the twins have appeared and the pole figure show that the twins have disappeared.

3.3 OXYGEN REMOVAL AT OTHER GRAINS

In the second investigations the effect of the oxygen removal within the polycrystalline films on (100) orientated YSZ was studied. In these metal films mainly 30° and 60° grain boundaries occurs. Figure 6 show the SEM images and the corresponding pole figures of such a platinum film before and after polarisation. Hole formation is not visible in the SEM investigations, but the metal film is waved and the grain boundaries are widened after polarisation. The pole figures show no significant difference due to polarisation, only a broadening of the reflexes, probably a result of the waved metal surface.

3.4 OXIDATION DURING OXYGEN REMOVAL

Thirdly, a possible oxidation in the case of all prepared and polarised platinum and palladium films was studied by measuring the phase composition before and after polarisation by X-ray scans. For platinum films no oxidation due to polarisation is measureable. On first observation, this is not in agreement with the investigations of Opitz and Fleig [4] or Foti et al. [5], but in the X-ray scan only platinum oxide species of some nm-sizes would be visible and not the species, which had been suggested by the authors.

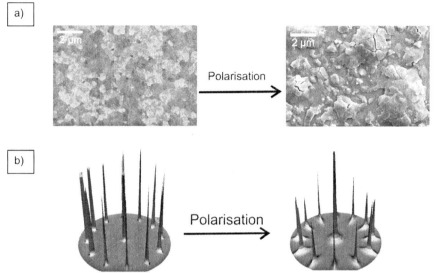

Figure 6. a) SEM images and b) {111} platinum pole figures of a platinum film before (left) and after (right) polarisation. Within the platinum film before polarisation a grain structure is visible in the SEM. The pole figure show that 30° and 60° occur. After polarisation the platinum film is waved and the grain boundaries are widened. In the pole figure the same reflexes are visible, but they are broadened – as a result of the waving of the platinum film.

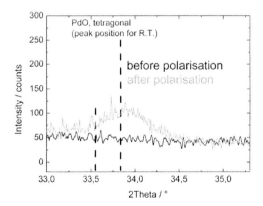

Figure 7. Cut-outs around the (002) and (101) PdO reflexes of the XRD-patterns of the palladium film before and after polarisation at 400 °C in air. After polarisation the PdO reflexes show a small intensity, indicating small oxidation of the palladium film.

In the case of palladium, reflexes of palladium oxide were always measureable after polarisation, as shown in figure 7 for the palladium film from figure 5, although the PdO-reflexes had only very low intensities. This indicates only a very thin (few nm-size) PdO layer. Moreover, within SEM investigations an electrical isolating film on palladium could be recognized after polarisation. Thus, the oxidation takes place not only at the interface between palladium and YSZ, but also at the surface of the palladium.

3.5 SUMMARY

By pulse-laser deposition platinum and palladium films with a low energy (111) surface were formed on YSZ. Single-crystalline metal films, metal films with only twin grains and polycrystalline metal films with mainly 30° and 60° grain boundaries on YSZ with different microstructure were prepared.

By electrical polarisation oxygen leaves the YSZ mainly at the triple phase boundary metal/YSZ/gas phase, but the investigations show that oxygen removal occurs also at the 30° and the 60° grain boundaries: Twin grains (only 60° grain boundary angles) within the platinum and palladium films are the location for hole formation due to bubble formation and cracking during oxygen removal - other grain boundaries are widened due to oxygen removal.

Twins are a special type of a high-angle grain boundary and they can be described by a twinning-plane between the twin grains or by a 60° rotation around the surface normal (compare pole figure measurements). In the investigated platinum and palladium films the twinning-planes are perpendicular to the surface and these twins are called contact twins. In the ideal case of a coherent twin the boundary is parallel to the twinning plane. If the twin boundary is not exactly parallel to the twinning plane, it is an incoherent or semicoherent twin boundary. At a coherent twin boundary the lattice is not widened, because the atoms in the boundary are essentially in undistorted positions. At a incoherent or semicoherent twin boundary the atoms do not fit perfectly into each grain and therefore, the lattice is widened here. The nearly round twins – which are visible in figure 5a - cannot be separated only by coherent twin boundaries. Therefore, also incoherent or semicoherent twin boundaries have to exist here and the platinum and the palladium lattices are widened here.

The 30° grain boundaries in the metal films on (100) YSZ are high-angle grain boundaries with a large disturbed region.

Accordingly, the hole formation within twins and the grain boundary widening due to oxygen removal are very plausible results, since the crystal lattice is widened at these grain boundaries and, therefore, the energy for removing metal atoms is less than for removing it from a lattice position.

4. CONCLUSIONS

On the basis of our investigations we conclude that twin grain boundaries and other grain boundaries are the location for oxygen removal during anodic polarisation. Therefore, the twins within the prepared platinum and palladium films have at least a partially incoherent or semicoherent twin boundary with a widened lattice. Accordingly, holes are formed within the twin positions and other grain boundaries are widened during anodic polarisation of Pt/YSZ or Pd/YSZ. Palladium films are also oxidised during polarisation at 400 °C in air.

ACKNOWLEDGEMENTS

The author thank the DFG (Deutsche Forschungsgemeinschaft) for financial foundation in the projects BE 3170/3-1, H. Pöpken (Justus-Liebig-University Giessen, Germany) for the preparation of the metal films and J.W. Ashford for revising the manuscript.

REFERENCES

[1] S. B. Adler, Chem. Rev. **104** (2004) 4791-4843.

[2] E. Mutoro, B. Luerßen, S. Günther, J. Janek, Solid State Ionics **179** (2008) 1214-1218.

[3] T. Ryll, H. Galinski, L. Schlagenhauf, P. Elser, J.L.M. Rupp, A. Biberle-Hutter, L.J. Gauckler, Adv. Funct. Mater. **21** (2011) 565–572.

[4] A.K. Opitz, J. Fleig, Solid State Ionics **181** (2010) 684-693.

[5] G. Foti, A. Jaccoud, C. Falgairette, C. Comninellis, J. Electroceram **23** (2009)   175-79.

[6] A. Schwartz, L.L. Holbrook, H. Wise, J. Catalysis **21** (1971) 199-207.

[7] E. Mutoro, S. Günther, B. Luerßen, I. Valov, J. Janek, Solid State Ionics **179** (2008) 1835-1848.

[8] H. Pöpke, E. Mutoro, B. Luerßen, J. Janek, Solid State Ionics **189** (2011) 56-62.

[9] G. Beck, H. Fischer, E. Mutoro, V. Srot, K. Petrikowski, E. Tchernychova, M. Wuttig, M. Rühle, B. Luerßen, J. Janek, Solid State Ionics **178** (2007) 327-337.

[10] G. Beck, H. Pöpke, B. Luerßen, J. Janek, Journal of Crystal Growth **322** (2011) 95–102

[11] G. Beck, J. Janek, to be published.

[12] D.A. Porter, K.E. Easterling, Second Edition 1992, Chapman&Hall London, Phase Transformation in Metals and Alloys.

[13] H.-J. Bargel, G. Schulze, 10. Edition 2008, Springer Verlag Berlin Heidelberg, Werkstoffkunde.

ELECTRICAL AND MORPHOLOGICAL CHARACTERIZATION OF MONOCULTURES AND CO-CULTURES OF SHEWANELLA PUTREFACIENS AND SHEWANELLA ONEIDENSIS IN A MICROBIAL FUEL CELL

K. E. Larson[1] & M. C. Shaw[2],*
[1]Biology Department
[2]Physics Department
60 West Olsen Road, #3750
California Lutheran University, USA
* Corresponding author

ABSTRACT
    The diminishing supply of fossil fuels has led to a global effort to develop novel renewable energy conversion strategies including microbial fuel cells (MFCs). MFCs operate by producing electricity via the breakdown of organic matter by living bacteria. The purpose of the present study was to design and commission an MFC and electrical characterization apparatus for the purpose of measuring and analyzing the electrical output of single cultures of the Shewanella Oneidensis and Shewanella Putrefaciens bacterial species, as well as the potential synergistic interactions of co-cultures of the two species. Here we present the design of a novel, low-cost and disposable MFC aimed at rapid screening of assays in MFC experiments. Furthermore, we found that co-cultures exhibited a steady-state open circuit voltage of $0.27V \pm 0.02V$, compared to $0.16V \pm 0.04V$ for the monocultures, thus indicating potential synergistic interactions between the two species. Finally, high-magnification scanning electron micrographs of the bacteria morphologies are presented as confirmation of the presence of bacteria. We conclude that these results demonstrate the feasibility of co-cultured MFCs comprising Shewanella Oneidensis and Shewanella Putrefaciens bacteria for enhanced output voltage.

INTRODUCTION & BACKGROUND
Renewable Energy
    Renewable energy sources are steadily becoming more prevalent in our fossil fuel-dependent society. The rising costs of energy and the detrimental effects that common energy sources have on the environment are just two of the factors leading the world toward investing more time and resources into developing and modifying renewable energy sources [1]. Furthermore, since every individual on the planet uses energy, and the world population increases by nearly a billion people per year, the rate of energy consumption will continue to rise for the foreseeable future. As this demand for energy continues to rise with the steadily increasing population, other factors, such as the cost of the production of various forms of energy, also begin to influence the world's movement toward the further development of renewable energy sources. Indeed, experts have predicted that oil prices could so much as double by the end of the 21st century [2]. This is primarily due to the rising costs of oil, coal and gas extraction, as well as the rising costs of cleaning the waste produced by these energy sources [2]. These increased prices have led to the cost of crude oil rising from $20 a barrel to $140 per barrel in a time period of 20 years [1]. Conversely, as an example of the volatility in oil prices, in a time period of just a single year, the cost dropped from the $140 per barrel to only $60 per

barrel [1]. The increasing cost and volatility in oil prices has led many to view it as an increasingly unreliable source of energy.

Not only can the energy sources of oil, gas and coal be seen as costly and unreliable, but they can also have negative effects on the environment and on human health. For example, the carbon dioxide emissions from burning fossil fuels continue to have adverse affects on our environment and radioactive waste disposal continues to pose a hazardous risk to human health [1].

Microbial Fuel Cells

Working to generate novel sources of renewable energy that can be cost efficient and safe to work with is thus critically important in mitigating these threats. Microbial Fuel Cells (MFCs) represent one such source of renewable energy [3]. An MFC is a biological system that uses bacteria to generate electrical power through the degradation of organic matter via normal biological activity (Fig. 1 [3]).

One of the primary applications of the MFC system is for wastewater treatment [3], where the widespread implementation of MFCs could lead to a reduction in the cost of operating wastewater treatment plants while simultaneously producing electricity. Studies conducted by Bruce Logan have shown that not only is wastewater treatment with MFCs possible, but also that many advantages could result from implementing such a system (4). For example, using microbial fuel cells to treat wastewater saves power and leads to a much lower production of solids, therefore making it a more efficient system [4]. Logan points out that the energy input required for the treatment of wastewater represents 50% of the cost of the entire system [4], a fact that strongly supports research investigating MFCs as an alternative method for wastewater treatment. Microbial fuel

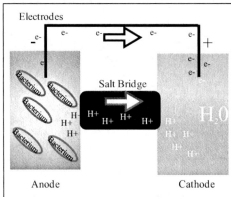

Fig. 1. Schematic of a typical Microbial Fuel Cell (MFC), consisting of four components: The anaerobic anode chamber containing the bacteria; the cathode chamber containing a saline solution, the salt/agar bridge and the electrodes.

cells use a culture of bacteria, which can use the organic matter present in the wastewater as an energy source to produce electrons and protons. This allows for the generation of electricity which can be harvested and then utilized to power other processes. Implementing such a system could lead to wastewater treatment plants becoming completely self-sustainable.

The MFCs utilized in our study have four main components (Fig. 1). The first is the anode chamber, which contains the bacterial broth solution and must maintain an anaerobic environment. The second component is the cathode chamber. This chamber houses a conductive saline solution. The third component is an agar-salt bridge which serves to separate the differing chemical environments of the cathode and anode chambers, while concurrently allowing for the passage of protons in order to maintain electrical charge neutrality throughout the system. The fourth component is the electrodes, which serve to transfer electrons out of and into the MFC and thus allow current flow.

Many prior investigations [5], including the present study, utilize platinum as the electrode material. Although high in cost, its chemical inertness is a major experimental advantage. In such cases, a platinum wire is inserted into both the anode and the cathode chambers, with the platinum wire in the anode chamber utilized by bacteria as the final electron acceptor. The wires are then connected to a data acquisition system to measure the electrical characteristics during an experiment.

Not all bacteria are able to use an inorganic metal, such as platinum, as an electron acceptor. Only certain species, known as *exoelectrogens* [3], have been found to possess this specific ability. Exoelectrogenic bacteria are able to transfer their electrons outside of their body to the electrode. They do so because they are able to generate more ATP by donating their electrons to the electrode than by donating their electrons to the organic molecules present in their environment, the normal process of non-exoelectrogenic bacteria in the equivalent environment. Also, if oxygen was present in the anode chamber, even exoelectrogenic bacteria would utilize oxygen as the final electron acceptor rather than the electrode because of the even greater amount of ATP production that is possible when oxygen is utilized. This is why the anode chamber must maintain an anaerobic environment; there must be no oxygen present so that the bacteria do not transfer their electrons to oxygen molecules instead of to the electrode thus diminishing the voltage production [6].

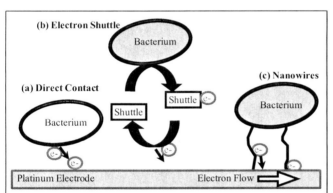

Three methods of electron transfer to the electrode have been identified: (i) direct contact, (ii) electron shuttles, and (iii) microbial nanowires (Fig. 2 (a-c)). Each will now be briefly described. The direct contact method (Fig. 2 (a)) involves the bacteria directly adhering to the platinum electrode and using outer membrane proteins, such as

Fig. 2. Three mechanisms of electron transfer in Microbial Fuel cells: (a) Direct Contact; (b) Electron Shuttles and (c) Nanowires. Shewanella Oneidensis is known to employ microbial nanowires and electron shuttles whereas Shewanella Putrefaciens is known to utilize the method of direct contact.

cytochromes, to transfer electrons to the platinum wire. In contrast, bacteria that use the electron shuttle method (Fig. 2 (b)) generate molecules, typically riboflavins, which diffuse to the electrode carrying the electrons. Once at the electrode, the shuttles oxidize and release the electrons to the electrode. They then diffuse back to the bacteria where they continue the cycle.

Finally, microbial nanowires (Fig. 2 (c)) are biological appendages that specific bacteria can produce. They are generally pili or flagella that have the ability to connect to the electrode and transfer electrons from the bacteria to the electrode. Of these three methods, some bacteria are known to utilize a single electron transfer method [7] while others have been found to utilize two or even all three of these mechanisms [7].

The primary goal of our study was to determine whether there was a measurable difference in the output voltage of MFCs comprising monocultures vs. co-cultures of individual bacterial species known to implement fundamentally different mechanisms of electron transport to the electrodes, as schematically illustrated in Fig. 3. Specifically we hypothesized that the co-culture would function in one of two possible ways: (i) the co-culture would act similarly to

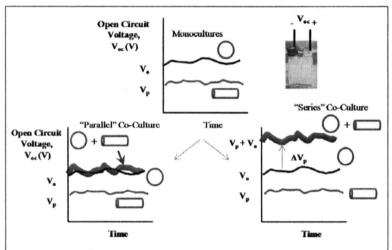

Fig. 3. Schematic diagram illustrating the goal of the present study: to determine whether: (i) the co-culture would act similarly to potential sources connected in parallel (a), thus producing an open-circuit voltage equal to the greatest voltage produced by either of the monocultures; or, (ii) the co-culture would act similarly to potential sources connected in series (b), and produce an open-circuit voltage that was equal to the simple sum of the voltages produced by the monocultures.

potential sources connected in *parallel* (Fig. 3 (a)), thus producing an open-circuit voltage equal to the greatest voltage produced by either of the monocultures; or, (ii) the co-culture would act similarly to potential sources connected in *series* (Fig. 3 (b)), and produce an open-circuit voltage that was equal to the simple sum of the voltages produced by the monocultures of the two species used in this study.

We selected the exoelectrogenic species Shewanella Oneidensis and Shewanella Putrefaciens as the bacteria implemented in the monocultured and co-cultured MFCs studied in the present investigation. These two species were chosen owing to the fact that Shewanella Oneidensis is known to use the electron transport methods of microbial nanowires and electron shuttles [8] whereas Shewanella Putrefaciens is known to utilize the method of direct contact [5].

EXPERIMENTAL

The organization of the experimental portion of this investigation is illustrated schematically in Fig. 4. As a summary, the experiments began with a freeze-dried pellet of bacteria, which was rehydrated, incubated and electrically characterized in the microbial fuel cell. Monocultures and co-cultures were investigated. Following electrical characterization, specimens were filtered and examined at high magnifications in a scanning electron microscope (SEM). Tryptic soy agar plates were also inoculated and examined optically to confirm the presence of the bacteria. Each of these steps is described in detail below.

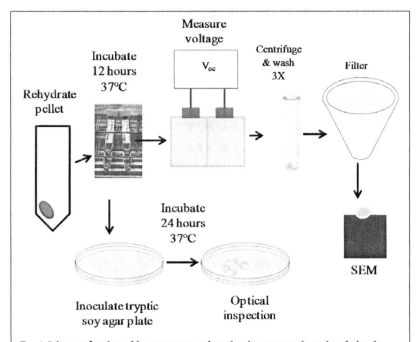

Fig. 4. Schematic flowchart of the experiments performed in the present study, as described in the text. A freeze-dried pellet of bacteria was rehydrated, incubated and electrically characterized in the microbial fuel cell. Monocultures and co-cultures were investigated. Following electrical characterization, specimens were filtered and examined at high magnifications in a scanning electron microscope. Tryptic soy agar plates were also inoculated and examined optically to confirm the presence of the bacteria.

Experimental – Transitory Microbial Fuel Cell

A novel design of a simple, low-cost, disposable microbial fuel cell was developed and commissioned. The design utilized two 15 mL culture flasks, with each flask functioning as one of the two MFC chambers. A circular aperture approximately 6 cm in diameter was introduced into the same side of each flask. The flasks were then glued together with airtight, waterproof silicon glue so that the apertures of each flask were aligned, and would thus allow unobstructed passage of electrolytes between the MFC chambers (Fig. 5). One flask was then designated as the anode chamber and one as the cathode chamber. A hole was drilled into the cap of the anode chamber through which the platinum wire electrode was inserted. The open space surrounding the point of insertion of the electrode into the cap was sealed with airtight glue. The other platinum electrode was placed in the cathode chamber, which was left open to the ambient environment.

The salt-agar bridge was prepared using 15 grams of agar (Acros Organics) and 300 mL of water. After heating the agar solution to boiling, 3 grams of salt (NaCl) were added to the mixture. The salt-agar solution was then poured into the MFC design so that it filled both chambers until only 5 mL of volume remained at the top of each chamber. The agar was allowed to cool and harden for 12 hours at 5° C. The 5 mL bacterial broth solution was then introduced into the corresponding 5 mL volume of the anode chamber, and 5 mL of a conductive salt solution was inserted into the cathode chamber.

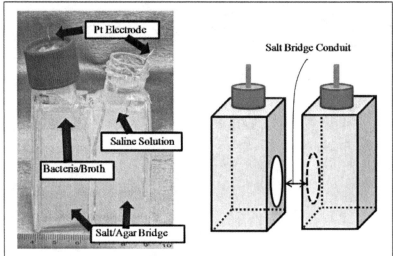

Fig. 5. Image and diagram of the transitory Microbial Fuel Cell developed during the present investigation. The design comprised two 15 mL cell culture flasks into which two 6 cm diameter apertures were introduced. The flasks were glued together and filled with salt/agar solution which formed the salt bridge when cooled. Platinum electrode wires were inserted into the anode and cathode chambers. The anode chamber was sealed to provide an anaerobic environment.

EXPERIMENTAL – Biological

Specimens of the bacterial species of Shewanella Oneidensis (Cat. # 700550) and Shewanella Putrefaciens (Cat. # 8071) were ordered from The American Type Culture Collection (ATCC) and were received in the form of freeze dried pellets. The pellets were rehydrated following the recommended protocol from ATCC with 5 mL of tryptic soy broth (TSB). Aseptic techniques were utilized during all bacteria culturing to prevent contamination. The resultant bacteria solution was re-mixed three times, placed in a 15 mL centrifuge tube and incubated at 37°C for 12 hours in an ambient air environment.

In addition to the TSB specimens, tryptic soy agar (TSA) plates were inoculated with the solution of blank TSB (negative control) and bacteria in order to confirm the presence of bacteria.

EXPERIMENTAL – Electrical

Following incubation, the bacteria solution was transferred from the centrifuge tube to the transitory microbial fuel cell apparatus. The MFC anode chamber was immediately sealed and electrical measurements were immediately recorded. Namely, the multimeter was attached to the platinum electrodes of each chamber of the MFC and the open-circuit voltage of the MFCs was measured using a Keithley 2700 digital multimeter with 0.1 microvolt resolution for a total of 30 minutes (1800 seconds). The voltage was recorded every ten seconds for the first five minutes and then every minute after that for the next 25 minutes.

EXPERIMENTAL – Scanning Electron Microscopy

After the MFC experiment was complete, in select cases, the bacteria solution was transferred to a 15 mL centrifuge tube. It was then centrifuged for three minutes at 5000 rpm. The supernatant was then removed, the resulting pellet was washed with deionized (DI) water and the process was repeated a total of three times. Following the final washing, the bacterial pellet was re-suspended in DI water.

The bacterial suspension was filtered through a Nalgene 0.45 μm micropore vacuum filter. The filter paper was sputter-coated with gold to enhance electrical conductivity prior to filtration. The resultant bacteria/filter paper was dried for 24 hours in an airtight, controlled humidity container, and then analyzed in a JEOL JSM-6390LV scanning electron microscope (SEM) to image the bacteria [9].

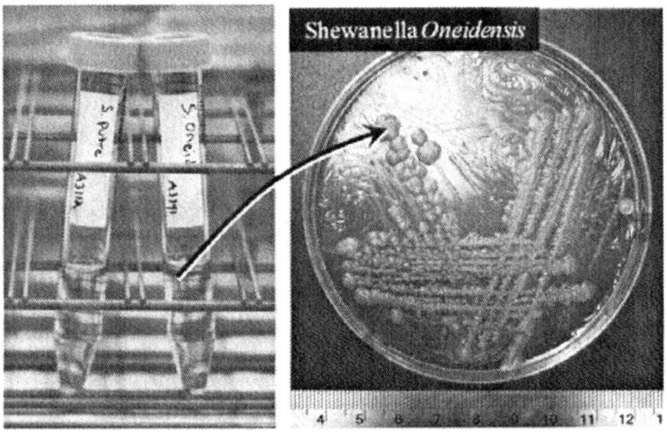

Fig. 6. Images of the cell cultures and the inoculated tryptic soy agar plates following incubation. The presence of bacteria was evident, and confirmed by the color and surface appearance in both cases of Shewanella Oneidensis and Shewanella Putrefaciens. Scale marker in cm.

Results
Results – Biological

The TSA plate for bacterial species exhibited successful proliferation of bacteria after incubation at 37°C for 24 hours in an ambient air environment (Fig. 6). The specific bacteria were indicated by the presence of opaque colonies that varied in color from pale pink to white for both species of bacteria. These results confirmed the presence of the bacteria in the bacterial broth solution.

RESULTS– Scanning Electron Microscopy

The samples of the monocultures Shewanella Oneidensis and Shewanella Putrefaciens that were analyzed in the SEM were imaged at magnifications between 100X – 5000X. The resulting micrographs clearly indicated the presence of bacteria as well as the distinct morphological differences between the two species (Fig. 7 (a) and (b)). Specifically, the Shewanella Oneidensis specimen exhibited a generally spherical morphology (Fig. 7 (a)), and the Shewanella Putrefaciens specimen exhibited a generally rod-like morphology (Fig. 7 (b)).

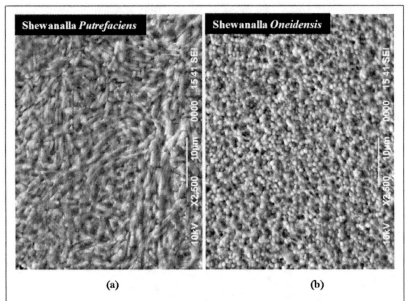

(a)   **(b)**

Fig. 7. Representative high-magnification (2500X) scanning electron micrographs of the filtered specimens of Shewanella Putrefaciens (a) and Shewanella Oneidensis (b). The rod-like morphology of the Shewanella Putrefaciens was clearly visible, as was the spherical morphology of the Shewanella Oneidensis specimen.

RESULTS – Electrical

A plot of the open-circuit voltage as a function of time from the monocultures of Shewanella Putrefaciens and Shewanella Oneidensis over a 30 minute time interval is shown in Fig. 8. Each data point on this plot represents data from three different trials, and the error bars were calculated using the data from each of the three trials. Also shown in Fig. 8 is the open-circuit voltage generated from co-cultures of S. Oneidensis and S. Putrefaciens over the same time interval.

Although S. Oneidensis produced a slightly higher average open-circuit voltage than S. Putrefaciens, there is not a significant difference between the open-circuit voltages produced by the monocultures. In contrast, the co-culture of S. Oneidensis and S. Putrefaciens produced an open-circuit voltage that was substantially higher than that produced by either monoculture (Fig. 8). Furthermore, to compare the open-circuit voltages at different time points, Fig. 9 shows plots of the open-circuit voltage produced by the monoculture of S. Oneidensis, the monoculture of S. Putrefaciens and the co-culture of both species at the specific time intervals of 10 seconds, 100 seconds and 1800 seconds. It is evident that the open-circuit voltages produced by the monocultures are lower than those produced by the co-culture over the two orders of time intervals investigated.

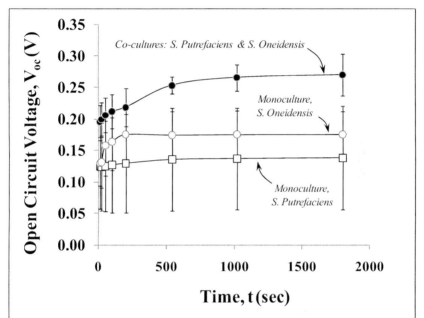

Fig. 8. Experimental measurements of the open-circuit voltage as a function of time from the monocultures of Shewanella Putrefaciens and Shewanella Oniednensis over a 30 minute time interval. Each point represents the average of three different trials, and the error bars were calculated using the data from each of the three trials. Also shown is the open-circuit voltage generated from co-cultures of S. Oneidensis and S. Putrefaciens over the same time interval.

DISCUSSION

Two distinct possibilities have been identified to rationalize the results obtained from this investigation. The first is that *new and/or different electron transport mechanisms are utilized by the co-cultured bacteria that are inoperable when monocultured.* For example, interactions among the bacteria could lead to the activation of different electron transport mechanisms that are inaccessible when in monoculture. A 2010 study by Bouhenni et al. [10] shows that outer membrane c cytochromes are a leading factor enabling the successful transfer of electrons to the electrode, thus indicating that the direct electron transfer mechanism used primarily by

Shewanella Putrefaciens could be having a substantial impact on the Shewanella Oneidensis in the co-culture [10]. The second possibility is that *biochemical byproducts are produced in the co-culture that affect the ability of the bacteria species to transfer electrons to the electrode.* A 2008 study conducted by Wall [11] indicated that the presence of various molecules can impact the voltage production of an MFC. Thus in the present investigation, the co-cultured bacteria species could produce different byproducts, thus influencing the behavior of the species in the MFC. The byproducts could lead to an increase in pH which has been shown to influence the efficiency of an MFC. In both cases, the significantly higher open-circuit voltage of the co-culture has significant implications for the implementation of microbial fuel cells.

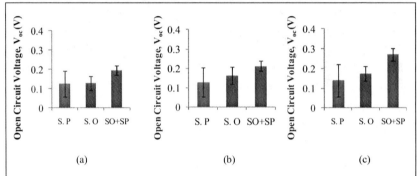

(a) (b) (c)

Fig. 9. Experimental measurements of the open-circuit voltage produced by monocultures of S. Oneidensis and S. Putrefaciens and the co-culture of both species at the specific time intervals of 10 seconds (a), 100 seconds(b) and 1800 seconds (c). Each point represents the average of three different trials, and the error bars were calculated using the data from each of the three trials.

The next steps in our investigation are to analyze the pH of the monocultures and co-cultures as a function of time, in order to explore the effect that pH has on the open-circuit voltage. We also intend to investigate the actual electron transport mechanisms used in the monoculture and the co-culture. This will be accomplished by SEM and transmission electron microscopy (TEM), which can reveal the presence and density of microbial nanowires. Furthermore, we intend to conduct quantitative protein analyses to investigate to what degree direct electron transport mechanisms are being used. Finally, we intend to quantify the byproducts of broth fermentation via quantitative chemical analysis to discern whether byproducts are having a significant effect on the voltage production.

CONCLUSIONS
We have developed and commissioned a novel, low-cost, disposable MFC design intended for single-use, rapid screening assays of MFC design variants. We have utilized the MFC to

determine the open-circuit voltage (V_{oc}) of monocultures and co-cultures of Shewanella Oneidensis and Shewanella Putrefaciens bacterial species over a time interval of 30 minutes. These two species were selected owing to their different mechanisms of electron transport to the anode electrode. We have found that the monocultured MFCs exhibit a V_{oc} of approximately $0.12 - 0.17$ V that stabilizes after a period of approximately $3 - 4$ minutes. In contrast, the co-cultured MFCs exhibit a V_{oc} of approximately $0.25 - 0.27$ V that stabilizes after a period of approximately $4 - 5$ minutes. We interpret these results to indicate a significant interaction between the bacterial species resulting in a different, potentially new mechanism of electron transport and/or different or higher concentrations of biochemical byproducts.

REFERENCES

1. Dicker, Daniel. Oil's Endless Bid: Taming the Unreliable Price of Oil to Secure the Economy. Hoboken, N.J: John Wiley & Sons, 2011.

2. Nersesian, Roy L. Energy for the 21st Century: A Comprehensive Guide to Conventional and Alternative Sources. Armonk, N.Y: M.E. Sharpe, 2007.

3. Logan, Bruce E. Microbial Fuel Cells. Hoboken, N.J: Wiley-Interscience, 2008.

4. Ahn, Y, and B.E Logan. "Effectiveness of Domestic Wastewater Treatment Using Microbial Fuel Cells at Ambient and Mesophilic Temperatures." *Bioresource Technology*. 101.2 (2010): 469-475.

5. Kim, H J, H S. Park, M S. Hyun, I S. Chang, M Kim, and B H. Kim. "A Mediator-Less Microbial Fuel Cell Using a Metal Reducing Bacterium, Shewanella Putrefaciens." *Enzyme and Microbial Technology*. 30.2 (2002): 145-152.

6. Biffinger, J. C., Ray, R., Little, B., & Ringeisen, B. R. (January 01, 2007). Diversifying biological fuel cell designs by use of nanoporous filters. Environmental Science & Technology, 41, 4, 1444-9.

7. Torres, CésarI, AndrewKato Marcus, Hyung-Sool Lee, Prathap Parameswaran, Rosa Krajmalnik-Brown, and BruceE Rittmann. "A Kinetic Perspective on Extracellular Electron Transfer by Anode-Respiring Bacteria." *Fems Microbiology Reviews*. 93.1 (2010): 3.

8. Jiang, Xiaocheng, Jinsong Hu, Lisa A. Fitzgerald, Justin C. Biffinger, Ping Xie, Bradley R. Ringeisen, and Charles M. Lieber. Probing Electron Transfer Mechanisms in Shewanella Oneidensis Mr-1 Using a Nanoelectrode Platform and Single-Cell Imaging. Ft. Belvoir: Defense Technical Information Center, 2010.

9. Ray, R, S Lizewski, L A. Fitzgerald, B Little, and B R. Ringeisen. Methods for Imaging Shewanella Oneidensis Mr-1 Nanofilaments. Ft. Belvoir: Defense Technical Information Center, 2010.

10. Bouhenni, Rachida A, Gary J. Vora, Justin C. Biffinger, Sheetal Shirodkar, Ken Brockman, Ricky Ray, Peter Wu, Brandy J. Johnson, Eulandria M. Biddle, Matthew J. Marshall, Lisa A. Fitzgerald, Jim K. Fredrickson, Alexander S. Beliaev, Bradley R. Ringeisen, and Daad A. Saffarini. The Role of Shewanella Oneidensis Mr-1 Outer Surface Structures in Extracellular Electron Transfer. Ft. Belvoir: Defense Technical Information Center, 2010.

11. Wall, Judy D, Caroline S. Harwood, and A L. Demain. Bioenergy. Washington, D.C: ASM Press, 2008.

Biomass

REFRACTORY CERAMIC LINING SELECTION AND TROUBLESHOOTING IN THERMAL BIOMASS OPERATIONS

Dana G. Goski, Timothy M. Green, Dominic J. Loiacona
Allied Mineral Products Inc., 2700 Scioto Parkway, Columbus, OH, 43221 USA
www.alliedmineral.com

ABSTRACT: There are innumerous creative alternative energy solutions incorporating high temperature processes and biomass emerging in the marketplace. These range from the small scale home heating level to the industrial large scale plants generating heat energy or alternative fuel gases. One of the limiting factors to the thermal efficiency and cost of these operations is the performance of the refractory ceramic containment materials. Different sources of biomass can contribute to different failure mechanisms in refractory linings. The authors will present a refractory selection and troubleshooting overview relating to differing biomass characteristics and operational conditions.

INTRODUCTION TO REFRACTORIES

Refractories are highly engineered ceramic materials that are designed to withstand conditions that would be damaging to other types of materials. They are chemically and physically stable in environments containing extremely high temperatures and corrosive chemicals. In the case of biomass thermal processes, refractories line the interiors of equipment such as boilers and gasifiers in order to contain the heat produced by these processes. Because such a wide variety of biomass fuel sources are available, a wide variety of refractory materials exists in order to withstand the various types of chemical and physical attack presented by these different fuel sources. Therefore, careful selection of refractory products for use in thermal biomass operations can maximize efficiency and lifespan of the equipment.

The market for refractories is very diverse in terms of the form in which the material can be purchased. Monolithic refractories come in many forms, including mixed, poured, gunited, shotcrete pumped, and dry rammed products. In addition to monolithics, bricks, mortars, and fiber insulation are commonly available, as well as precast and prefired specialty shapes.

Refractories are comprised of a wide variety of materials, ranging from many different forms of silica, alumina, magnesia, cements, clays, various aluminosilicates, spinels, zircon, and silicon carbide. Some of these materials are naturally occurring minerals, while others are manufactured materials. The refractory's function and environment dictate the form it should take, as well as the materials it will be engineered from.

BIOMASS SOURCES

There is a diverse variety of biomass sources available for thermal processing, and each biomass source presents its own unique refractory challenges based on its own unique chemistry. For example, rice straw tends to produce much more ash than other biomass sources [1], which increases the need for refractory and furnace design that can withstand high levels of ash, as well as the harmful components of the ash itself. Almond hulls have a higher level of potassium oxide, which increases the need for a refractory that is alkali-resistant [1]. Hybrid poplar wood and wheat straw ash contain high levels of sulfur, which could lead to an acidic environment the refractory will need to withstand and contain [1]. Wheat straw contains high levels of chlorine, alkalis, and sulfur, so refractory used in a wheat straw biomass process would have to combat a host of alkali and acidity issues [1]. Tables 1 and 2 provide an overview of these chemical attributes as percentage weight of ash after thermal processing, as well as typical ash percentage and chlorine content before thermal processing.

Table 1. Examples of Diverse Biomass Chemistry I

	Furniture Wood Waste [1]	Hybrid Poplar [1]	Switchgrass [1]	Rice Straw [1]	Almond Hulls [1]	Wheat Straw [1]
Ash %	3.61	2.70	8.97	20.34	6.13	9.55
Chlorine %	<0.01	0.04	0.19	0.51	0.02	1.79
Chemical Composition of Ash (% wt of ash)						
SiO_2	57.62	0.88	65.18	80.15	9.28	37.06
Al_2O_3	12.23	0.31	4.51	1.46	2.00	2.66
TiO_2	0.50	0.16	0.24	0.06	0.05	0.17
Fe_2O_3	5.63	0.57	2.03	0.85	0.76	0.84
CaO	13.89	44.40	5.60	2.03	8.07	4.91
MgO	3.28	4.32	3.00	2.11	3.31	2.55
Na_2O	2.36	0.23	0.58	0.91	0.87	9.74
K_2O	3.77	20.08	11.60	8.51	52.90	21.70
SO_3	1.00	3.95	0.44	1.22	0.34	4.44
P_2O_5	0.50	0.15	4.50	1.68	5.10	2.04
CO_2		19.52			20.12	
Undetermined/ Unreported	-0.78	5.43	2.32	1.02	-2.89	14.32
Total %	100.00	100.00	100.00	100.00	100.00	100.00

Table 2. Examples of Diverse Biomass Chemistry II

	Urban Wood [1]	Corn and Canola Straw [2]	Corn Residues and Silage [2]	Pellets from Corn Residues [2]	Hay Pellets [2]	Dried Sewage Sludge [3]
Ash %	5.54	N/A	N/A	N/A	N/A	N/A
Chlorine %	0.06	N/A	N/A	N/A	N/A	N/A
Chemical Composition of Ash (% wt of ash)						
SiO_2	55.12	53.4	44.7	31.8	28.4	22.4
Al_2O_3	12.49	3.0	3.0	2.4	0.9	9.0
TiO_2	0.72	0.1	0.2	0.2	0.1	0.9
Fe_2O_3	4.51	0.9	1.4	1.1	0.5	24.6
CaO	13.53	14.5	11.5	12.8	7.0	3.5
MgO	2.93	4.0	6.0	6.1	6.6	2.8
Na_2O	3.19	1.0	0.5	0.4	0.2	4.9
K_2O	4.78	17.5	15.1	18.2	27.3	2.2
SO_3	1.92	N/A	N/A	N/A	N/A	N/A
P_2O_5	0.88	N/A	N/A	N/A	N/A	19.3
CO_2		N/A	N/A	N/A	N/A	N/A
Undetermined/ Unreported	-0.07	5.6	17.6	27.0	29.0	10.4
Total %	100.00	100.0	100.0	100.0	100.0	100.0

REFRACTORY ISSUES RELATED TO THE BIOMASS SOURCE
    Biomass sources are so varied that refractories can experience a variety of failure conditions based on the biomass source and their processing conditions.

Alkali Attack
    One of the most common issues is alkali attack of the refractory, in which sodium and potassium vapors or ashes will infiltrate the refractory, and condense in the pores of the material. This occurs at temperatures between 750-1200 °C [4]. The alkali elements then react with the refractory, forming phases such as kalsilite, carnegieite, leucite, or β-aluminas. Volume expansion can be as high as 55.5% in the (K_2O, Na_2O)·Al_2O_3·SiO_2 systems, with the greatest volume expansion attributed to the formation of kalsilite and carnegieite [5]. In the case of β-alumina formation, up to approximately 25% volume expansion may be observed, based on crystallographic unit cell volumes [6] [7]. This causes significant volume expansion and stress on the refractory, eventually causing partial or total failure of the material as it cracks and spalls off. Examples of these mechanisms are shown in Figures 1 through 3.

		$K_2O·Al_2O_3·6SiO_2$	Orthoclase (1)
Aluminosilicates	K_2O or Na_2O $\longrightarrow$	$K_2O·Al_2O_3·4SiO_2$	Leucite (2)
(Mullite, Fireclay, Chamotte)		$K_2O·Al_2O_3·2SiO_2$	Kalsilite (3)
		$Na_2O·Al_2O_3·2SiO_2$	Carnegieite (4)

Alumina	K_2O or Na_2O $\longrightarrow$	$Na_2O·11Al_2O_3$	Na β-Alumina (5)
		$K_2O·11Al_2O_3$	K β-Alumina (6)

    As seen in the example reactions, alkali attack is enhanced in cases with elevated or excess alumina in the refractory components. In cases where service temperatures are low enough to allow flexibility in choice of refractory linings, a material with less alumina is preferred to reduce alkali attack. Thus, a flint/chamotte based refractory can be expected to be more resistant to alkali attack than a mullite based refractory, and further, than an alumina based refractory. In cases where service temperatures disallow use of flint/chamotte refractories, use of additives to the base refractory composition such as SiC or Zr-containing phases may be beneficial.
    In cases where silica (fused silica in particular) is a suitable refractory, it may actually be the best material for resisting alkali attack. Silica on the refractory hot face readily reacts with alkali to form potassium and/or sodium silicates, or phases such as kalsilite. The formation of these compounds on the hot face actually forms a protective barrier, protecting the rest of the refractory from further attack. However, a balance of the material selection process is required, if a high level of ash is present, it will tend to accumulate on the silica surface.
    In addition to the volume expansion associated with the formation of these phases, there is also an issue of glassy slags forming as these alkali-alumina-silica phases melt at higher temperatures. Glassy slags can act as collection points for particulate material generated during the biomass conversion process. The particulate matter impacts onto the viscous surface, and in worst case scenarios will lead to high rates of ash buildup and fouling [8]. An example of this can be seen in Figure 4.

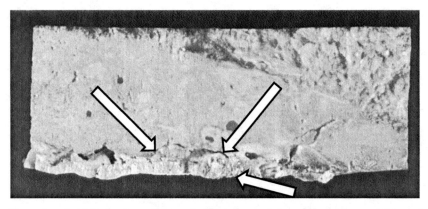

Figure 1. A high alumina refractory hot face exposed to sodium vapor, reacting and subsequently expanding, creating volume expansion and stress, finally generating a cracking plane in the refractory which will eventually spall off.

Figure 2. Lab-controlled alkali test on refractory, illustrating cracking due to volume expansion caused by alkali attack. Test is performed with various Na and K salts at high temperatures, to mimic operating conditions in the biomass unit.

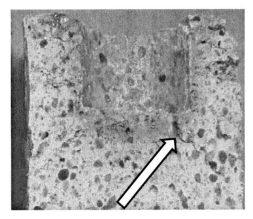

Figure 3. Alkali cup test post mortem exhibiting the glassy reaction phase when the Na and K salts were contained and a large crack where destructive expansive phases developed.

Figure 4. Buildup can be caused by either particulate matter transported in the gas, or vapors condensing on surfaces. This region of a woody biomass unit (red oak) was not in direct contact with the fuel. This buildup was mainly potassium-based ash components.

Acidic Attack, Sulfur, Chlorine, and Phosphorus

Depending on the chemical composition of the biomass, acidic attack on the refractory can occur. Chlorine can form hydrogen chloride gas or hydrochloric acid, given an appropriate source of hydrogen. Sulfur-containing compounds will typically oxidize to SO_2 gas during combustion of the biomass, which can further react with water vapor (if the biomass feedstock contains significant moisture) to form sulfuric acid. Acid will attack the calcium aluminate bond in typical refractory castables, so it is important to use a low-cement or no-cement castable in high-acidity conditions. Hydrochloric acid will also corrode steel and brick anchor supports. Sulfur can also oxidize and react with alkali to form sulfate compounds, further weakening the refractory.

In addition to acidity concerns, the presence of chlorine can exacerbate alkali movement in the gaseous state, causing alkali to penetrate the refractory at a faster pace than it normally would [1]. If a high degree of phosphorus is present, it can react with silica and alkali to form deposits of glassy slag with a low melting point.

Another important consideration with acid-producing biomass is that liquid-state acids are much more damaging than gaseous-state acids. Damage to the refractory and to the entirety of the boiler metal containment structure can be minimalized if temperatures are kept high enough that water never has a chance to condense, and acids are kept in the gaseous state.

Carbon Monoxide Disintegration

Incomplete combustion of biomass can be caused either by the biomass source, or the process or equipment used to process it, and will lead to excessive production of carbon monoxide gas. This is harmful to a refractory with a high degree of iron or iron oxide, where CO vapor transports into the refractory, and is catalyzed by iron and its oxides via the Boudouard reaction into carbon dioxide and elemental carbon [9].

$$2CO \xrightarrow{\text{Fe, Fe Oxides}} CO_2 + C \qquad (7)$$

The carbon is deposited around the iron source, and continues to expand until the surrounding structure is disrupted, eventually causing cracking and failure of the refractory. The more reduced valence state the iron in the refractory is, the stronger a catalyst it is, and the worse the effect on the refractory is (i.e. metallic iron is more damaging than FeO, which is more damaging than Fe_2O_3). The temperature range this tends to occur in is 430°C to 560°C. Because this is a relatively low temperature for these types of systems, this is not typically an issue that occurs on the surface of the refractory lining. It is more likely to affect the deeper layers of the refractory, causing total failure. This issue can be easily avoided by using a refractory with low iron content, typically less than 1.0% Fe_2O_3.

Figure 5, Refractory before (left) and after (right) carbon monoxide test (ASTM #C 288-87), performed at 500°C for 100 hours, with 95% CO.

GENERAL REFRACTORY SELECTION GUIDELINES FOR BIOMASS PROCESSING

A major point to remember in design of vessels for biomass reactions is that the refractories are consumables. In many cases choice of the 'best' refractory involves a number of trade-offs to provide maximum service life. The ultimate purpose of the refractory is to protect the process vessel from high temperature and/or environmental factors that would bring the process out of service.

The physical form of the vessel plays a major control on the physical form of the refractory used. Refractory materials are variously available as bricks and cast shapes, castables, plastics, mortars, and dry vibratables. They may be installed by hand lay-up, casting, gunning, shotcreting, ramming, troweling and other methods. In general, the shape of the refractory and installation method is chosen to provide the fastest and/or more cost effective way to get the vessel on line.

In general terms, the first constraint on refractory selection is maximum service temperature – the highest temperature to which the refractory is likely to be exposed. An analysis of the process should consider whether the temperature is likely to be constant, or will be intermittent. In both cases, the service temperature of the refractory should be sufficient to resist alteration at the highest process temperature.

For biomass processes, the next consideration is process chemistry. Is the process likely to have an environment that is high in alkali, high in CO, or other components that will react with the refractory lining? It is important to remember that multiple issues could be a factor to some degree, and that an ideal situation is rarely present. A simplified example decision chain is presented in Table 3.

Table 3. Simplified Refractory Decision Guide

Alkali environment: Flint-base > Mullite-base > Alumina-base
High temperature (Flint not refractory enough): Mullite + SiC > Alumina > SiC
High oxidation (SiC oxidizes): Mullite + Zr-phase > Alumina + Zr-phase
Hydrogen environment:
Low temperature: Controlled by other process factors
High temperature: Minimize silica in refractory (high alumina systems)
Carbon Monoxide environment: Minimize Fe-containing aggregates in refractory
Thermal cycling: Fused silica > Low alumina > High alumina
Acidic Environment:
Fused silica > Flint > Mullite
Reduce or eliminate cement

SUMMARY

The biomass source and the processing conditions will determine what the best choice of refractory is for the particular application. Often times, several detrimental conditions to the refractory are present in a system, and can have a compounding effect on one another, so many similar refractory products may be tested before optimal conditions are achieved. For this reason, it is extremely important to have good process control, as well as knowledge of the chemical composition of the biomass, in order to choose the right refractory.

WORKS CITED

1. Miles, T. *Alkali Deposits Found in Biomass Power Plants.* s.l. : National Renewable Energy Laboratory, 1996.

2. M. Henek, N. Pavkova. *Development of Castables for Biomass Furnaces.* Prumyslova keramika, Rejec-Jestrebi, Czech Republic : s.n.

3. *The Influence of Biomass Co-Combustion on Boiler Fouling and Efficiency.* Pronobis, M. s.l. : Fuel, 2006, Vol. 86, pp. 474-480.

4. A.P. Green Industries, INC. Destructive Forces on Refractories in a Waste Incineration Environment. [Online] www.banksengineering.com.

5. *Potassium Vapor Attack in Refractories of the Alumina-Silica System.* Luis Scudeller, Elson Longo, Jose Varela. 5, s.l. : Journal of the American Ceramic Society, 1990, Vol. 73, pp. 1413-1416.

6. *Stuctural Determination of Single-Crystal K Beta-Alumina and Cobalt-Doped Beta-Alumina.* Remeika, J.P., Dernier, P.D. s.l. : Journal of Solid State Chemistry, 1976, Vol. 17, pp. 245-253.

7. *A Structural Investigation of Alpha-Al2O3 at 2170K.* Iwai, S.I., Marumo, F. s.l. : Acta Crystallographica B, 1982, Vol. 36, pp. 228-230.

8. Carmen Bartolome, Indelson Ramos, Antonia Gil. *Ash Deposition in Co-Firing using Cynara Biomass Residues with Coal in a PF Pilot Plant.* Zaragoza, Spain : CIRCE (Centre of Research for Energy Resources and Consumption).

9. *Evolution of in Situ Refractories in the 20th Century.* William E. Lee, Robert E. Moore. s.l. : J. Am. Ceram. Soc., 1998, pp. 1385-1410.

COMPARISON OF TWO STAGE MESOPHILIC AND THERMOPHILIC ANAEROBIC
DIGESTION OF OFMSW

Gayathri Ram Mohan[1], Patrick Dube[1], Alex MacFarlene[2], Pratap Pullammanappallil[1]

[1]Department of Agricultural and Biological Engineering
University of Florida
Gainesville, FL 32611-0570

[2] Harvest Power Waltham MA
221 Crescent St., Suite 402
Waltham, MA 02453

ABSTRACT
Concerns about rising energy costs have greatly increased the awareness of renewable fuels and
alternative energy sources on a global scale. Anaerobic digestion of municipal solid wastes has
the advantage of being able to produce a biofuel in the form of biogas, as well as reduce the
volume of incoming waste streams. In a single stage system, biological and chemical reactions
categorized into four stages namely, hydrolysis, acidogenesis, acetogenesis and methanogenesis,
occur within a single reactor. In a two-stage system biogas production is maximized by carrying
out hydrolysis, acidogenesis and acetogenesis in the preliminary solids digester and then
transferring the liquid containing the solubilized and acidified compounds from the solids
digester to a second reactor, an anaerobic filter that is predominant in methanogens, for
methanogenesis. In this paper, the performance of a two-stage system was compared at
mesophilic (37ºC) and thermophilic (55ºC) temperatures. A synthetic source separated waste
mix based on residential waste characteristics that included food waste, yard waste and paper
waste was used for digestion. In the two-stage system, 5L solids digester was sequenced with a
12 L packed bed anaerobic filter reactor. During the course of each run soluble COD, pH, gas
composition and production, and nutrient content were monitored on a daily basis.

INTRODUCTION
    On average, 4.5 pounds of municipal solid waste (MSW) is generated per person per day in
the United States and over 249.9 million tons per year (EPA Factsheet, 2010). Average MSW
composition: paper (28.5%), yard trimmings (13.4%), food scraps (13.9%), wood (6.4%), rubber,
leather and textiles (8.4%), plastics (12.4%), metals (9%) and glass (4.6%)(EPA Factsheet, 2010)
.
    Typically MSW is managed by landfilling, incineration or composting (Daskalopoulos, et al.
1997). Landfills aim to dispose of the waste while at the same time inefficiently capturing
landfill gas (LFG) that is produced when the waste degrades anaerobically (Morrissey and
Browne, 2004). LFG contains predominantly methane and is converted to electricity. LFG can
also be upgraded to renewable natural gas which can be compressed for use as a vehicular fuel or
injected into the natural gas pipeline. Another way to manage disposal of municipal solid waste
is composting where microorganisms turn the organic matter in the waste into compost which
can land applied (Daskalopoulos, et al. 1998).
    However anaerobic digestion offers an advantage over landfilling and composting as a way
to reduce MSW waste while producing a valuable end product (Kiely, et al. 1997). Anaerobic
digestion is a microbial process that converts organic matter into biogas, a mixture of methane

47

(CH_4) and carbon dioxide (CO_2) under anaerobic conditions. Unlike landfills, which are inefficient in capturing methane, a greenhouse gas, anaerobic digestion is carried out in engineered systems under controlled conditions and aims to maximize energy production from the waste. Being a microbial process like composting, anaerobic digestion only converts the organic fraction of MSW, namely the food, paper and yard wastes.

Two types of anaerobic digestion process configurations are frequently employed for OFMSW (organic fraction of municipal solid waste): 1) a high-solid single-stage process (CIWMB, 2008) and 2) a hybrid two-stage process (CIWMB, 2008). In a single-stage system, waste is loaded into a vessel (reactor) and then flushed with inoculum. Digester liquid is recirculated, the solids are degraded and biogas is produced. Although this configuration is cheaper and can be easier to implement, a possible drawback is microbial inhibition caused by a drop in pH due to accumulation of volatile organic acid intermediates during the digestion process. One approach to overcoming this drawback is to operate a two-stage configuration which separates the process into preacidification step followed by methanogenesis in a separate vessel. In the first stage (or reactor), the solids are loaded and then flushed with inoculum rich in a mixed culture of microorganisms that initiates acidogenesis. The digester liquid from the solids digester is then transferred to the second stage (reactor) in which the acidified influent is converted to biogas. Simultaneously, the same amount of liquid from the second reactor is transferred to the solids digester (first reactor) in a process known as sequencing (Chynoweth, 1991). Sequencing is continued until the pH of the anaerobic filter is at neutral 7, at which transferring will be stopped. It is in the second stage that methanogenesis takes place and much of the biogas is generated due to the high concentration and specificity of methanogens (Braber, 1995). The second stage could be an anaerobic filter (reactor) (Polematidis, et al, 2010), upflow anaerobic sludge blanket reactor (Schmidt, 1996), or a packed bed reactor containing residue that has been previously digested (Chynoweth, 1992).

Anaerobic digestion is typically carried out at two different temperature regimes, mesophilic (28-40°C) and thermophilic (50-57°C) (Koppar and Pullammanappallil, 2008). Currently two stage processes are operated at a mesophilic temperature of 37 °C. However, thermophilic digestion has a number of advantages over mesophilic digestion in that rates of degradation and biogasification are much faster and according to EPA 503 regulations, biosolids put through thermophilic digestion process qualify as Class A solids (Gray, et al., 2008). However, in mesophilic digestion, these solids have to be dried further in order to qualify as Class A biosolids. Also, thermophilic digestion leads to higher volatile solids destruction than mesophilic digestion. However, in spite of these advantages, commercial thermophilic digesters are not preferred due to the additional costs associated with special equipment and additional energy requirement to maintain the temperature at 55°C. In addition thermophilic digestion is considered to be inherently unstable tending to accumulate solubilized organic matter. There are some reports that thermophilic digestion can lead greater solids degradation (Kayhanian and Rich 1995)

The aim of this work is to evaluate and compare the performance of mesophilic and thermophilic two stage anaerobic digestion systems for biogasification of organic fraction of municipal solid wastes.

METHODS
Feedstock

A synthetic feedstock was created to provide a consistent results from trial to trial. The synthetic waste mix formulated had twice the amount of yard waste, 1/3 the amount of paper and over 3 times as much food waste as compared to the EPA's MSW characteristics. The higher amount of organic matter in the synthetic waste mix was due to the substitution of non-organics such as rubber, wood, plastics, metals and glass. The mixture was formulated based on analysis of typical OFMSW data and is shown in Table 2. The yard waste portion consisted of 30% of the total weight and came from fresh and dead trees and shrubs from the green waste collected by University of Florida's (UF) Physical Plant Division. The shrubbery was shredded to uniform size before it was added to the mixture. Food waste was formulated based upon an audit performed on UF's Broward dining hall on campus. It consisted of cooked and uncooked vegetables (onion, potato, carrot) and fruit (lemon, orange, grape, watermelon, pineapple, apple), carbohydrate (macaroni and cheese, potato roll, tortilla) and meat (chicken) which were cut to a consistent size and mixed. The remaining part of the solid waste was 10% of paper which was shredded office paper, newspaper and cardboard. There was a final 10% water (weight basis) added to obtain the desired level of moisture observed in MSW. The waste was portioned into 1 kg amounts in freezer bags and stored in a -20 °C chamber until used.

Anaerobic Digester

Two solids digesters each with a total capacity of 5L and working volume of 4L were constructed by modifying pyrex glass bottles. The digesters were built to a height of 16" and an inner diameter of 2.4". A lid, with an outer diameter of 3.8" was constructed using pyrex glass to seal the digesters and provide anaerobic conditions. The lid was clamped to the body with an O-Ring made of rubber which was used to provide a good seal between the digester body and the lid. Three ports were provided at the top of the lid, one for gas outlet and the other two for withdrawing samples. A liquid outlet port was constructed at the bottom of the digester.

An anaerobic filter reactor with a total volume of 18L was constructed using pyrex glass. The working volume was adjustable and can be modified between 6.5 L and 12L. The reactor has a cylindrical body with an inner diameter of 11" and a conical bottom sealed with a cup that is attached to the bottom using a clamp and an O-Ring. A liquid outlet port was constructed at the bottom of the lid. The body of the reactor was wrapped with a heating tape to achieve uniform temperature distribution (55 °C) and reduce temperature fluctuations. 1" thick insulation was provided on the outside of the reactor assembly to prevent any heat loss (Polematidis, Koppar et al. 2010).

Mesophilic and thermophilic digestion for the solids reactor took place in a temperature controlled incubating chamber. The anaerobic filter's temperature was maintained by heating tape surrounded by insulation and controlled by a Campbell Scientific CR-10 measurement and control system. The first run was inoculated with 3 L of inoculum obtained from previous trials that had digested similar wastes and after digestion was complete, the reactor was drained and the inoculum was stored mesophilic and thermophilic conditions, respectively. A similar method was employed for the anaerobic filter inoculum. Subsequent runs were inoculated with the digestate from the previous trial. Prior to start of the trials, the anaerobic filter handling capacity was tested by varying the amount of leachate transferred during the sequencing process. Sequencing between the solids digester and anaerobic filter was carried out by transferring between 0.5 L and 1 L on a daily basis.

A peristaltic pump was used to recirculate the liquid in the anaerobic filter reactor to obtain good mixing but no recirculation was utilized in the solids digester. A thermocouple was used to measure the temperature of the digestate and on wall of the anaerobic filter. A CR10 data logger was used to collect data. A positive displacement gas meter made of a PVC U tube filled with anti freeze solution was used to measure the volume of biogas produced. This set up was made using a solid state time delay, a float switch, a counter and a solenoid valve. The gas meter was calibrated to determine the milliliters of gas required to make a click on the counter. An Orion Bench top pH meter was used to measure the pH.

Experimental Protocol

Bags of synthetic feedstock were removed prior to experiment to thaw. The experiments at a particular temperature were run sequentially using the same digesters. Experiments at the mesophilic and thermophilic temperatures were conducted in parallel.

Analysis

Total Solids (TS) was measured gravimetrically using Standard Methods (APHA, 1992) procedure. A sample of waste mixture was dried overnight in a 105C oven. Percent total solids was calculated from the difference between dry and wet weight measurements. The dried sample was then ashed in a 55 °C furnace for 2-3 hours and the weight was measured to determine the percent volatile solids. The composition of biogas produced from the anaerobic digesters was analyzed using a gas chromatograph (Fisher Gas Practitioner, Model 1200). The gas chromatograph was calibrated using a gas mixture with $N_2:CH_4:CO_2$ in the ratio 25:45:30. Cumulative biogas yield measured over the length of the experiment was multiplied by the methane percent to obtain cumulative methane yield. The methane yield obtained from these calculations was normalized by the kilogram volatile solids loaded in the solids digester and reported at standard temperature and pressure (STP). Soluble chemical oxygen demand (sCOD) analysis was performed on the percolate which is the solids digester effluent samples using a HACH COD kit (150-1500ppm range). Samples were centrifuged and filtered using a 0.45 um Whatman filter paper and diluted to obtain readings in the desired range. Prepared samples were added to the COD vials and treated in a COD digester for 2 hours following which the COD's were measured using a colorimeter (HACH DR/890 Colorimeter).

The performance of the anaerobic digesters was evaluated by fitting the cumulative methane production data to the modified Gompertz equation. The Gompertz equation describes cumulative methane production from batch digesters assuming that methane production is a function of bacterial growth.

The modified Gompertz equation is presented below:

$$M = P \times \exp\{ -\exp[(Rm \times e)/P \times (\lambda - t) + 1] \}.$$

where M is the cumulative methane production, m^3 (kg VS)$^{-1}$ at any time t, P is the methane yield potential, m^3 (kg VS)-1, Rm is the maximum methane production rate, m^3 (kg VS)$^{-1}$ d^{-1}, λ is the duration of lag phase, and t is the time (days) at which cumulative methane production M is calculated. The parameters P, λ and Rm were estimated for each of the data sets by using the 'Solver' feature in MS-Excel. The value of parameters which minimized the sum of the square of errors between fit and experimental data were determined.

RESULTS

Figure 1 shows that the pH profiles in the solids digester were similar for mesophilic and thermophilic temperatures. The pH was continued to be recorded until the end of experiments but is not shown past day 18 for mesophilic trials as it remain constant. Mesophilic trials started out at a pH of 7 and dropped to a low of 5.22 by day 4 due to acidification of the waste. By day 17 Mesophilic Trial 1 was at a pH of 7.2 and by day 13 Mesophilic Trial 2 was at a pH of 7.3 and remained between 7.3 and 7.8 for the rest of the experiments. In Thermophilic Trial 1, the pH was at 8.15 before sequencing was initiated. However, the pH dropped to 4.63, 2 days after the transfer was begun. By day 9, it had increased to approximately 7.36 where it remained thereafter for rest of the run. In Thermophilic Trial 2, the pH took a similar path, beginning at 7.11, dropping to 4.5 after 2 days, then increasing to 7.1 by day 9 and remaining stable.

pH in the anaerobic filter was also monitored (data not shown). During Mesophilic Trial 1, the pH remained around 6.96 throughout the experiment and during Mesophilic Trial 2, the pH remained at about 7.72 for the entire duration of the trial. In the thermophilic trials, the pH fluctuated between 7.2 and 8.61 in Trial 1 and 7 and 7.64 in Trial 2. These pH ranges were appropriate for anaerobic digestion indicating that the anaerobic filter reactors were able to handle the degradation of acidified liquid transferred from the solids digester without being overloaded.

Soluble chemical oxygen demand was measured in both the solids digester and anaerobic filter and the solids digester values are shown in Figure 2. Mesophilic Trial 1 showed a maximum release of sCOD at day 4 of 16,080 mg/L and had dropped to 2,140 mg/L by day 7 in the solids digester. At day 2, Thermophilic Trial 1 released a maximum 19,600 mg/L and had dropped to 2,050 by day 10. Thermophilic Trial 2 had a high of 38,550 mg/L released and was finished by day 9. The anaerobic filters for both trials were steady between 1,000 and 2,000 mg/L throughout the trials.

Once methane was being produced, the biogas in Mesophilic Trial 1 had a maximum methane percent of 49.52% in the solids digester as seen in Figure 3 and between 46-53.3% in the anaerobic filter. In Mesophilic Trial 2, the methane percent was 56.85 % in the solids digester and peaked at 68% before leveling out at 47.2% in the anaerobic filter. Thermophilic Trial 1 had a maximum methane percent of 55.0% in the solids digester and between 56.57 and 67.23% in the anaerobic filter. In Thermophilic Trial 2, the solids digester had a high of 56.77% methane recorded and ranged between 64.11 and 70.66% in the anaerobic filter.

Biochemical methane potential of the waste mixture was determined by running two stage anaerobic digestion systems at mesophilic and thermophilic conditions. Total methane yield, (calculated as the volume of methane produced at STP conditions per kg volatile solids of the waste mix loaded into solids digester) was determined by summing the individual methane yields from the solids digester and the anaerobic filter. Under mesophilic conditions, the solids digester produced 125 and 158 L CH4 at STP (Kg VS)$^{-1}$ in Trials 1 and 2 respectively. While, under thermophilic conditions the solids digester produced 110 and 100 L CH4 at STP (Kg VS)$^{-1}$ in Trials 1 and 2 respectively. The anaerobic filter made 65 and 37 L CH4 STP (Kg VS)$^{-1}$ in Trials 1 and 2 under mesophilic conditions and 55 and 85 L CH4 at STP (Kg VS)$^{-1}$ in Trials 1 and 2 under thermophilic conditions respectively. Figure 4 (ii) shows the total methane yield profiles for mesophilic and thermophilic anaerobic digestion trials. For mesophilic Trials 1 and 2 the total methane yields were 190 and 195 L CH4@ STP (Kg VS)$^{-1}$ and for thermophilic Trials 1 and 2 it was 155 and 195 L CH4 at STP (Kg VS)$^{-1}$ respectively. The average biochemical methane potential of waste mix was previously determined to be ~185 L CH4 at STP (Kg VS)$^{-1}$ based on long term experiments conducted in a single stage digester at optimal conditions of inoculation and nutrient supply. Table 1 shows the duration of time required to reach 90% of the ultimate methane potential in each trial. Figure 4 (i) is a plot of the fraction of total methane yield in the mesophilic and thermophilic digestion trials which shows how quickly digesters reach their 90% of ultimate methane yield. Under thermophilic conditions, 90% of the ultimate methane yield was reached in ~17-20 days, while under mesophilic conditions; it took ~34-38 days. Total solids reduction was measured for both thermophilic and mesophilic conditions and was found to be 42% for thermophilic and 40.5% for mesophilic digestion.

DISCUSSION

Total and Volatile solids were measured to ensure uniformity in feedstock characteristics loaded in the anaerobic digesters. Simulated waste mixtures used as feedstock for the anaerobic digestion trials were characterized with average TS of 35.3% and a VS of 96.2%. Variation in

TS, VS percentages was maintained within ±10%. These average TS, VS percentages were used in the total methane yield calculations from both single and two stage anaerobic digestion trials.

Overall, pH trends for both mesophilic and thermophilic digestions were similar. As seen in Figure 1, in the solids digester an initial drop in pH was observed for all trials as volatile organic acids are being formed in the initial phase of digestion. As methanogenesis starts and sequencing moves solubilized waste from the solids digester to the filter, the pH begins to increase and levels out at 7-8 for both thermophilic and mesophilic digestion. Figure 1 shows that pH gets back to neutral much quicker under thermophilic conditions than mesophilic conditions. With mesophilic digestion, it takes on an average 12 days to reach a pH of 7 while under thermophilic, it only takes 8 days. sCOD release in the solids digester was higher in the thermophilic trials (average 29,075 mg/L) than the mesophilic trials (17,580 mg/L), but the final methane yields were similar meaning lingering organic matter was left remaining in the thermophilic digester.

Maximum methane percent in the thermophilic trials was higher than in mesophilic and reached peaks in the thermophilic solids digester more quickly. Thermophilic conditions promote rapid solids digestion which is evident from higher methane percent values during the digestion process. . A higher methane percent means a higher exothermic output when the gas is burned to be converted to energy, meaning that it is more valuable. Since pH profiles in thermophilic digestion show that pH reaches back to neutral sooner, establishing methanogenesis is favored at a earlier stage than in mesophilic digestion. This aids in reaching higher methane percent n the digesters. Peak values in the solids digester were reached in 11 days under thermophilic digestion and 16 days with mesophilic digestion.

The average biochemical methane potential of the simulated waste mixture was determined to be 185 L CH4 at STP (Kg VS)$^{-1}$. The total methane yield under thermophilic and mesophilic conditions seem to be similar values lying in the range of 190-195 L CH4 at STP Kg VS^{-1}. Thermophilic Trial 1 had a lower yield of 155 L CH4 (Kg VS)$^{-1}$ than the second thermophilic trial and the mesophilic trials carried out on the waste mix. This might have resulted from using inoculum that was not well acclimatized. However, the Thermophilic Trial 2 that was conducted using well acclimatized inoculum from Trial 1 showed results similar to mesophilic digestion. Table 1 shows the total methane yield from two stage anaerobic digestion process along with the percent total methane yield from solids digester and the anaerobic filter. Mesophilic and thermophilic anaerobic digestion runs showed that higher percentage of total methane yield resulted from the solids digester (~60%) while the filter produced the remaining ~40%. Mesophilic Trial 2 also showed a similar trend, except that it produced approximately 80% of the methane yield from the solids digester and remaining 20% coming from the filter. The plot of fraction of total methane produced from the digesters versus time period clearly shows the duration of time required to obtain 90% of the ultimate methane yield. It took about 17-20 days for thermophilic digestion to produce 90% of the ultimate methane yield, while under mesophilic conditions it took twice the time period, ~34-38 days to produce 90% of the methane yield. This shows that thermophilic digestion has much higher methane production rates and can be completed in half the time required to digest the waste mix under mesophilic conditions.

CONCLUSION

This study has shown that thermophilic digestion could be favored for the biogasification of municipal solid waste. When compared to mesophilic digestion, similar yields with regard to pH, sCOD, percent and total methane yield ~185 L CH4 at STP Kg VS-1 are achieved. But, mesophilic digestion takes about twice as long (34-28 days) as thermophilic digestion (17-20 days) to achieve 90% of the ultimate methane yield. The result of being able to complete digestion in half the time means doubling the amount of waste that can be collected and processed over the year in an existing facility.

Based on our results, we can conclude that, if GICON switches its operation from mesophilic to thermophilic digestion a number of added benefits will be obtained including:

1. Treating more waste results in a greater amount of methane gas produced, up to double the amount when compared to mesophilic digestion within the given time frame of one complete mesophilic digestion trial which would help double the revenue obtained from shipping out waste from client sites to the digester facility.

2. Thermophilic digestion produces biogas with a higher percentage of methane, hence it has a higher heating value when used as a fuel to produce energy.

3. No negative issues with equipment were observed during the trials. Facilities operating at mesophilic conditions should be able to switch to thermophilic conditions without adverse side effects assuming equipments are made of compatible material for operation in either temperature regimes .

4. Class A biosolids are produced as a result of thermophilic digestion as defined by EPA 503 regulations, thus eliminating the need for added costs associated with post-processing to inactivate the pathogens in the waste mixture that would be required when using mesophilic digestion.

5. Two stage digestion presented no issues in regard to pH control. Sequencing between the reactors prevented inhibition due to low pH in the solids digester and the anaerobic filter was able to handle the incoming sCOD loads that were transferred on an everyday basis, thus eliminating the need for addition of caustic soda for pH control.

ACKNOWLEDGEMENTS

The authors wish to acknowledge the financial assistance provided by Harvest Power, Waltham, MA for this research project.

Table 1: Results

	Total Methane Yield at STP (L CH4/kg VS)	Yield from Solids Digester (% Total Yield)	Yield from Anaerobic Filter (% Total Yield)	Time to reach 90% Total Methane Yield (days)
Mesophilic Trial 1	190.0	63.1	36.9	38
Mesophilic Trial 2	202.5	81.0	19.0	34
Thermophilic Trial 1	170.3	67.0	33.0	20
Thermophilic Trial 2	180.8	61.7	38.3	17

Table 2 Feedstock

Component	Amount (g)
Uncooked Vegetables/Fruit	1500
Meat	150
Carbohydrate	750
Cooked Vegetables/Fruit	600
Yard Waste	1800
Recycled Paper	150
Office Paper	150
Cardboard	300
Water	600
Total	6000

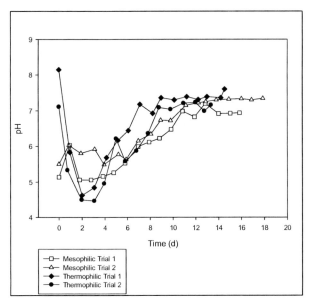

Figure 1: pH in Solids

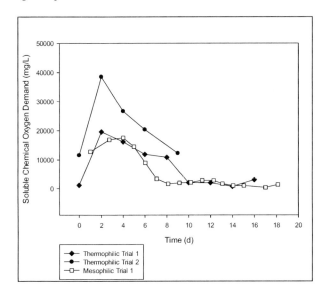

Figure 2: Soluble Chemical Oxygen Demand in Solids

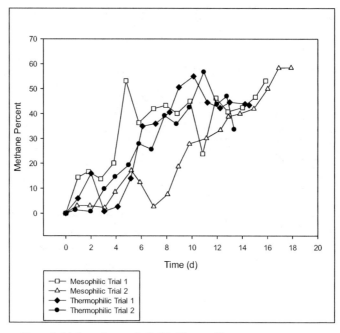

Figure 3: Methane Percent in Solids Digester Biogas

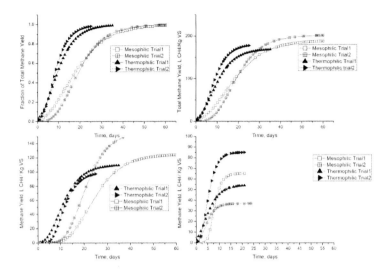

Figure 4 (i) Fraction of Methane produced from Anaerobic Digesters, (ii) Total Methane yield from two stage Anaerobic Digestion, (iii) Methane yield from Solids Digester, (iv) Methane yield from Anaerobic Filter

Table 2. Average of Gompertz Parameter values used to fit the experimental data plotted in Figure 4

	P Ultimate Methane Production (L CH4/Kg VS)	λ Lag (Days)	Rm Maximum Methane rate (L CH4/Kg VS/Day)
Thermophilic Solids Digester	$105.3 \pm 5\%$	5.07	8.35
Thermophilic Filter	$69.8 \pm 20\%$	0.97	8.99
Mesophilic Solids Digester	$141.5 \pm 11\%$	12.7	7.36
Mesophilic Filter	$51.2 \pm 24\%$	1.23	8.49

REFERENCES

American Public Health Association (APHA), American Water Works Association (AWWA), & Water Environment Federation (WEF). (1992). *Standard Methods for the Examination of Water and Wastewater: 18th Edition.* Washington, DC: American Public Health Association.

Braber, K. (1995). Anaerobic digestion of municipal solid waste: A modern waste disposal option on the verge of breakthrough. *Biomass & bioenergy, 9*(1-5), 365-376.

California Integrated Waste Management Board (CIWMB). 2008. Current anaerobic digestion technologies used for treatment of municipal organic solid waste.

Chynoweth, D. P. (1991). A novel process for anaerobic composting of municipal solid waste. *Applied biochemistry and biotechnology, 28-29*(1), 421-432.

Chynoweth, D. P., Owens, J., Okeefe, D., & Earle, J. F. K. (1992). Sequential Batch Anaerobic Composting of the Organic Fraction of Municipal Solid-Waste. *Water science and technology, 25*(7), 327-339.

Daskalopoulos, E., Badr, O., & Probert, S. D. (1997). Economic and Environmental Evaluations of Waste Treatment and Disposal Technologies for Municipal Solid Waste. *Applied energy, 58*(4), 209-255.

Daskalopoulos, E., Badr, O., & Probert, S. D. (1998). An integrated approach to municipal solid waste management.*Resources, conservation and recycling, 24*(1), 33-50.

Gray (Gabb), D.M.D., Suto, P.J., and Chien, M.H. (2008) Green Energy from Food Wastes at Wastewater Treatment Plant. *BioCycle, 49(1),* 53-58.

Environmental Protection Agency. 2010. Municipal solid waste generation, recycling, and disposal in the United States: facts and figures for 2010.

Kayhanian, M. and D. Rich (1995). Pilot-scale high solids thermophilic anaerobic digestion of municipal solid waste with an emphasis on nutrient requirements. Biomass & Bioenergy 8(6): 433-444.

Kiely, G., Tayfur, G., Dolan, C., & Tanji, K. (1997). Physical and mathematical modelling of anaerobic digestion of organic wastes. *Water research (Oxford), 31*(3), 534-540.

Koppar, A., & Pullammanappallil, P. (2008). Single-stage, batch, leach-bed, thermophilic anaerobic digestion of spent sugar beet pulp. *Bioresource technology, 99*(8), 2831-2839.

Morrissey, A. J., & Browne, J. (2004). Waste management models and their application to sustainable waste management. *Waste management (Elmsford), 24*(3), 297-308.

Polematidis, I., Koppar, A., & Pullammanappallil, P. (2010). Biogasification potential of desugarized molasses from sugarbeet processing plants. Journal of Sugar Beet Research, 47(3/4), 89-104. doi: 10.5274/jsbr.47.3.89

Schmidt, J. E. (1996). Granular sludge formation in upflow anaerobic sludge blanket (UASB) reactors. *Biotechnology and bioengineering, 49*(3), 229-246.

BIOGASIFICATION OF MARINE ALGAE *NANNOCHLOROPSIS OCULATA*

Samriddhi Buxy[1], Robert Diltz[1,2], Pratap Pullammanappallil[1]

[1] Agricultural and Biological Engineering Department, University of Florida
[2] Air Force Research Laboratories, Tyndall Air Force Base, Florida

ABSTRACT
    Bio Methane potential of marine microalgae *Nannochloropsis oculata* is determined in a 5-L batch digestion setup. *N. oculata* is grown in an open raceway pond at 25 degree C, with 1% CO2 and 99% air supplied. *N. oculata* was grown for 2-3 weeks to final concentration of 600-800 mg/L, and then harvested in a 30-gallon batch by adding base and concentrating algae to 3.15% volatile solids. The average Bio Methane potential of *N. oculata* is determined as 204.3 LCH$_4$/kg VS with the total digestion time of 52-66 days. Anaerobic microbes adapts to the saline conditions, displaying reduced lag for methane production with inoculum reuse.
    Anaerobic digestion eliminates the need of dewatering, extraction or economical separation, which is required in other biofuel production from algae. *N. oculata* is not rich in oil but contains predominantly cellulose and other carbohydrates, which makes it a good feedstock for anaerobic digestion. *N. oculata* grown in an open raceway pond, harvested and then digested anaerobically as presented in following study has the potential to supply energy need (electric or natural gas) of one household with utilization of 1.83 acre of land.

INTRODUCTION

    With the gaining interest in biofuels, algae are looked upon as a potential biomass resource for biofuels production. According to the International Energy Agency (IEA), by the year 2020, the source for a quarter of biofuels produced will be algae[1]. Algae are autotrophic microorganisms utilizing a non-fossil carbon source for growth by fixing carbon dioxide from atmosphere[2-3]. Use of algae eliminates the need for terrestrial energy crops and hence will not impact availability of land for food and feed. Algae have higher growth rates, about ten times more productivity than other terrestrial crops[4]. Several species of algae can be grown in saline or brackish water obviating the need for precious fresh water resources. Recent developments in photo-bioreactor designs have enabled algae to be grown at much higher rates in compact areas[5], with high composition of lipids and carbohydrates.
    Algae are ideal feedstocks for biofuel production due to its rich composition of lipids and carbohydrates. Ethanol can be obtained from carbohydrates by fermentation[6]. Lipids and triglycerides can be utilized to produce biodiesel and alkanes[7]. Algae biomass can be directly combusted for energy or thermochemically gasified to produce syngas[8]. Further research has shown the viability of genetic modification of algae to secrete biofuels like ethanol[9], long-chain hydrocarbons[10] and hydrogen[11] in appreciable quantities.
    Biogasification (or anaerobic digestion) is a biochemical process that converts organic matter to biogas (a mixture of methane, 50-70%, and carbon dioxide) under anaerobic conditions. The biogas can be used as a replacement for natural gas or it can be converted to electricity. The process is mediated by a mixed undefined culture of microorganisms at near ambient conditions. The advantages are that algal slurries can be processed without dewatering and that sterile conditions need not be maintained for operating the fermenter. The anaerobic digestion process will also mineralize the organic nitrogen and phosphorous, and these nutrients can be recycled for algae growth[12]

Aquatic biomass – macrophytes[13-14], micro and macro algae, have all been tested as feedstock for biogasification. Previous studies have shown that macro algae like *Ulva lactuca, Gracillaria vermiculophylla, Saccharina latissima* etc. can be anaerobically digested producing methane at yields ranging from 0.1-0.3 LCH$_4$/g VS[15]. Methane yields of microalgae like *Spirulina* (fresh water), *Dunaliella* (saline water), and *Chlorella vulgaris* (fresh water) ranged from 0.3-0.4 L CH$_4$/g VS[16-18], whereas other micro algae like *Tetraselmis sp* (marine), *Spirulina maxima* (fresh water), mixed culture of *Chlorella* and *Scenedesmus* (fresh water) produced a lower methane yield ranging from 0.09-0.136 L CH$_4$/g VS[19-20] when these were codigested with other feedstocks.

In the present study, the methane potential of the saline microalgae *Nannochloropsis oculata* is assessed. *N. oculata* is primarily made up of carbohydrates (all polysaccharides with a 60% glucose content), proteins, amino acids, fatty acids and unsaturated alcohols[21]. Since this alga is not known for accumulating lipids, methane potential of *N. oculata* is seen as a standard for any oil-free biomass. The marine algae species were grown in open raceways. Harvested algal cells were anaerobically digested in laboratory scale digesters to determine the extent of degradation, rate of degradation, methane production potential. The study throws light on the overall steps that are required to set up a biogasification system from algae and the potential of biogasification of *N. oculata* to support energy requirements for a household.

MATERIALS AND METHODS

Algae growth and harvesting

Nannochloropsis oculata Culture Utex LB2164 (UTEX Culture Collection, Austin, TX) was grown at Tyndall Air Force Base, Florida at the Air Force Research Laboratory, Deployed Energy Research facility (AFRL/RXQE). 10 ml of *N. oculata* sample was diluted weekly to a final volume of 14 L and grown indoors at a constant temperature of 25 °C. A 125 W Compace Fluorescent Bulb grow light was turned on for 14 hours daily for the growth culture. When the algae reached a maximum concentration of about 800-1000 mg/L in 14 L vessels (which took about a month) the contents were transferred to a raceway pond in an outdoor greenhouse. The raceway pond had a working volume of 0.19 m^3 (50 gallons) and a base area of 0.63 m^2 (6.78 sq ft). The depth of the water in the raceway was 0.3 m (~1 ft). In the raceway *N. oculata* was grown using natural light at ambient temperatures. A 1% CO$_2$/99 % air mixture was sparged for 12 hours daily and the contents of pond were kept circulating using a Maxi-jet 1200 aquarium pump. Culturing time from initial inoculation to full growth is between 2-3 weeks, depending on seasonal conditions. Final concentrations were around 600-800 mg/L. Algae were harvested in 30 gallon batches, with the remaining suspension being refilled with seawater to a volume of 50 gallons. Harvesting was done by transferring 30 gallons of algae from the tank into a settling hopper and adjusting the pH to a value of 10 using sodium hydroxide and allowing the suspension to settle overnight. The supernatant liquid was decanted and settled algae drained into a collection vessel. The settled algae were filtered using a cheese cloth and the resulting paste was shipped overnight in coolers to the Agricultural and Biological Engineering Department, University of Florida. Upon receipt, the paste was stored at 5 °C. The paste was apportioned into two 900 mL aliquots. The first was labeled 'Paste 1' and the second 'Paste 2' and were used as feedstock for anaerobic digestion.

Anaerobic digestion

A 5 L anaerobic digester was constructed by modifying a Pyrex glass jar. The digester was provided with ports for venting and sampling biogas, and for liquid withdrawal. The biogas production from the digester was measured using a positive displacement gas meter. Details of the set up can be found in a previous study[22]. The digester contents were stirred using a magnetic stirrer at 300 rpm. The digester was placed in a chamber maintained at 55 °C.

Two experiments were conducted using Paste 1 and 2 respectively. In experiment 1 900 mL of paste was mixed with 3.1 L of thermophilic digester inocula. The inoculum was mixed liquor collected from a thermophilic digester that was being operated on a sugarbeet residuals feedsctock for over two years. When biogas production ceased at the end of experiment 1, the digester was opened, 900 mL of mixed liquor removed and 900 mL of Paste 2 was added. During the digestion process daily volumetric biogas production was monitored as well as the biogas composition. Liquid samples were collected periodically and analyzed for pH and soluble chemical oxygen demand (sCOD).

Analysis

The algae paste was analyzed for dry matter, volatile solids and ash content. Dry matter and ash analysis is done by conventional standard method of drying the solids[23]. Total solids (TS) were determined after drying the wet sample overnight at 105°C. The dried sample was burned at 550°C in a muffle furnace for 2 h to determine the Volatile solids (VS) content and the ash-free dry weight.

Gas composition (CH_4, CO_2) was measured using a gas chromatograph (model 1200 gas partitioner, Fisher Scientific, Inc.) equipped with a thermal conductivity detector. The gas chromatograph was calibrated with an external standard containing N_2, CH_4, and CO_2 in 25:45:30 volume ratios. Leachate samples were analyzed for pH, soluble chemical oxygen demand (sCOD). sCOD of leachate was measured by colorimetric method by HACH COD measuring kit after filtering the samples through a 0.22 micron filter paper.

RESULTS AND DISCUSSION

Dry matter content of the paste was 7% (w/w) of which only 45% was volatile (the rest 55% being ash). The high ash content was due to two reasons; 1) salts from the seawater used for growth of the algae and 2) addition of sodium hydroxide for settling the algae. The seawater alone would have contributed to 46.5 % ash content (assuming a salinity of 3.5%). The volatile solids content is an indication of organic matter; hence the degradable matter in the feedstock was only 3.15 % (w/w) of the mass harvested. Table 1 lists the volatile solids content of biomass that has been used as feedstocks for biogasification. Aquatic biomass has a range of 5.1 to 11.2% (w/w) [24] degradable matter; sargassum shows higher volatile solids content of 26% (w/w) [24]. Range of degradable matter in plant biomass feedstock varies from as low as 8.4% w/w (Napier grass) [24] to 89.9% w/w (Poplar) [24], but contains high lignin content, which is not desirable for methane production. Biomass from wastes like sugarbeet tailings[25], pressed spent sugar beet pulp[26], Municipal solid waste[27] etc. also are potential feedstock for biogasification as the volatiles range from 12.75-26.25% (w/w). Therefore, it should be noted that the degradable matter in the *N. oculata* paste is lower than biomass feedstocks but it is higher than organic matter in numerous wastewater effluent (0.5 – 2 % w/w) that have been anaerobically digested.

The cumulative methane yield (L/kg VS) from the experiments is shown in Fig 1. A lag of 8 days in methane production can be seen initially during the digestion of Paste 1. This is the adaptation period of inoculum probably for the new feedstock, and since the mixed liquor was reused for anaerobic digestion of Paste 2 there was no appreciable lag phase before initiation of methane production. The methane yield from the digestion of Paste 1 was observed to be 204 L/kg VS and was obtained in 52 days and from Paste 2 the methane yield was 204.6 L/kg VS which was obtained in 66 days. The total time taken for complete anaerobic digestion of *N. oculata* appeared to fall within a range of 52-66 days. Methane production exhibited a bi-phasic profile. During digestion of paste 1 after the lag phase methane production rate (as measured by the slope of the cumulative methane plot) increased remaining more or less constant until day 32 after which it decreased appreciably until day 46. A similar slow methane production rate period beginning on day 28 was seen during digestion of Paste 2 as well. It is possible that from day 28 or 32 onwards slowly degradable portions of algae are beginning to be degraded.

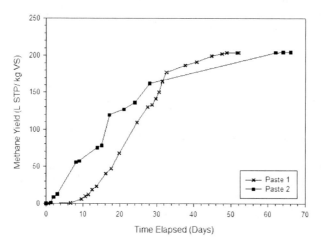

Figure 1. Bio Methane Potential of N.oculata

Soluble Chemical Oxygen Demand (sCOD) measures the amount of dissolved organic matter in the digester mixed liquor. Mineralization of organic matter to biogas proceeds through a sequence of steps in an anaerobic digester. The first step is enzymatic hydrolysis (or solubilization) of particulates or macromolecules followed by acidification and then methanogenesis. sCOD is a lumped measurement of solubilized monomers and organic acids. sCOD concentration is shown in Figure 2. The inoculum used for digestion contained some dissolved organic matter as seen by the initial sCOD concentration of 4 – 5 g/L. The sCOD rapidly increases when the paste is incubated peaking at 24.6 g/l during digestion of Paste1 by day 13 and by day 14 during digestion of paste 2. Since sCOD increased quickly it appears that the inoculum was able to solubilize and perhaps acidify the feedstock. The lag time seen during digestion of paste 1 was due to the delay in degrading the solubilized components. This could be an adaptation of methanogens to saline conditions. Usually, saline conditions are inhibitory for the mixed microbes culture population of the anaerobic digester but they can be adapted to the saline conditions. Unadapted anaerobic digester microbial cultures can tolerate concentration upto 15 g/L (chlorine) without exhibiting toxicity effects[28]. The salt adaptability can be achieved by gradual feeding of the digester with salt in a batch process or at increased hydraulic retention time (HRT)[28]. No study has been found in the previous literature about the mixed anaerobic microbial culture adaptability to saline microalgae culture.

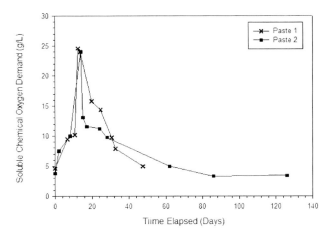

Figure 2. sCOD from the digestion of N.oculata

After peaking, the sCOD concentration continued to decrease as the solubilized components were fermented to biogas. By day 30 the sCOD concentration was 10 g/L after which the remaining sCOD was slowly degraded which is in agreement with the slow methane production rate observed after around day 30 in both experiments. The residual sCOD at the end of digestion of paste 1 and 2 was around 3-4 g/ L, similar to that in the inoculum.

These experiments indicated that it is feasible to anaerobically digest a saline microalgae species. Since mixed microbial culture of anaerobic digestion is adaptable to the saline conditions, the need for fresh water microalgae species can be avoided. The hydrolysis, solubilization and acidification of marine microalgae were independent of the adaptation time of the methanogenic microorganisms. This can facilitate a two-stage continuous digester operation where the solubilization and acidification process is separated from methanogenesis. This type of a digester design is commonly employed commercially for biogasification of terrestrial biomass and wastes. The solubilized and acidified mixed liquor from the first stage is fed to a second reactor where methane is produced. Various species of algae have differences in morphological structure, encapsulation behavior and cell wall rigidity[29]. Therefore different species exhibit different methane yields when anaerobically digested. The methane yield obtained from anaerobic digestion of *N. oculata* here was lower than that obtained from anaerobic digestion of *Spirulina, Dunaliella,* and *Chlorella vulgaris*[16-18], but higher than methane yields from *Tetraselmis sp* and mixed culture of *Chlorella* and *scenedesmus*[19-20]. Table 1 lists the methane yields of other biomass feedstocks. The methane yield from pure cellulose is also tabulated to serve as a benchmark for theoretical methane yield if the biomass predominantly contains carbohydrates. In comparison to cellulose, *N. oculata* produced about 60% of the theoretical methane yield. Other aquatic biomass like kelp, sargassum has relatively higher yield of about 75% of theoretical methane yield[24]. This could be due to the difference in cellular structure. Pressed spent sugar beet pulp, sugarbeet tailings, MSW has a low volatile solids content but due to high sugar content, it makes a good feedstock for biogas production[25-27]. Different pretreatment methods like sonication and acid hydrolysis[30] can also be employed to increase the methane yield.

Table 1. Organic matter and methane potential of some feedstocks used for biogasification

Feedstock	Volatile solids* (% total weight)	Methane Yield (L@STP Kg^{-1} Volatile Solids)	Reference
Kelp (*Macrocystis*)	7.74	390 - 410	Chynoweth 2002
Kelp (*Laminaria*)	11.20	260 - 280	Chynoweth 2002
Sargassum	26.00	260 - 380	Chynoweth 2002
Sorghum	32.55	260 - 390	Chynoweth 2002
Napiergrass	8.43	190 - 340	Chynoweth 2002
Poplar	89.90	230 – 320	Chynoweth 2002
Water hyacinth	5.12	190 - 320	Chynoweth 2002
Sugarcane	41.28	230 - 300	Chynoweth 2002
Willow	24.12	130 - 300	Chynoweth 2002
Municipal solid waste	26.25	200 – 220	Chynoweth, et.al., 1993
Sugar beet tailings	12.75	275 – 320	Liu et.al., 2008
Pressed spent sugar beet pulp	21.5	302 – 374	Koppar and Pullammanappallil, 2007
Cellulose powder (Avicel)	94.7	370	Cho and Park, 1995

Currently ethanol and biodiesel dominate the US biofuel market[31]. Algae are primarily cultivated for its ability to secrete or accumulate lipids. The lipids are extracted and converted to biodiesel. *N. oculata* is not rich in oil but contains predominantly cellulose and other carbohydrates. This species can be utilized for bioenergy via anaerobic digestion, combustion or gasification processes. Biodiesel production and other thermochemical processes demand drying and/or oil extraction, which are energy intensive unit operations and require large capital investment. But anaerobic digestion does not require drying of feedstock, which can be processed in a slurry form. So the net energy through anaerobic digestion would be higher than that obtained by converting algae lipids into biodiesel or thermochemically processing algal biomass by combustion or gasification.

It was envisaged that an *N. oculata* growth and harvesting system could be coupled with an anaerobic digester to produce biofuel in the form of biogas. The algae would be grown using seawater thereby avoiding use of freshwater resources. The biogas can be used directly as a substitute for natural gas of converted to electricity. Based on *N. oculata* growth observed in the pilot raceways and the methane yield from digestion of this alga, an analysis was carried out to estimate energy production and land requirements. Currently the algae harvesting rate from the raceways are 9.64 g afdw/m^2/d. Note that afdw (ash free dry weight) is the same as volatile solids content. An often cited study for algae growth has yielded a much higher productivity of 50 g afdw/m^2/d for *Platyomonas sp*[32]. The algae biomass yield obtained in this study was only about 20% of the productivity attainable. Optimization of growth conditions for *N. oculata* may improve its productivity. Using the methane yield value of 204 L/kg VS for anaerobic digestion of *N. oculata*, the annual energy output from a facility that grows the algae and subsequently digests it would be 27 MJ/m^2/year. The area occupied (or footprint) of the digester(s) would be far less than the land area required for growing the algae. If the methane produced from this facility is converted to electricity, the electrical energy output would be 2.25 KWH$_e$/m^2/year assuming that the efficiency of converting thermal energy to electrical energy is 30%. The household electrical energy and natural gas consumption in the city of Gainesville, FL for the year 2011 was 12,100 KWH/year and 1,344.2 m^3/year respectively[33]. If the algae biogasification facility were to supply the entire electrical energy requirements for a household, the land area required would be 5377 m^2 (=1.33 acres). If in addition, the facility were to supply the natural gas needs, then an additional 1880 m^2 (=0.5 acres) would be needed. In other words 1.83 acres of land could supply all the energy needs of a household in Florida. If the algae productivities were improved then land requirement would be further reduced. At 50 g afdw/m^2/d algae productivity, the land requirement would only be about 0.4 acres.

CONCLUSIONS

The marine algae species, *N. oculata* was grown in pilot scale raceways producing 9.6 g afdw/m^2/year. Settled *N. oculata* was successfully biogasified as a sole feedstock in an anaerobic digester without any pretreatment or further processing. The methane yield from anaerobic digestion was 204 L STP/kg VS. The proposed system of growing, harvesting and anaerobically digesting *N. oculata* has the potential to produce 27 MJ of thermal energy/m^2/year or 2.25 KWH$_e$/m^2/year. A land area of 1.83 acres for growing the algae in open ponds could supply the electricity and natural gas needs of a typical household in Florida. Land requirements would be much lower if the algae productivity can be increased by optimization of its growth conditions.

* Volatile solids content as % weight of feedstock harvested or collected

REFERENCES
1. Available at http://www.iea.org/
2. L. Rodolfi, G.C. Zittelli, N. Bassi, G. Padovani, N. Biondi, G. Bonini and M.R. Tredici, "Microalgae for oil: Strain selection, induction of lipid synthesis and outdoor mass cultivation in a low-cost photobioreactor", Biotechnology and Bioengineering, 102[1] 100-112 (2009)
3. B. Wang, Y. Li, N. Wu and C. Q. La, "CO2 bio-mitigation using microalga", Appl. Microbiology and Biotechnology, 79[5] 707-718 (2008)
4. Available at http://www.globalccsinstitute.com/publications/accelerating-uptake-ccs-industrial-use-captured-carbon-dioxide/online/28516
5. E. Molina, J. Ferna´ ndez, F.G. Acie´n and Y. Chisti, "Tubular photobioreactor design for algal culture", Journal of Biotechnology, 92[2] 113-131 (2001)
6. S. M. Lee, J. H. Lee, "Ethanol fermentation for main sugar components of brown-algae using various yeasts", Journal of Industrial and Engineering Chemistry, 18[1] 16-18 (2012)

7. Q. Hu, M. Sommerfeld, E. Jarvis, M. Ghirardi, M. Posewitz, M. Seibert and A. Darzins, "Microalgal triacylglycerols as feedstocks for biofuel production: perspectives and advances", The plant Journal, 54[4] 621-663 (2008)
8. S. Wegeberg, Algae biomass for Bioenergy in Denmark, Available at http://www.bio4bio.dk/~/media/Bio4bio/publications/Review_of_algae_biomass_for_ene rgy_SW_CF_April2010.ashx
9. Available at http://www.algenolbiofuels.com/
10. R. Radakovits, R.E. Jinkerson, A. Darzins and M.C. Posewitz, "Genetic Engineering of Algae for Enhanced Biofuel Production", Eukaryotic Cell, 9[4] 486-501 (2010)
11. K. Srirangan, M. E. Pyne and C. P. Chou, "Biochemical and genetic engineering strategies to enhance hydrogen production in photosynthetic algae and cyanobacteria", Bioresource Technology, 102[18] 8589-8604 (2010)
12. B. Sialve, N.Bernet and O. Bernard, "Anaerobic digestion of microalgae as a necessary step to make microalgal biodiesel sustainable", Biotechnology Advances, 27[4] 409-416 (2009)
13. V. Singhal and J.P.N Rai, "Biogas production from water hyacinth and channel grass used for phytoremediation of industrial effluents", Bioresource Technology, 86[3] 221-225 (2003)
14. D.P. Chynoweth, D.A. Dolenc, S. Ghosh, M.P. Henry, D.E. Jerger and V.J. Srivastava, "Kinetics and advanced digester design for anaerobic digestion of water-hyacinth eichhornia-crassipes and primary sludge", Proceedings of the fourth symposium on biotechnology in energy production and conservation, Gatlinburg, Tennessee, May 11-14 1982
15. H.B. Nielsen and S. Heiske, "Anaerobic digestion of macroalgae: methane potentials, pre-treatment, inhibition and co-digestion", Water science and technology 64[8] 1723 (2011)
16. P.H. Chen, "Factors influencing methane fermentation of micro-algae", PhD thesis, University of California, Berkeley, CA, USA (1987)
17. E.P. Sánchez Hernández and L. Travieso Córdoba, "Anaerobic digestion of *Chlorella vulgaris* for energy production", Resources, Conservation and Recycling, 9[1-2] 127-132 (1993)
18. C.M. Asinari Di San Marzano, A. Legros, H.P. Naveau and E.J. Nyns, " Biomethanation of the marine algae Tetraselmis", International Journal of Sustainable Energy, 1 263-272 (1982)
19. C.G. Golueke, W.J. Oswald and H.B. Gotaas, "Anaerobic digestion of algae", Applied Microbiology, 5[1] 47-55 (1957)
20. H.W. Yen and D.E. Brune, "Anaerobic co-digestion of algal sludge and waste paper to produce methane", Bioresource Technology, 98[1] 130-134 (2007)
21. M.R. Brown, "The amino-acid and sugar composition of 16 species of microalgae used in mariculture", Journal of Experimental Marine Biology and Ecology, 145[1] 79-99 (1991)
22. A.Koppar and P.Pullammanappallil, "Single-stage, batch, leach-bed, thermophilic anaerobic digestion of spent sugar beet pulp", Bioresource Technology, 99[8] 2831-2839 (2008)
23. A.E. Greenberg, L. S. Clescerl, and A. D. Eaton, "StandardMethods for the Examination of Water and Wastewater", Washington. D.C.: American Public Health Association, 18th edition, (1992)
24. D.P. Chynoweth, "Review of biomethane from marine biomass", A report prepared for Tokyo Gas Company, 2002.

25. W. Liu, P.C. Pullammanappallil, D.P. Chynoweth and A.A. Teixeira, "Thermophillic Anaerobic Digestion of Sugar Beet Tailings", Transactions of the ASABE, 51[2] 615-621 (2008)
26. I. Polematidis, A. Koppar and P. Pullammanappallil, " Biogasification potential of desugarized molasses from sugarbeet processing plants", Journal of Sugar Beet Research, 47[3/4] 89-104 (2010)

27. D.P. Chynoweth, C.E. Turick, J.M. Owens, D.E. Jerger and M.W. Peck, "Biochemical methane potential of biomass and waste feedstocks", Biomass and Bioenergy, 5[1] 95-111 (1993)
28. R. Mendez, F. Omil, M. Soto and J.M. Lema, "Pilot plant studies on the anaerobic treatment of different wastewaters from a fish-canning factory", Water Science and Technology, 25[1] 37-44 (1992)
29. J. C. Rooke, C. Meunier, A. Léonard and B.L. Su, "Energy from photobioreactors: Bioencapsulation of photosynthetically active molecules, organelles, and whole cells within biologically inert matrice", Pure Appl. Chem., 80[11] 2345–2376 (2008)
30. K.Sander and G.S. Murthy, "Enzymatic Degradation of Microalgal Cell Walls", Presentation at ASABE Annual International Meeting, June 21-June 24, 2009
31. Available at < http://www.emerging-markets.com/PDF/Biodiesel2020Study.pdf

32. J. Sheehan, T. Dunahay, J. Benemann and P. Roessle, "A Look Back at the U.S. Department of Energy's Aquatic Species Program—Biodiesel from Algae", National Renewable Energy Laboratory report, July 1998
33. Available at <https://www.gru.com/>

A PRELIMINARY STUDY OF AN INNOVATIVE BIOMASS WASTE AEROBIC DEGRADATION SYSTEM FOR HOT WATER HEATING

Haorong Li and Daihong Yu
University of Nebraska-Lincoln,
Omaha, NE, USA

Yanshun Yu
Nanjing University of Science & Technology
Nanjing, Jiangsu, China

ABSTRACT

As energy prices, environmental pollution, and waste generation rates continue to rise, the need for green renewable energy and alternative waste treatment method is becoming critical. The U.S. alone produces over 250 million tons of municipal solid waste per year, two-thirds of which is organic waste. However, most of this valuable energy resource is directly disposed through landfills and incineration, which causes enormous environmental damage. Only 7% is naturally decomposed to minimize environmental effects and produce fertilizer. Traditional composting technology is too slow to be adopted prevalently. In addition, massive green heat generated from bio-degradation is neglected and fully discharged to the atmosphere. In this study, a prototype biomass waste aerobic degradation system for hot water heating is developed to reclaim free energy in a fast and effective manner. Preliminary laboratory tests using various biomass wastes demonstrate that, on average, a sustainable net heat generation rate of about 14.6 W/kg of biomass waste on a wet basis can be produced. It is estimated that adopting such an aerobic degradation system only with household biomass waste can adequately support daily domestic hot water needs in the U.S. The study concludes that this cost-effective system could produce significant ecological, environmental, economic and social benefits.

INTRODUCTION

As energy prices, environmental pollution, and waste generation rates continue to rise worldwide, so will the desperate need for green renewable energy and alternative waste treatment solutions. In particular, as estimated by the U.S. Environmental Protection Agency (EPA) and Department of Energy (DOE), with less than 5% of the world's population, the U.S. ranks as the world's largest producer of solid waste[1], and consumes 22% of the world's primary energy[2]. In 2009, more than 250 million tons of municipal solid waste (MSW) was generated[1], over two-thirds of which was organic material. Organic MSW composed of food scraps, yard trimmings, wood waste, and paper and paperboard products is known as free and valuable energy resources. However, only 7% is composted to produce fertilizers[3]. The majority part of organic MSW in the U.S. is dumped in landfills or incinerated directly every year. Sustainable management of biomass waste as a renewable energy source becomes an important issue in the U.S.

In general, the state-of-the-art methods of biomass MSW treatment include landfills, incineration, and composting. However, these technologies either have serious adverse effects to the environment, agriculture and/or human health, or have not recycled the useful materials, or have not reclaimed the renewable energy.

As the current primary waste disposal method worldwide, landfills (in Figure 1) buried up to 54.3% of biomass MSW in 2009 in the U.S.[1]. According to a study conducted by Leak Location Services, Inc. in 2000[4], 82% of surveyed landfill cells had leaks while 41% had a leak area of more than 1 square foot. Agricultural and ecological problems (e.g., soil compaction and erosion, increased irrigation, planting, and re-vegetation requirements, reduced flora and fauna health and vigor), and consequent human health risks (e.g., cancer, birth defects, genetic mutations) are potentially serious due to heavily polluted leachate with massive, harmful, organic and inorganic compounds. Globally, the rotting waste in landfills continuously emits a significant source of methane, which is a potent greenhouse gas with 21 times the global warming potential of carbon dioxide[5]. Moreover, the U.S. had 3,536 active landfills and over 10,000 old municipal landfills in 1995[6]. Although these landfills occupy only a small percentage of the total land in the U.S., public concern over possible ground water contamination and odors from landfills makes finding new sites difficult. In many areas worldwide, landfill space is running out, and a landfill shortage crisis is looming within the next 10 years.

As estimated by the U.S. EPA, over one-fifth of biomass MSW in 2010[7] was also managed through incineration (in Figure 1), which uses thermal treatment at very high temperatures to wastes in specially designed furnaces. The volume of waste can be reduced by up to 90% and the weight of the waste by up to 60%[8] and therefore incineration is falsely represented as a highly effective method of organic waste reduction. Overall, it provides little to no benefit to the entire biosphere. First, incineration does not eliminate the waste, but rather changes the form of waste into hazardous smoke, gases and ash, all of which severely impact the environment, agricultural production and human health. For instance, the globally spread hazardous contaminations are a major source of 210 different dioxin compounds, plus mercury, cadmium, nitrous oxide, hydrogen chloride, sulfuric acid, and fluorides. The highly toxic dioxins could cause acid rain which destroys vegetation, wildlife, rivers, soils and even architecture. Then, the leftover harmful ash from incinerator also has to be buried in landfills and consequently exacerbates water pollution. Second, the primary and secondary costs of incineration are high and enormous energy and resources are utilized for the construction and operation of such facilities, even if they are developed in conjunction with energy recovery.

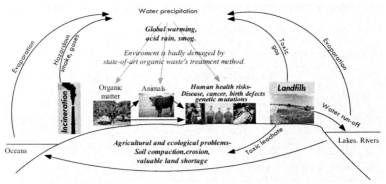

Figure 1. Organic waste treatment methods: landfills and incineration.

Thus, it can be seen that neither landfills nor incineration is a sustainable biomass waste management method that prevents enormous environmental damage from being inflicted on water, air and soil quality. On the other hand, composting is an environmental-friendly method to handle biomass solid waste by minimizing environmental effects including neutralizing carbon, significantly reducing sulfur, nitrogen oxides, and methane emissions, and also delivers organic fertilizer for better agricultural productivity and food production. In addition, composting conserves landfill space and reduces disposal cost. However, the current biomass waste aerobic bio-degradation technology remains in a stage of infancy such that:

    1)   The bio-degradation rate is too slow to be effective.

Usually, it takes up to a year (including maturation phase) for the microorganisms to compost garden clippings aerobically. As a matter of fact, aerobic bio-degradation is an exothermic process that could be controlled on the basis of temperature feedback, but the present technologies neglect the direct control of the temperatures, airflow rates, and water availability[9]. The optimized biomass waste composting process is overlooked and not exploited to help to speed up the bio-degradation rate. As a result, to handle a huge amount of biomass waste, an additional vast area of land is required for traditional aerobic degradation technology.

2)   It is used only for producing fertilizer purposes.

A significant amount of green thermal energy generated from the composting process is neglected and fully discharged to the atmosphere. As shown in Figure 2, the most significant energy driven process on Earth with respect to the ecosystems is that of photosynthesis, which converts light into a form of potential energy held in the chemical bonds within organic matter. The natural biological process, aerobic bio-degradation, then transforms putrescible organic matter into CO_2, H_2O and complex metastable compounds (e.g., humic substances) by microorganisms. Thus, massive green heat generated from the bio-degradation is wasted.

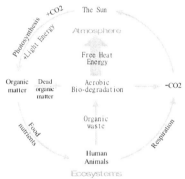

Figure 2. The wasted free heat energy from aerobic bio-degradation.

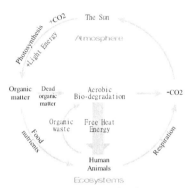

Figure 3. Reclaiming the free heat energy from aerobic bio-degradation.

There are large amounts of heat contained in biomass wastes. For instance, the gross calorific values of kitchen, paper, and garden wastes are 15.7GJ/t, 17.4GJ/t, and 16.1GJ/t, respectively[10]. In the U.S. there were about 32.9 million tons of yard wastes in 2008[11], representing about 200 trillion Btu of thermal heat energy, which can be good energy resources for domestic hot water heating. As estimated by the DOE, the hot water heating need for an average household in the U.S. is 1491.8 kWh/year (4.1 kWh/day), which accounts for 14~25% of the total U.S. energy consumption in residential buildings[12]. That is, despite the additional massive production of biomass wastes during the agriculture, industry, and transportation processes without reliable estimations nationwide, the potential biomass energy from yard waste alones in the U.S. can adequately support the annual domestic hot water heating needs for more than 28% (about 28.7 million) of families[13], if about 73% of the theoretical energy value of the biomass waste is recovered from an optimized control of composting[14]. However, no attempt is made to reclaim such green heat within biomass wastes, which are good renewable energy sources.

In this study, as shown in Figure 3, to reclaim the free green heat energy from the biomass waste in a fast and effective manner, a prototype biomass waste aerobic degradation system is proposed and used for hot water heating. The system with two chambers, a biomass waste aerobic degradation chamber and an air-to-water heat exchange chamber, is developed using modular industrialization design method. Preliminary laboratory tests with multiple types of biomass wastes (e.g., grass clipping, leaves, sawdust, sludge) for the aerobic degradation chamber demonstrates that, it can generate adequate bio-heat to support the daily domestic hot water needs in the U.S., by just disposing the daily biomass waste produced in an average household. Adopting this cost-effective biomass waste aerobic degradation system is promising in which a variety of benefits can be achieved, including providing significant free green renewable energy, improving organic waste management, rapid delivery of healthy and fertile soil, and minimizing the impacts of environmental pollution and human health.

The study begins by addressing the modular design of a prototype biomass waste aerobic degradation system for hot water heating with two chambers, a bio-degradation chamber and a heat exchange chamber. Then, laboratory experiments for the bio-degradation chamber are carefully conducted for investigating the novel system. Experimental results reveal that the bio-degradation chamber can effectively accelerate the bio-degradation process under controlled conditions (e.g., mechanical ventilation, moisture, temperature) to rapidly break down the organic wastes. The study concludes this novel biomass waste treatment system for hot water heating could produce significant ecological, environmental, economic and social benefits.

SYSTEM DESIGN

As shown in Figure 4, modular design of a prototype biomass waste aerobic degradation system for hot water heating has two chambers: a biomass waste aerobic degradation chamber and an air-to-water heat exchange chamber. To ensure the system for scalability in practice without sacrificing performance, modular industrialization design methods are adopted in the laboratory. Organic feedstock is loaded from the heat exchange chamber which will sit on the top of the bio-degradation chamber and easily unloaded through the feedstock discharge port at the bottom of the bio-degradation chamber.

Mechanical ventilation with controlled temperature and airflow rates evenly supplies oxygen to microbes through an air supply griller from the bottom of the bio-degradation chamber. To maximize the ventilation reception, four inner sub-chambers with additional individual air supply grillers are built. To observe the feedstock volume reduction rate, multi-visible windows are added on each sub-chamber. The disposal capacity (m) of the testing prototype aerobic degradation chamber is 92.4 kg.

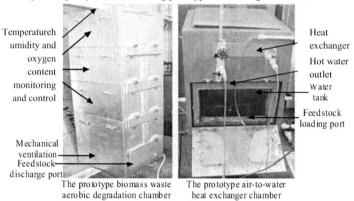

Temperature humidity and oxygen content monitoring and control

Mechanical ventilation
Feedstock discharge port

Heat exchanger
Hot water outlet
Water tank
Feedstock loading port

The prototype biomass waste aerobic degradation chamber   The prototype air-to-water heat exchanger chamber

Figure 4. A prototype biomass waste aerobic degradation system for hot water heating.

Innovatively, the bio-heat energy contained in the hot air exhausted from the top of bio-degradation chamber is recovered using an air-to-water heat exchanger. Real-time monitoring and control of inner and/or exhaust air temperatures, humidity, and oxygen content is applied to ensure optimal built-environments for microorganisms. In addition, external multi-layer (2~3) thermal insulations are adapted to improve energy conversion and energy efficiency by maintaining high temperatures in the bio-degradation chamber and significantly reducing heat loss of both bio-degradation chamber and air-to-water heat exchange chamber.

The heat exchange chamber includes two parts: a water tank and an air-to-water heat exchanger. The water tank sits right upon the bio-degradation chamber and the air-to-water heat exchanger is crafted on the top of the water tank. The hot air exhausted from the bio-degradation chamber is connected to the inlet of the airside of the heat exchanger. Temperatures of the heated water can be controlled by the flow rate at the waterside of the heat exchanger.

SYSTEM PERFORMANCE INVESTIGATION

The air-to-water heat exchanger has been tested previously in the laboratory that it could reclaim more than 80% of heat energy. Hence, in this study, system performance of the biomass waste aerobic degradation chamber is investigated with carefully designed and conducted experiments.

Experimental Settings

Experimental settings for organic feedstock and mechanical ventilation are specified in Table 1:

1) Organic feedstock

Carbon: Nitrogen (C:N) ratio Different organic waste has different C:N ratio, e.g., grass clippings' C:N ratio is about 20:1, leaves' 60:1, sawdust's 325:1, and sludge's 10:1. Microbes can only thrive under proper C:N ratios. A high C: N ratio leads to slow decomposition and the reaction stops when useable C is used up. However, if the C:N ratio is too low, a lot of ammonia which is toxic to microbes will be generated and the decomposition will slow down. Consequently, a mixture of grass clippings, leaves, sawdust, and sludge with a C:N ratio range of 25~35 is used in this study.

2) Mechanical ventilation

Supply air flow rate $\dot{V}$ Aerobic biodegradation consumes large amounts of oxygen for microorganisms to transform organic matters into CO_2, H_2O and release the free green energy as shown in Equation (1). Therefore, to maximize the composting speed, constant mechanical ventilation to ensure oxygen content of exhaust air higher than 6~10% and so achieve fast composting is set up here with $\dot{V}$ about 5.5 cfm.

$$C_6H_{12}O_6 + 6O_2 + Microorganisms \xrightarrow{\dot{V}} 6CO_2 + 6H_2O + thermal\ energy \qquad (1)$$

where $C_6H_{12}O_6$ models organic matter such as glucose.

Supply air temperature T_{sa} Temperature has a self-limiting effect on microbial activity and thus the rate of degradation of organic materials. Above 40 °C is the best condition for activating thermophilic microbes, and mesophilic microbes are active in the temperature range of 10 ~ 40 °C. Hence, in this study, T_{sa} is designed in a range of 30 ~40 °C and 32 °C on average.

Supply air relative humidity ϕ_{sa} Moisture is essential to all living organisms. Moisture is lost through evaporation during the aerobic composting process. Thus, to maintain liquid rich conditions, supply air is constantly humidified and ϕ_{sa} is stabilized at around 75%.

Table 1. Experimental design of the biomass waste aerobic degradation chamber

Organic feedstock		
Biomass waste types	Disposal capacity m (kg)	Mixture C:N ratio
Grass clipping, leaves, sawdust, sludge	92.4	25~35
Mechanical ventilation		
Supply airflow rate $\dot{V}$ (cfm)	Supply air temperature T_{sa} (°C)	Supply air humidity ratio ϕ_{sa}
5.5	32	0.75

Experimental results

In the laboratory, instant readings of supply air and exhaust air temperatures (T_{ea}) of the biomass waste aerobic degradation chamber are sampled every 5 minutes to represent the corresponding test results. In addition, the exhaust air humidity ratio (ϕ_{ea}) is maintained at nearly 100% as measured during the testing.

With the knowns of T_{sa}, T_{ea}, ϕ_{sa}, ϕ_{ea}, and $\dot{V}$, the heat flow rate $\dot{Q}_{H,total}$ of the prototype aerobic degradation chamber can be obtained as

$$\dot{Q}_{H,total} = \frac{\dot{V}}{v}(h_{ea} - h_{sa}) = \frac{\dot{V}}{v}[f(T_{ea},\phi_{ea}) - f(T_{sa},\phi_{sa})] \tag{2}$$

where h_{ea}, h_{sa} = enthalpy of exhaust air and supply air of the aerobic degradation chamber in kJ/kg and v = specific volume of air in m³/kg.

Therefore, the heat flow rate per unit mass of the biomass waste on a wet basis $\dot{Q}_H$ of this prototype aerobic degradation chamber can be achieved as

$$\dot{Q}_H = \frac{\dot{Q}_{H,total}}{m} \tag{3}$$

Experimental results of T_{sa}, T_{ea}, $\dot{Q}_{H,total}$ and $\dot{Q}_H$ are plotted in Figure 5. The horizontal axis represents the reaction duration in hours and the vertical axes represent air temperatures in Celsius degrees and heat flow rate in Watt (W) or Watt/kg (W/kg).

Temperature and humidity control

Observed from the acquired data in Figure 5, after the reaction occurred in less than 8 hours, the exhaust air temperature rose to 75°C with a temperature difference higher than 40°C. Within 18 hours, 85.4°C of peak exhaust air temperature could be achieved. T_{sa} was maintained above 75 °C for more than 3 days.

Additional laboratory test also proves that, optimum supply air temperature during the composting is within a range of 30~40 °C. If T_{sa} is controlled to be higher (e.g., >45 °C), instead, the exhaust air temperature will be lower than 75 °C. That means, it will recover much less energy than that of controlling T_{sa} in a range of 30~40 °C. But, if T_{sa} is controlled to be lower (e.g., 15~25 °C), the bio-degradation will get slow. It may take days to reach 60 °C for the exhaust air temperature. Similarly, if ϕ_{sa} is low, as the dry air expands through the feedstock, the water content decreases and so the bio-degradation slows rapidly.

Net energy generation rate

When a temperature difference was higher than 40°C as shown in Figure 5, the corresponding $\dot{Q}_H$ was 15.1 W/kg. At the peak condition, $\dot{Q}_H$ topped 24.3 W/kg, $\dot{Q}_{H,total}$ was 2242W. On average, it obtained a sustainable heat flow rate of 18.2 W/kg and daily energy generation rate of 40.4kWh/day.

In this study, the small amount of forced air was directly created by utilizing an air-compression system, which had been readily available in the building (e.g., used for laboratory cleaning purposes). It had been estimated that the bio-degradation chamber consumed about 1.5~2.5 W for the pressure loss through the chamber. It is only 0.02 ~0.03 W/kg of biomass waste. Meanwhile, by utilizing thick insulation materials for both chambers, the heat loss was lower than 20%. Innovatively, we also re-used this "lost" energy by placing the chambers in a small space and warming up this space (about 30 °C) for the two chambers.

In sum, in terms of this testing, the net energy generation rate of this prototype biomass waste aerobic degradation system is about 14.6 W/kg and daily net energy generation rate is 32.3kWh/day. It is estimated that it can support the daily domestic hot water needs (about 4.1 kWh/day) for more than 8 households[1,12]. In the meantime, the daily biomass waste generation of each household is 4.0 kg in the U.S. By just using the daily generated biomass waste of U.S. households, the prototype biomass waste aerobic degradation system can reclaim an adequate amount of thermal heat for hot water heating at home. The experiment demonstrates the biomass waste aerobic degradation system for hot water heating can rapidly dispose of biomass waste and also reclaim the free green energy effectively.

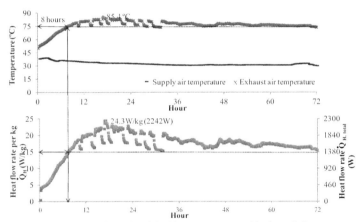

Figure 5. Experimental results of the biomass waste aerobic degradation system.

CONCLUSIONS AND DISCUSSIONS

Due to continued rapid population growth and economic development, the demand for renewable green energy and the amount of biomass waste treatment worldwide, particularly in the U. S., has increased dramatically in the past 100 years. As valuable energy resources, massive organic wastes have been long ignored and randomly buried in landfills or incinerated directly. However, neither landfills nor incineration is a sustainable biomass waste management method. On the other hand, current aerobic biodegradation technology that relies mainly on the natural aerobic biological processes is also too slow to be applied prevalently, and significant amounts of high-grade heat released during aerobic degradation are ignored and wasted.

In this study, an innovative prototype biomass waste-to-renewable energy system from speedy aerobic degradation process is developed for hot water heating. Laboratory testing with multiple types of biomass waste is carefully conducted and demonstrates that, on average, a sustainable net heat generation rate of about 14.6 W/kg of biomass waste on a wet basis can be produced. As estimated, adopting this biomass waste treatment system would be able to achieve self-sufficiency in domestic hot water heating by just using the household produced biomass waste.

The study concludes that this novel biomass waste treatment system for hot water heating is economical and could result in significant ecological, environmental, economic and social benefits, such as

- *Providing another option for using free green renewable energy resources,*
- *Producing renewable heat energy and thus decreasing natural gas or electricity usage for domestic hot water heating,*
- *Improving organic waste management and saving the government expense of waste treatment,*
- *Decreasing the dangerous air and water emissions which severely impact environmental pollution and human health, and*
- *Rapidly delivering healthy and fertile soil for agriculture and avoiding the extensive use of harmful chemical fertilizer and insecticides.*

Immediately after this study, integer performance of the novel biomass waste aerobic degradation system including the bio-degradation chamber and heat exchange chamber will be tested and investigated in a separate article.

NOMENCLATURE

h_{ea} = Enthalpy of exhaust air, kJ/kg

h_{sa} = Enthalpy of supply air, kJ/kg

m = Disposal capacity, kg

$\dot{Q}_H$ = Heat flow rate per unit mass of the biomass waste on a wet basis, W/kg

$\dot{Q}_{H,total}$ = Heat flow rate of the prototype aerobic degradation chamber, W

v = Specific volume of air, m^3/kg

$\dot{V}$ = Supply air flow rate, cfm

T_{sa} = Supply air temperature, °C

T_{ea} = Exhaust air temperature, °C

ϕ_{sa} = Supply air humidity ratio

ϕ_{ea} = Exhaust air humidity ratio

REFERENCES

[1]Environmental Protection Agency (EPA), 2009. Sustainable materials management: the road ahead, available at <www.epa.gov/osw/inforesources/pubs/vision2.pdf> (accessed on March 21, 2012).

[2]Department of Energy (DOE), 2009. Building data book, available at <http://buildings databook.eren.doe.gov/> (accessed on March 21, 2012).

[3]Rob van Haaren, Nickolas Themelis and Nora Goldstein. 2010. The State Of Garbage in America. BioCycle, 51(10).

[4]Leak Location Services, Inc. (LLSI). 2000. Landfill surveys, available at < http://www.llsi.com/land fills. htm> (accessed on March 21, 2012).

[5]Environmental Protection Agency (EPA), 2009b. Basic Information about Food Waste: the road ahead, available at < http://www.epa.gov/osw/conserve/materials/organics/food/fd-basic.htm> (accessed on March 21, 2012).

[6]D.L. Rockwood, C.V. Naidu, D.R. Carter, M. Rahmani, T.A. Spriggs, C. Lin, G.R. Alker, J.G. Isebrands and S.A. Segrest. 2004. Short-rotation woody crops and phytoremediation: Opportunities for agroforestry? Agroforestry System 2004, 61:51-63.

[7]Environmental Protection Agency (EPA), 2011. Combustion: the road ahead, available at < http://www.epa.gov/osw/nonhaz/municipal/combustion.htm> (accessed on March 21, 2012).

[8]A.V. Bridgwater and K. Lidgren Van Nostrand Reinhold. 1981. Household waste management in Europe: Economics and techniques, New York, 1981.

[9]Ekinci, K., Keener, H.M. and Akbolat, D., 2006. Effects of feedstock, airflow rate, and recirculation ratio on performance of composting systems with air recirculation, Bioresource Technology, 97.

[10]Parfitt, J., 2002. Analysis of Household Waste Composition and Factors Driving Waste Increases, Waste & Resources Action Programme (WRAP), U.K.

[11]Environmental Protection Agency (EPA), 2008. Municipal solid waste generation, recycling, and disposal in the United States: facts and figures for 2008, available at < http://www.epa.gov/epawaste/nonhaz/municipal/msw99.htm> (accessed on March 21, 2012).

[12]Department of Energy (DOE), 2011. Water Heating, available at < http://www. energy savers.gov/your_home/water_heating/index.cfm/mytopic=12760> (accessed on March 21, 2012).

[13]United States Census Bureau. 2010. Current Population Survey, available at < http://www.census.gov /main/www/cprs.html> (accessed on March 21, 2012).

[14]Lekic, S, 2005. Possibilities of heat recovery from waste composting process. M.Phil. Dissertation, University of Cambridge, U.K.

MULTI-ENERGY OPTIMIZATION PROCESS: BIODIESEL PRODUCTION THROUGH
ULTRASOUND AND MICROWAVES

S. Getty, M. Kropf Ph.D, B. Tittmann Ph.D , The Pennsylvania State University

ABSTRACT
The an ever increasing demand for fuels and new energy sources is forcing a growing
need for new, innovative concepts of alternative and renewable energy sources. To meet this
demand and become competitive with fossil fuels, industry must increase production of
renewable energy while decreasing energy and costs of producing them. One such source of
renewable energy, which has been changing rapidly in North America, is biodiesel. A novel
processing approach developed at The Pennsylvania State University leverages high intensity,
focused ultrasound and microwave heating to improve the material and energy efficiencies of the
production of biodiesel. Technical approaches towards technology scale-up will be discussed
along with the validation studies at industrial scales.

INTRODUCTION
Fossil fuels have been the means for power and transportation since the industrial
revolution. With high energy and power densities as well as the ability to be stored relatively
easily, they have a high attraction for power sources. One major downside of fossil fuel is the
inability to be reused once depleted. Scares of these depletions can be seen through history, as in
the late 1970's during the first oil crisis. With the growth of other countries, such as India and
China, there is concern about when these sources will run out. Increasing fuel prices are the first
evidence in the decline of fossil fuels.
As fossil fuels are depleted there is more push to develop alternative sources of energy
which can easily be replenished. Some of these sources include technologies such as solar power,
wind power, battery technologies, and biofuels. All of these are considered "green technologies"
for their environmentally friendly production and the ability to be replenished. Biodiesel is one
such biofuel which is created chemically using plant oils, recycled cooking oils and animal fats
making it a renewable energy source [1]. Biodiesel has a lower energy density than that of
petroleum diesel, 118,296 Btu/gal vs. 129,500 Btu/gal respectively. Currently biodiesel is
typically blended with petroleum diesel at a 20 to 80% ratio, also known as B20 [2]. Likewise,
diesel engines tend to have a higher efficiency, about 40%, than that of gasoline engines' 30%.
Diesel engines, while being heavier than spark-ignition engines, have significant torque
advantages and are easily converted to run on B20, which is the competition-mandated fuel for
diesel engines. There has been evidence of reduced maintenance schedules in biodiesel fleets
based on the lubricity properties. However, diesel engines have greater NOx and particulate

emissions which are harmful to the environment [3] [4]. With advances in nano technology, these emissions are being better controlled and reduced.

CURRENT BIODIESEL PRODUCTION

Biodiesel is made through the chemical process of transesterification where the glycerin is separated from the triglyceride, fat or vegetable oils. The product from the chemical process is methyl esters and glycerin [5]. To separate the glycerin, the oil or fat is combined with an alcohol, typically methanol, at a ratio of 1:3 stoichiometrically. Figure 1 is a visualization of the transesterification process for biodiesel using triglycerides and methanol.

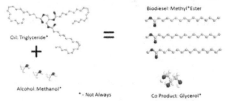

Figure 1. Biodiesel reaction with triglycerides and methanol to create glycerol and methyl-esters

Biodiesel is typically a batch-production process, which involves large quantities all at once in a single chamber. For the reaction to occur the oil and alcohol must be in close proximity. Since oil and methanol are immiscible fluids, much like oil and water, they do not distribute between each other to help the reaction. Shear mixing via an impeller is used to create a mixture of the fluid to help the reaction. One the feed stocks are mixed the system is heated to supply the energy to create the reaction through resistance coils or flame. Due to limitation of mixing process typically there is a requirement for higher levels of alcohol to stimulate the reaction toward the product, 1:6 or 1:9 ratios. Likewise, a catalyst is typically used to reduce the amount of energy needed to break the bonds to create the chemical reaction and to further enhance the mixing of the two components as an emulsifying agent. Both the increase in alcohol used and addition of the catalyst cause the cost of the product to increase. With the use of a catalyst there is also the problem with the containment of the catalyst in the glycerol once the reaction is complete and the need for removal downstream before it can be sold.

METHODS

While batch reactions are simple and yield a sale quality product, they also leave room for large loss if the reaction is not properly executed. Penn State has developed a piped system which runs through the use of ultrasound and microwave technology for a continuous flow system. The ultrasound provides the mixing for the system while the electromagnetic radiation supplies the heat necessary to stimulate reaction [6]. With better mixing and higher temperatures the system can be moved toward being more stoichiometric and better rates of conversion. It has been proven that with an increase mixing rate the time to reach conversion of methyl esters decreases as temperature is held constant at 50°C. Therefore as the mixing is optimized the rate of conversion can be increased. Conversely, if the mixing rate is held constant and the temperature is increased than the rate of conversion can be increased as temperature is increased [7].

The concept was first tested through bench scale testing in order to confirm theoretical concepts of superheated fluids could produce fuel. Ultrasound was deemed a viable source for mixing as technology was researched. Microwaves have proven to be useful for creating superheating with polar molecules. With bench scale testing concepts were proven to be capable of producing results. Positive bench scale results lead to the development of a pilot scale system. Since commercial components are limited for these applications many products are being developed through Penn State. The most necessary development concept for this application has been in ultrasound. Simulations have proven to be the most beneficial part in developing adequate ultrasonic emulsifiers to allow more design shapes and sizes while cutting cost in machining components.

Ultrasound

Penn State is currently using ultrasound to create a stable cavitation within the mixture of the methanol and oil which results in a semi-stable emulsion of micrometer sized methanol bubbles in oil as a lyophobic emulsion. While ultrasound is not a new concept to be utilized in the production of biodiesel in an industrial application, it is typically used to supplement the mixing process in the biodiesel system [8]. Penn State is using ultrasound to replace shear mixing typically used in a conventional biodiesel plant and reduce the action of catalyst towards emulsification. The ultrasound uses the concept of cavitation to create small shockwaves within the fluid. These shockwaves are a result of micro-jetting which allows for better mixing and smaller micro, methanol bubbles dispersed within the oil. Figure 2 visualized the three step process in the creation of the emulsion via ultrasonic cavitation.

1) Immiscible Solution   2) Ultrasonic Cavitation

3) Emulsion Complete

Figure 2. Emulsion creation by micro-jetting caused by cavitation

Controlling stable cavitation is critical to creating a well distributed emulsion and decreasing reaction time. Shear mixing is capable of producing micro-sized methanol droplets; however the distribution of the methanol bubbles greatly varies in size. When comparing the size of the droplets in relation to the ultrasonic frequency applied the distribution is more concentrated. In Figure 3, the shear mixing has a large distribution of droplet size. Then the ultrasound is applied at a frequency of 22 kHz the distribution is more isolated in the 4 micrometer size range. When an ultrasonic frequency is shifted to 44 kHz the distribution not only becomes more concentrated, but it also decreases in average size.

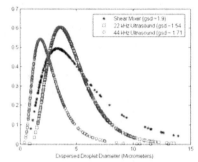

Figure 3. Comparison between droplet size dispersion in relation to mixing type [9].

An even distribution of size allows for a more complete reaction as well as a more stable emulsion. Figure 4 are images of the micro bubbles and their distribution. In the conventional shear mixing scenario the size is not as uniform and there are larger agglomerations of methanol. At an ultrasonic frequency applied at 44 kHz the micro bubbles are much smaller and overall well distributed. While there are no agglomerations present in the current ultrasonic mixing image, if the emulsion were left over time the like methanol particles would join and grow in size. With the adaption of ultrasonic mixing, a more even emulsion can be created in order to help further decrease the time necessary to achieve the reaction.

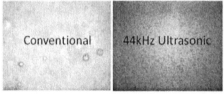

Figure 4. Visual comparison between conventional shear mixing and ultrasonic mixing at 44 kHz.

Microwaves

Microwave heating can be used to replace conventional heating methods in the biodiesel process. In consideration of the components of the transesterification process, some of the reagents are microwave reactive, while other are microwave transparent [10]. Figure 5 shows that both the ionic and polar materials are reactive to electromagnetic radiation, albeit at different frequency regimes. This means that the catalyst and alcohol molecules can be excited by specific frequencies of microwave. Particularly the alcohol has a specific dipolar relaxation peak that can be isolated. While alcohol and catalyst ions are microwave reactive, oil is non-reactive.

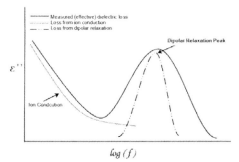

Figure 5. Visual of materials in biodiesel transesterification reactive to electromagnetic radiation [9].

Through a dielectric analysis the electromagnetic radiation excitation frequency can be determined for different materials. Figure 6 is the measured excitation frequencies for the alcohol and catalyst. With the known excitation frequency for the alcohol, a single mode microwave can be used to isolate the frequency to stimulate heating and the reaction.

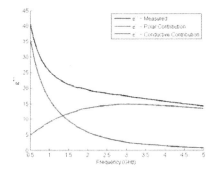

Figure 6. Measured excitation frequencies for ion and polar molecules [6].

Since oil is non-reactive with microwaves only the alcohol would be heated. This would cause the alcohol to boil off and not allow for the reaction to occur. As the micro bubbles of methanol are decreased in size and more evenly dispersed in an emulsion the system begins to look like a single material rather than two immiscible fluids. This allows for the microwaves to stimulate the transesterification reaction. As microwaves are applied to the emulsion the normal phase equilibrium graph correlating temperature and pressure and dictating boiling points is overcome by the direct input of energy. The result is the ability to reach a superheated (fluid above boiling temperature at a given pressure) conditions and ultimately faster reaction rates [9]. This allows for faster reaction times and a decrease in catalyst usage with a push to zero catalyst reaction.

RESULTS

The development of a continuous flow production design required the use of commercially available devices as well as new innovative technologies. While there are several commercial ultrasonic emulsifiers, there is much to be improved in design and capabilities especially in the development of a novel biofuels production concept. In order to prove this concept, simulations in Comsol Multi-Physics were performed to determine the proper resonance of the sonotrode shapes and dimensions. Figure 7 is a Comsol simulation showing the maximum displacement of the sonotrode being located at the center point of the working face. For this sonotrode the resonant frequency was determined to be 38.4 kHz. This was later verified with the working sonotrode using a LabVIEW program to determine the resonant frequency of the sonotrode.

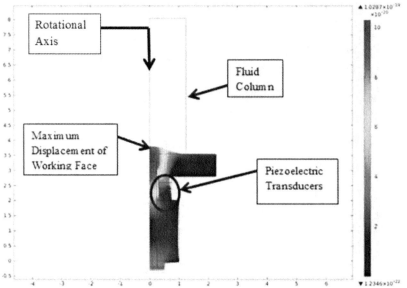

Figure 7. Comsol simulation showing maximum displacement of 1.03×10^{-19} in at resonant frequency of 38.4 kHz.

The ultrasound is produced by a piezoelectric transducer which converts electrical energy into mechanical energy. An electrical pulse is applied to the transducer to develop a continuous wave. The maximum displacement occurs at the resonant frequency resulting in a wave of ultrasonic energy. Using LabVIEW software the frequency is controlled at the resonant frequency of the structure to produce the highest efficiency of conversion between electrical and mechanical energy, measured by the change of RMS voltage absorbed divided by the initial voltage. Figure 8 shows a resonant frequency scan of a sonotrode design similar to Figure 7. The largest resonant frequency peak is at 38.2 kHz, validating the simulation. Using the forward and reflected power from the amplifier, the software can monitor the difference and scan for the highest efficiency. Likewise, the software can monitor if the efficiency changes more than 5% and scan for a new resonant frequency, which ensures optimal output in dynamic environments. Shifts in resonant frequency can be due to multiple issues, but the most common to piezoelectric materials is heat from the conversion of electrical to mechanical energy. In this way, the resonance frequency is adjusted to compensate for heating which prevents decreasing electro-mechanical coupling efficiency which would otherwise result in more excess heating ultimately destroying the elements.

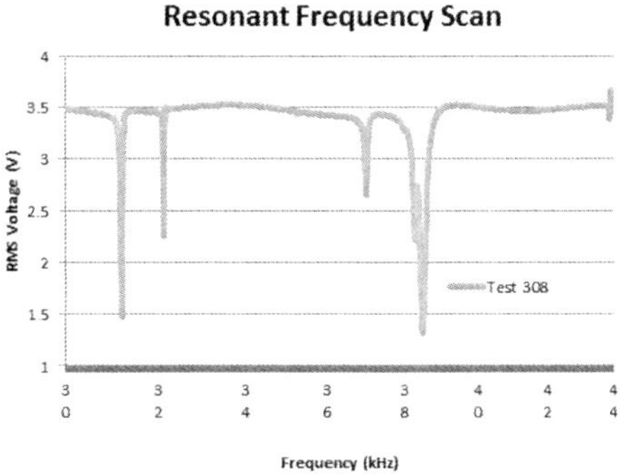

Resonant Frequency Scan

Figure 8. Resonant frequencies scan with maximum displacement peak at 38.2 kHz

Upon an initial proof of concept through bench scale testing, Penn State used ultrasound and electromagnetic technology to develop an automated pilot scale testing system via a single mode microwave and multiple ultrasonic sonotrodes. Figure 9 is an image of the automated pilot scale testing apparatus able to achieve fuel within ASTM specifications at rates above 10 gallons per hour (GPH). The pilot system is automated using LabVIEW to not only control the ultrasound and microwaves, but also monitor temperature and pressure.

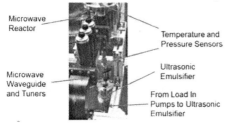

Figure 9. Automated pilot testing apparatus

Penn State transitioned from a pilot scale reactor at 10 GPH flow rate towards a 60 GPH system. The scale up proved to be fairly linear in terms of both the ultrasound and microwave technology. With a better understanding of cavitation from simulation and bench top testing the ultrasonic emulsifier was able to be scaled to accommodate an increase in flow rate. The current work on ultrasound involves scaling from a system initially using 2 ultrasonic sonotrodes to a system which combines static mixing and 4 ultrasonic transducer stacks. Pressure and temperature sensors were fitted before and after the single mode microwave wave guide to monitor the changes. The microwave reaction chamber was designed as a thick walled, alumina pipe fitted for the wave guide opening capable of handling pressures and temperatures up to and exceeding the super critical reaction range. The byproduct once reacted is sent to a glycerin separation tank where it is agitated when sent to the tank then allowed to settle over time.

Using fryer oil, containing high levels of free fatty acids (FFA), provided by the Penn State campus fryers, the pilot plant is capable of creating fuel within ASTM specifications and with lower catalyst and power consumptions. The single mode magnetron functions using 1 kW of power and the ultrasound uses roughly 60 W of power to generate the necessary energy to create a semi-stable emulsion. At 10 GPH, the pilot reactor produced fuel with a conversion efficiency of 0.99806 (as measured by total and free glycerin remaining through gas chromatography), well within biodiesel ASTM specifications, using under 0.2% catalyst compared to the industry standard of 0.5% plus an amount correlated to the additional FFA.

CONCLUSIONS AND FUTURE WORK
Penn State was capable of producing biodiesel through the use of ultrasound technology for semi-stable cavitation and single mode electromagnetic radiation at a single excitation frequency isolating the dipolar relaxation point of methanol. With the successful construction and conversion of an automated bench top reaction system, Penn State moved forward to build a pilot scale reactor capable of producing fuel within ASTM specification at a rate of 10 GPH with minimal energy consumption to run the plant and decreased catalyst content.

With the success of a pilot plant, Penn State is moving forward to produce a plant towards capable of 10 gallons per minute (GPM). Using the anticipation of linear scaling of both the ultrasound and microwave technology, Penn State is working on developing new ultrasonic emulsifiers to meet the increased flow rate. The new system is in development to continue the decrease in percent catalyst being used with the anticipation of reaching a reaction similar to that of the supercritical reaction with zero catalyst.

Since biodiesel and the catalyst used are corrosive materials it is becoming more important to consider materials being used in the new system. Another concern as the system moves toward full automation at a commercial capacity is deterioration of the working face of the ultrasound due to pitting caused by cavitation. Currently there have been no issues due to material failure due to corrosion or pitting, however these become larger issues with the progression to commercialization. As the technology reaches capabilities of industrial flows and standards materials must be changed from aluminums to stainless steels to reduce reactions with biodiesel and catalyst. The ultrasonic emulsifiers will be used continuously; other issues arise such as pitting on the working face as well as cooling of the ultrasonic transducer stacks. With heating issues the piezoelectric transducers can fracture due to heating as well as begin to de-pole and lose efficiency. These are only a few of the material concerns. More issues will become prominent as scaling progresses.

Penn State is currently working on performing a techno-economic analysis on the system with considerations of power consumption and material cost. As part of the research, Penn State is also working on diversification of feedstock to even higher FFA content oils as well as looking into the application of the system to other refinement uses.

BIBLIOGRAPHY

1]   "About Biodiesel," National Bidiesel Foundation, [Online]. Available: www.biodiesel.org. [Accessed July 2011].

2]   "Energy Content," National Biodiesel Board, October 2005. [Online]. Available: http://biodiesel.org/pdf_files/fuelfactsheets/BTU_Content_Final_Oct2005.pdf. [Accessed March 2012].

3]   Shirk and et al., "Investigation of a Hydrogen-Assisted Combustion System for a Light-Duty Diesel Vehicle," *International Journal of Hydrogen Energy,* vol. 33, 2008.

4]   Chapman and et al., "Penn State FutureTruck Hybrid Electric Vehicle: Light-Duty Diesel Exhaust Emission Control System to meet ULEV Emissions Standard," *SAE,* 2005.

5]   "Biodiesel Basics," National Biodiesl Board, 2012. [Online]. Available: http://biodiesel.org/resources/biodiesel_basics/. [Accessed March 2012].

6]   M. M. Kropf, "MULTI-ENERGY OPTIMIZED PROCESSING: THE USE OF HIGH INTENSITY ULTRASONIC AND ELECTROMAGNETIC RADIATION FOR BIOFUEL PRODUCTION PROCESSES," PhD Dissertation, The Graduate School, The Pennsylvania State University, University Park, PA, 2008.

7]   H. Noureddini and D. Zhu, "Kinetics of Transesterification of Soybean Oi," *JAOCS,* vol. 74, no. 11, 1997.

8]   L. Yustianingsih, S. Zullaikah and Y. H. Ju, "Ultrasound assisted in situ production of biodiesel from rice bran," *Journal of the Energy Institute,* vol. 83, pp. 133-137, 2009.

9]   M. M. Kropf, "ULTRASONIC AND MICROWAVE METHODS FOR ENHANCING THE RATE OF A CHEMICAL REACTION AND APPARATUS FOR SUCH METHODS". United Stated Patent US 2009/0000941 A1, 26 June 2008.

10] C. Mazzocchia, A. Kaddouri, G. Modica and R. Nannicini, "Fast Synthesis of Biodiesel from Triglycerides in Presence of Microwaves," in *Advances in Microwave and Radio Frequency Processing: 8th International Conference in Microwave and High-Frequency Heating,* 2006.

CONTACT INFORMATION
Shawn C. Getty, The Pennsylvania State University, Department of Engineering Science and Mechanics, scg5022@psu.edu
Dr. Matthew M. Kropf, The Pennsylvania State University, Department on Engineering Science and Mechanics, mmk230@psu.edu
Dr. Bernhard R. Tittmann, The Pennsylvania State University, Department of Engineering Science and Mechanics, brtesm@engr.psu.edu

ACKNOWLEDGEMENTS
The contributions of the research, modeling, test, and results from Matt Verlinich, Ryan Johnson, Eric So and Jesse Rodgers are gratefully acknowledged. We would like to thank A E Resources Inc. for their cooperate sponsorship. A final thanks the Pennsylvania State University for their support.

EFFECTS OF FUEL GRADE ETHANOL ON PUMP STATION AND TERMINAL FACILITIES

Greg Quickel and John Beavers
Det Norske Veritas (U.S.A.) Inc.
5777 Frantz Road
Dublin, OH   43017-1886
Phone (614) 761-1214   Fax (614) 761-1633
John.Beavers@dnv.com

Feng Gui and Narasi Sridhar
DNV Research and Innovation
5777 Frantz Road
Dublin, OH   43017-1886
Phone (614) 761-1214   Fax (614) 761-1633
Narasi.Sridhar@dnv.com

ABSTRACT

There is significant interest within the North American pipeline industry in transporting fuel grade ethanol (FGE) as a result of the increased usage of ethanol as an oxygenating agent for gasoline and interest in ethanol as an alternative fuel. Several materials compatibility issues must be resolved before FGE can be safely transported in pipelines. Many of the issues relate to corrosion of pipeline steels have been addressed in previous research projects. However, there is relatively little information on the effects of ethanol and ethanol - gasoline blends on other components in pipeline systems, such as pumps, valves, screens, springs, and metering devices. The Pipeline Research Council International (PRCI) and the US Department of Transportation, Pipeline and Hazardous Materials Safety Administration (PHMSA) funded research to address these technical issues. The scope of the work includes a survey of knowledge and gaps, laboratory studies designed to close the gaps, and a guidelines document for the pipeline industry. This paper summarizes recent research results.

INTRODUCTION

Ethanol has been used for the last several years as an environmentally friendly alternative to methyl tertbutyl ether (MTBE), which is an oxygenate additive to gasoline, to increase octane levels, and to facilitate the combustion process. However, the need to find alternatives to imported oil and gas has spurred the increased use of ethanol as an alternative fuel source. Further, ethanol is being promoted as a potential trade-off for CO_2 emissions from the burning of fossil fuels since CO_2 is consumed by the plants used as the ethanol source. The Renewable Fuels Standard requires the gradual increase in the use of a variety of renewable fuels over the next two decades, in which ethanol will play a dominant role.[1] One approach is to increase the concentration of ethanol in the gasoline. Ethanol is blended into 70 percent of America's gasoline and the EPA has granted two partial waivers that would allow increasing the maximum ethanol concentration in gasoline from 10 to 15 volume %.[2] A second approach is to encourage the use of alternative fuel vehicles, such as those that use E-85 (85% ethanol–15% gasoline). The widespread use of ethanol will require efficient and reliable transportation from diverse ethanol producers to distribution terminals. Pipelines are, by far, the most cost-effective means of transporting large quantities of liquid hydrocarbons over long distances. For transporting ethanol, both existing pipeline infrastructure and new pipeline construction are being contemplated.

Ethanol is produced in bio-refineries and is transported to terminals, where it is blended with gasoline to produce the most commonly used blends; E-10 (10% ethanol – 90% gasoline) and E-85. Presently, rail and truck are the predominant means of transporting ethanol in North America. Barge transportation will become more important with increased shipments of ethanol from South America. In terms of the volumes transportable by the different modes, one barge load is roughly equivalent to 15 to 20 rail cars or 60-80 truckloads. In comparison, a 16-inch pipeline can transport an equivalent of 15 barges on a daily basis. Barge transport can benefit substantially by a pipeline delivery system.

Many of the issues related to corrosion of pipeline steels have been addressed in previous research projects. However, there is relatively little information on the effects of ethanol and ethanol – gasoline blends on other components in pipeline systems, such as pumps, valves, screens, springs, and metering devices. PRCI and PHMSA funded research to address these technical issues. The scope of the work includes a survey of knowledge and gaps, laboratory studies designed to close the gaps, and a guidelines document for the pipeline industry. This paper summarizes issues with stress corrosion cracking (SCC) of carbon steels and recent research results on other materials found in pipeline facilities.

CORROSON OF CARBON STEELS

Stress corrosion cracking (SCC) is the single, most significant corrosion issue for the carbon steels found in pipelines, tanks, terminal piping, pumps, and other components in FGE service. A survey of literature and service experience with SCC in FGE was published by the American Petroleum Institute (API) in 2003.[3] Documented SCC failures of equipment in user's storage and transportation facilities have dated back to the early 1990s. The majority of the cracking has been found at locations near welds where the primary stresses leading to SCC have been residual welding stresses. An example of SCC in terminal piping is shown in Figure 1. No cases of SCC were reported in ethanol manufacturer facilities, tanker trucks, railroad tanker cars, or following blending of the FGE with gasoline. One possible case was identified in a barge but the vast majority of occurrences of SCC were in the first major hold point (FGE distribution terminal) or in the subsequent end-user gasoline blending and distribution terminals.

The API survey did not pinpoint what causes ethanol SCC, but the failure history suggests that the SCC may be related to changes in the FGE as it moves through the distribution chain over a period of days, weeks, or months. These observations led to an industry-sponsored research program to identify the causative factors. A Roadmapping Workshop held in October 2007 identified a number of research gaps in safely transporting FGE via pipelines.[4]

In the last five years, considerable research has been conducted with respect to ethanol SCC in several laboratories to address these gaps. This research has led to an increased understanding of the causes of SCC in steel and ways to mitigate it.[5, 6] The factors that are important to SCC of carbon steel in ethanol are discussed below.

Tensile Stress

As is the case with SCC in general, the presence of a tensile stress, either from fabrication in the shop, installation the field, or operation is necessary for SCC to occur. Tensile stresses may arise from welding, gouges/dents, bending during installation, or improper support (especially for tank bottoms). Cyclic stresses/strains, for example through pumping or loading/unloading operations, can increase the SCC tendency.

Oxygen

The results of research on chemistry effects on SCC have demonstrated that FGE that meets applicable ASTM standards (Table 1) is a potent cracking agent in the presence of oxygen.[5] Figure 2 shows the effects of oxygen and ethanol concentration in ethanol-gasoline blends on the occurrence of SCC. These results were obtained from slow strain rate (SSR) SCC tests in which samples of a carbon steel (line pipe steel) were slowly strained to failure (initial strain rate = 10^{-6} sec^{-1}) in ethanol-gasoline blends prepared with simulated fuel grade ethanol (SFGE) and gasoline containing no additives. The SFGE was prepared with reagent grade ethanol and additives at the upper limits for the ASTM standard (Table 1). Following testing, the specimens were examined for evidence of cracking. Optical photographs of SSR specimens with and without cracking are shown in Figures 3 and 4. Figure 2 shows that SCC readily occurs in susceptible blends in the presence of air (21% oxygen) and in oxygen-nitrogen gas mixtures containing as little as 0.5% oxygen.

Ethanol Concentration

Figure 2 shows that SCC was observed in ethanol-gasoline blends containing more than 15% ethanol by volume. The results of longer-duration crack growth tests indicate that E-50 is the most aggressive blend, with higher crack growth rates than observed in FGE, which contains less than 5% gasoline.

Chloride

Chloride is an important accelerant of ethanol SCC of carbon steels.[7, 8] Figure 5 shows that very minor SCC occurred in SSR tests with an aerated FGE simulant containing no added chloride. The addition of even a small amount of chloride (5 ppm) resulted in SCC and the severity of cracking did not increase with further additions of chloride. The fracture mode also changes from predominantly intergranular to predominantly transgranular as the chloride concentration is increased from 0 to 40 ppm. An example of mixed mode cracking in a simulated FGE (SFGE) containing 5 ppm chloride is shown in Figure 6.

Water

Water can act as an inhibitor of ethanol SCC but the required concentration is above 4.5% (Figure 7), which is significantly higher than the 1% maximum water specified by ASTM D-4806 for automotive application, see Table 1. Within the ASTM limits, water does not appear to have a significant effect on SCC.[8] The inhibiting effects of high concentrations of water in ethanol explains why SCC has not been reported in hydrous ethanol, which contains approximately 6% ethanol. While the SCC susceptibility decreases with increasing water content, the general corrosion rate of carbon steels increases with increasing water content of the ethanol. Furthermore, in ethanol-gasoline blends, addition of water may cause phase separation leading to enhanced corrosion in the water phase.

Other Factors

Methanol, which is a contaminant in FGE, affects the SCC potency. Methanol additions, within the ASTM limits, increased SCC susceptibility in SSR tests in aerated FGE.[7] Several environmental factors have a minimal influence on SCC in FGE. These include acidity, within the typical range for FGE, a common general corrosion inhibitor to protect against corrosion in automotive components, and the denaturant.[7, 8] The source of the FGE also does not appear to be important. A double blind round robin study was performed to investigate the effect of ethanol source and feed stock (corn, sugar cane, or grapes) on ethanol SCC using the SSR technique. Io obvious trends were identified. SCC was observed in SSR tests with FGE produced using all of the feed stocks. The only two ethanol samples that did not show any SCC were low volume percent ethanol-gasoline blends and therefore were not expected to cause SCC.

The metallurgy of the steel also does not appear to be particularly important. Neither the grade of the steel or the weld microstructure has a major influence on ethanol SCC behavior. The weld metal was slightly less susceptible to SCC than the base metal or heat affected zone, but the effect was not large.[9] The primary effect of welding appears to be to introduce residual stresses. The cast microstructure, found in pumps and valve bodies, also appears to be somewhat less susceptible to SCC than the wrought microstructure, as shown in Figure 8.[9]

Mitigation of SCC

The removal of oxygen by chemical (hydrazine at 1000 ppm concentration), mechanical, or electrochemical methods all resulted in suppression of SCC in SSR tests in a SFGE.[5, 6, 10] True SCC inhibitors also have been shown to be effective. Several inhibitor packages, prepared by commercial suppliers, have exhibited SCC inhibition in both SSR tests and the more realistic crack growth tests.[10] Simple ammonium hydroxide also has been found to inhibit SCC, as demonstrated in SSR tests (Figure 4) and crack growth tests with pre-cracked specimens, see Figure 9.

The association of ethanol SCC with high residual or cyclic stresses suggests that methods to reduce these stresses, or introduce residual compressive stresses are likely to be effective methods to mitigate ethanol SCC. These methods include adequate support of tank floors, post weld heat treatment of welds in piping systems, and shot peening or grit blasting wetted surfaces. These mitigation methods are the subject of ongoing research.

CORROSION OF STAINLESS STEELS AND NON-FERROUS METALS

A survey of pipeline operators was performed to identify the common stainless steels and nonferrous metals contained in pump stations and other pipeline facilities that are or may be used for ethanol service. The results of the survey are summarized in Tables 2 and 3. This information shows that a variety of stainless steels and non-ferrous metals are used in these facilities. These include aluminum alloys for floating roofs in tanks, bronzes and stainless steels in pumps and other components and several nickel base alloys in meters. Type 300 austenitic stainless steels as well as ferritic, martensitic and heat treatable stainless steels are commonly used in these facilities.

A literature search was performed for these materials in ethanol and relatively little information was found. The available information is summarized in Table 4. A review of their performance in ethanol suggests that aluminum alloys are not compatible metallic materials in ethanol. Aluminum has exhibited pitting and SCC in ethanol. This observation is of considerable interest to the pipeline industry given the extensive use in tanks of floating roofs made of aluminum alloys.

With the exception of brasses and other copper alloys that contain significant concentrations of zinc, copper base alloys, nickel base alloys, and stainless steels are probably compatible in ethanol, but more testing is needed on SCC behavior given the limited information on this failure mode and the SCC experience with carbon steels.

Research is ongoing to address these issues. This includes electrochemical testing to evaluate the localized corrosion behavior of these alloys, SSR testing to asses SCC susceptibility, and long-term exposure of stressed and unstressed coupons to evaluate general and localized corrosion as well as SCC. Figures 10 and 11 summarize the results of corrosion rates measured from cyclic potentiodynamic polarization (CPP) tests performed on several alloys in SFGE and an E-50 ethanol gasoline blend. While the copper – 10% tin bronze exhibited the highest corrosion rate in the tests, all of the corrosion rates were low. However, the aluminum alloy, the bronze, and the nickel alloy exhibited evidence of susceptibility to pitting corrosion.

Given the excellent performance of ammonium hydroxide as an SCC inhibitor for carbon steel, CPP tests were performed to evaluate its effect on the corrosion behavior of these alloys. Figure 12 shows that ammonium hydroxide had no effect on the general corrosion behavior of these alloys in deaerated SFGE. Pitting corrosion also was not affected.

PERFORMANCE OF NON-METALLIC MATERIALS

The survey of pipeline operators included the identification of non-metallic materials in pump stations and other pipeline facilities that are or may be used for ethanol service. Results are summarized in Table 5. This table shows that a number of different types of elastomers and plastics are found in these facilities including Buna N (nitrile) rubber, Viton®, which is a fluroelastomer, TFE (Teflon), polyether ether ketone (PEEK), nylon, polyurethane, Armstrong TN 9004, and Fiberglass.

The majority of these materials is used in elastomeric seals and gaskets and may undergo volume swelling, softening, and permanent set upon exposure to ethanol-gasoline blends. Table 6 summarizes the results of the literature search performed on these materials in ethanol. This table shows that, with the exception of polyurethane, most of the materials would be expected to be compatible with ethanol. Information on the behavior of Armstrong TN 9004 and fiberglass was not found in the literature. An issue with fiberglass is that a number of different resins are used and the specific type present in the pipeline facilities was not known.

Furthermore, very little data are available on gasoline blends with ethanol. Even FGE contains around 3.5% gasoline. An investigation of a number of elastomeric materials in contact with ethanol-gasoline blends indicated that the volumetric swell increases with gasoline content of the blend and was higher for E-20 (20 percent ethanol in gasoline) than either pure gasoline or 95% ethanol blend. Typical data comparing swell in E-20 and gasoline are shown in Figure 13. Viton GF, Viton GFLT, and Teflon samples were determined to offer the best hardness retention and least volume swelling.

FUTURE RESEARCH

Exposure testing of three to six months is being started for a number of stainless steels, non-ferrous metals, and elastomers in one lot of corn based FGE, SFGE, and E-50 prepared with SFGE and gasoline. The metallic specimens will be stressed to evaluate SCC and will contain crevice washers to evaluate localized corrosion. Performance of the elastomers will be evaluated based on changes in weight, hardness, and dimensions.

Additional work on SCC inhibitors is needed. For example, ammonium hydroxide is a promising SCC inhibitor but its influence on downstream components has not been evaluated. Furthermore, the dosing rates and the best method for dosing need to be established. Additional work on the effect of ethanol-gasoline blends on dynamic seals also is needed.

ACKNOWLEDGMENTS

The recent research presented here has been sponsored by PRCI and PHMSA. A number of individual pipeline companies have supported this research.

REFERENCES
1. Environmental Protection Agency, 40 CFR Part 80, Regulation of Fuels and Fuel Additives: 2012 Renewable Fuels Standards, Federal Register, 77 (5), January 9, 2012; Rules and Regulations, p. 1320.

2.   Environmental Protection Agency, partial grant of clean air waiver application submitted by Growth Energy to increase the allowable ethanol content of gasoline to 15 percent; Decision of Administrator, Federal Register, 76 (17), January 26, 2011/Ïotices.

3.   API Technical Report 939-D, R. D. Kane and J. G. Maldonado, "Stress Corrosion Cracking of Carbon Steel in Fuel Grade Ethanol: Review and Survey," American Petroleum Institute, Washington, DC, September 2003.

4.   Energetics, Inc., Safe & Reliable Ethanol Transportation & Storage Technology Roadmapping Workshop, Dublin, Ohio; October 25 – 26, 2007.

5.   J. A. Beavers, F. Gui, and Į. Sridhar, "Eff ects of Environmental and Metallurgical Factors on the Stress Corrosion Cracking of Carbon Steel in Fuel-Grade Ethanol," Corrosion, 67 (2), 025005-1 – 15, 2011.

6.   F. Gui, Į. Sridhar, and J. A. Beavers, "Localized Corrosion of Carbon Steel and Its Implications on the Mechanism and Inhibition of Stress Corrosion Cracking In Fuel-Grade Ethanol," Corrosion, 66, (12), p. 125001, 2010.

7.   Į. Sridhar, K. Price, J. Bu ckingham, and J. Dante, "Stress Corrosion Cracking of Carbon Steel In Ethanol," Corrosion, 62 (8), p. 687 – 702, 2006.

8.   J. A. Beavers, M. P. H. Brongers , A. K. Agrawal, and F. A. Tallarida, "Prevention of Internal SCC in Ethanol Pipelines," ĮACE, Corrosion 2008 Conference & EXPO, Ïew Orleans, LA, Paper Ïumber 08153, March 2008.

9.   J. A. Beavers and Į. Sridhar, and C. Zamarin, "Effects of Steel Microstructure and Ethanol-Gasoline Blend Ratio on SCC of Ethanol Pipelines, ĮACE, Corrosion 2009 Conference & EXPO, Atlanta GA, Paper Ïumber 095465, March 2009.

10. J. A. Beavers, F. Gui, and Į. Sridhar, "Recent Progress in Understanding and Mitigating SCC of Ethanol Pipelines," Corrosion 2010, Conference & EXPO, San Antonio, TX, Paper Ïumber 10072, March 2010.

Table I.   Specification for Fuel Grade Ethanol.

Requirement	ASTM D4806 LIMITS	
	Minimum	Maximum
Ethanol, vol. %	92.1	–
Methanol, vol. %	–	0.5
Solvent-washed gum, mg/100 ml	–	5.0
Water content, vol. %	–	1.0
Denaturant content, vol. %	1.96	4.76
Inorganic chloride, ppm (mg/L)	–	5 (4)
Copper, mg/kg	–	0.1
Acidity (as Acetic Acid CH_3COOH), mass % (mg/L)	–	0.007 (56)
pH_e	6.5	9.0

Table II.   Summary of non-ferrous metals in pipeline facilities.

MATERIAL	MATERIAL DESCRIPTION	COMPONENT
7075 Aluminum	An Al-Zn-Mg-Cu precipitation hardenable alloy	Rotor hub in meter
3000 Series Aluminum Alloys	Work hardenable Mn-Al alloys	Floating roofs for storage tanks
6000 Series Aluminum Alloys	precipitation hardenable Al-Mg-Si alloys	Floating roofs for storage tanks
Bronze	Copper alloys, with tin, aluminum, phosphorus, manganese, or silicon.	Pump components
Ni 200	99% pure nickel alloy.	Rotor blade in meter
Hi mu 80	80% nickel-iron-molybdenum alloy.	Rim button in meter

Table III.   Summary of stainless steels in pipeline facilities.

MATERIAL	MATERIAL DESCRIPTION	APPLICATIONS
CA6NM	A hardenable cast Fe-Cr-Ni-Mo alloy	Pump components.
302SS	Cr-Ni austenitic stainless steel.	Valves and meters components.
303SS	Cr-Ni austenitic stainless steel designed for improved machinability.	Valve and meter components.
304SS	Cr-Ni austenitic stainless steel with lower carbon content than 302SS.	Valve and meter components.
316SS	Cr-Ni austenitic stainless steel with 2 to 3 % Mo	Pump, valve and meter components, throttle sleeve, throat sleeve, mechanical seal.
317SS	Cr-Ni-Mo austenitic stainless steel 316SS.	Stud in pump.
18-8	Stainless steels having approximately 18% Cr-8% Nil.	Pump, valve and meter components.
17-7 PH	A Cr-Ni-Al precipitation hardening stainless steel	Valve and meter components.
416	A heat treatable martensitic free-machining stainless steel	Screw and shaft for mechanism in meter. Shaft in pump.
420	Heat treatable Cr stainless steel.	Pump and meter Components. Casing ring. Impeller ring. Throttle bushing. Throat bushing. Throttle sleeve. Throat sleeve.
430F	A low carbon ferritic stainless steel that contains Mo.	Screw for rotor in meter .
440C	A hardenable martensitic stainless steel.	Meter components.
630/17-4 PH	Martensitic stainless steel that is capable of precipitation hardening	Valve and meter components.

Table IV.   Summary of results of literature survey on corrosion of stainless steels and non-ferrous metals in ethanol.

MATERIALS	COMPATIBILITY	COMMENTS
Al Alloys	Not compatible	Pitting and SCC
Cu Alloys	Bronze and higher Cu-base alloys probably compatible; Brass is probably not compatible	Severe corrosion of brass
Ni-base Alloys	Probably compatible	Ni plating recommended; SCC and pitting unknown
Stainless Steels	Probably compatible	No SCC found on 316L SS

Table V.   Summary of elastomers/plastics identified in pipeline facilities.

MATERIAL	MATERIAL DESCRIPTION	COMPONENTS AND APPLICATIONS IN PUMP STATIONS
Fiberglass	Fine fibers of glass used as a reinforcing agent for polymers	Sump tank.
Buna N (nitrile)	Copolymer of butadiene and acrylonitrile.	O-ring in meter.
Polyurethane	Can be categorized as a polymer or elastomer.	Sphere in prover.
TFE (Teflon)	A synthetic fluoropolymer of tetrafluoroethylene.	Body seal and stem packing in ball valve.
Polyether ether ketone (PEEK)	A high performance thermoplastic generally used with fiber reinforcements such as glass, carbon, or Kevlar.	Pump and meter components
Viton®	An FKM fluoroelastomer. There are various grades of Viton®.	Pump, valve, pressure switch, pressure transducer, and meter components.
Viton® B	A specific grade of Viton®.	Mechanical seal and seal in pump.
Nylon	Generic designation for a family of synthetic polymers known generically as polyamides	O-ring in surge relief flow valve.
Armstrong TN 9004	A heavy-duty, high-density material with fully cured nitrile butadiene rubber binder.	Casing gasket in pump bearing housing. Flange gasket in mainline pump.

Table VI.   Summary of results of literature survey on degradation of elastomers/plastics in ethanol.

MATERIALS	COMPATIBILITY	COMMENTS
Polyurethane	Not compatible	Known to degrade in FGE. High volume change in ethanol.
Nitrile (Buna N)	Probably compatible	Minimal volume change in FGE.
Nylon	Probably compatible	Successfully used in FGE.
Teflon	Compatible	Minimal volume change in ethanol.
PEEK	Compatible	Excellent resistance to ethanol and M-85.
Viton®	Compatible	Minimal volume change in FGE.

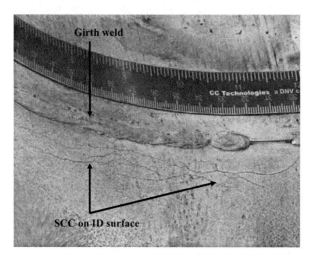

Figure I.   SCC observed in terminal piping containing FGE.

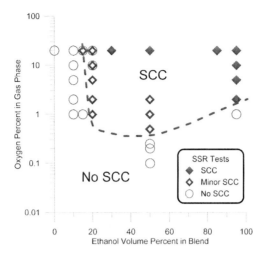

Figure II.   Effects of oxygen and ethanol concentration in ethanol-gasoline blends on the occurrence of SCC in slow strain rate tests of a carbon steel in simulated fuel grade ethanol [after 5].

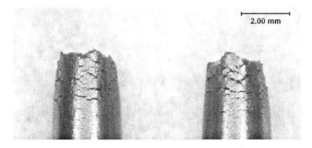

Figure III.   Optical photograph of gage section of X60 line pipe steel specimen tested in SFGE.

Figure IV.   Optical photograph of gage section of X60 line pipe steel specimen tested in SFGE containing an SCC inhibitor (500 ppm of 30% NH$_4$OH).

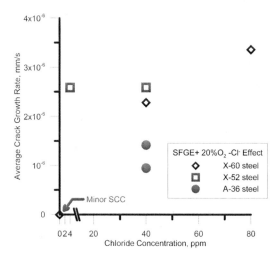

Figure V.   Average crack-growth rate from SSR tests as a function of chloride concentration added to aerated FGE simulant (after 5).

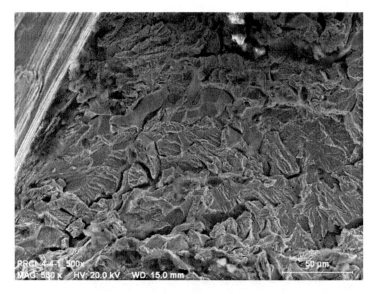

Figure VI.   Scanning electron microscope photograph of the fracture surface of a notched SSR specimen tested in SFGE containing 5 ppm chloride.

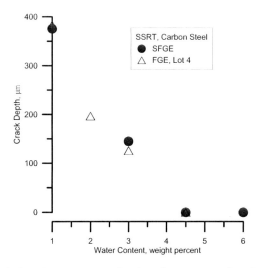

Figure VII.   Crack depth in SSR tests as a function of water content for a SFGE and one lot of corn based FGE [after 5].

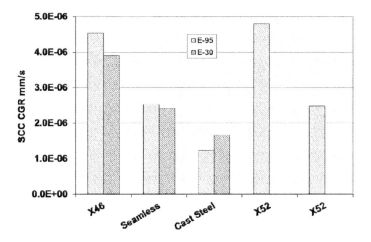

Figure VIII.   Crack growth rate (CGR) for notched tensile specimen of several steels tested in two SFGE-gasoline blends [after 9].

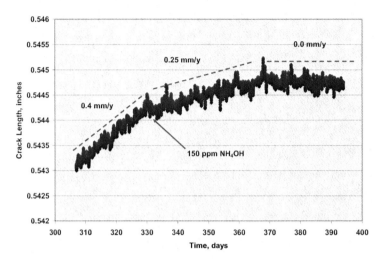

Figure IX.   Crack length and growth rate as a function of time for pre-cracked compact type specimen of an X-46 line pipe steel cyclically loaded in SFGE. 500 ppm of 30% NH₄OH added on Day 330 [after 10].

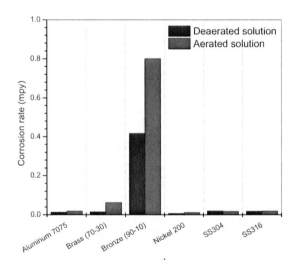

Figure X.   Corrosion rates, from cyclic potentiodynamic polarization tests, of alloys in aerated and deaerated SFGE (E-95).

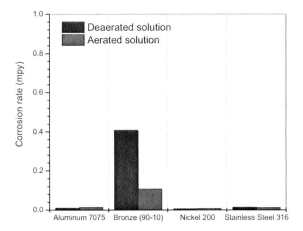

Figure XI.   Corrosion rates, from cyclic potentiodynamic polarization tests, of alloys in aerated and deaerated E-50 prepared with SFGE and gasoline.

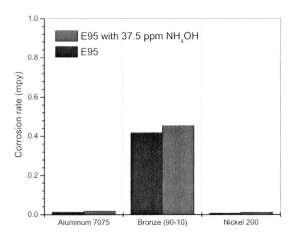

Figure XII.   Effect of 37.5 ppm NH₄OH on the corrosion rates, from CPP tests, of alloys in deaerated SFGE.

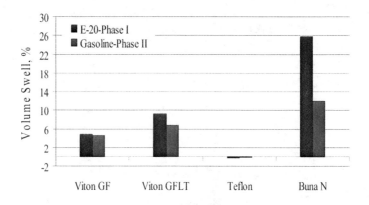

Figure XIII.   Effect of 28 days fuel transitions from E20 to Neat Gasoline on swelling of several elastomers.

DISTRIBUTED HYDROGEN GENERATION AND STORAGE FROM BIOMASS

Peter J. Schubert[a], Joseph Paganessi[b], Alan D. Wilks[b], Maureen Murray[b]
[a] Indiana University-Purdue University Indianapolis, Indiana, USA
[b] Packer Engineering, Inc., Naperville, Illinois, USA

ABSTRACT
    A new paradigm is presented for local generation and storage of hydrogen, in which biomass is the only feedstock. A novel thermochemical conversion method produces hydrogen gas which can then be separated. The mineral ash produced contains carbon and silica. With proper stoichiometric ratios, post-processing of this ash produces pure silicon. This silicon can be converted into a storage media for hydrogen suitable for stationary or portable applications. In this way, local plant matter can be used to renewably produce and store hydrogen for fuel cells, prime movers, or for heat. This paper presents experimental results and overall system metrics.

INTRODUCTION
    The US Energy Information Administration projects global energy use increasing 50% by 2030[1]. This is equivalent to installing 50 new mega-nuclear facilities every single year. That is unlikely to happen. It is believed by most experts that peak oil production has occurred globally[2] with the only remaining questions being how early will the plateau pass, and how quickly will be the decline in global production? With demand for petroleum products rising around the world, prices will rise and supplies will be limited. As concerns for resources and the environment intensify there is pressure to reduce energy production from coal, which will cause prices to rise. These factors speak to a softening of global economies and an increase in prices of most physical commodities, and in particular: energy.
    Renewable energy has been studied for decades yet non-hydro sources remain a minuscule portion of global energy production. As the impact of energy on economy, international relations, and human health becomes more and more significant, there will be increasing emphasis on developing alternatives. Two significant obstacles to renewable energy development are the lack of existing infrastructure (e.g. hydrogen), and the challenges of building large-scale installations (e.g. biomass). In addition to the technical hurdles for renewable energy, there exists within many US States considerable regulatory impediments and public opposition for new sources of energy[3]. Federal agencies heavily favor large-scale projects, several of which have famously flamed out. This situation is driven by the paradigm that large, centralized energy production with massive infrastructures provides the best way to address energy needs.
    The coming gap between traditional and renewable energy sources will likely last long enough and be economically painful enough to preclude large-scale, infrastructure-reliant solutions. Already in most of the world conditions preclude heavy investment in large, interconnected energy systems. Together these factors point to the need for distributed renewable energy where producers and users are local to one another. The purpose of this work is to outline a system which heavily favors local, distributed generation of energy and fuel.
    The technologies presented here provide a means for a farm operation (or other facility with access to biomass) to generate and store their own fuel renewably. One solution is for the farmer to own all the technology needed, and switch to self-generated fuel immediately. However, this

is capital intensive, and therefore will be less attractive in a protracted global recession. To address this, a system is presented which allows the farmer to start with a modest investment, then barter for additional energy generation and storage hardware and a portion of energy generation, using locally-produced raw materials for both as the medium of exchange.

METHOD

Figure 1 shows a schematic of a biomass hydrogen generation and storage system for local generation of fuels, electricity, and heat energy.

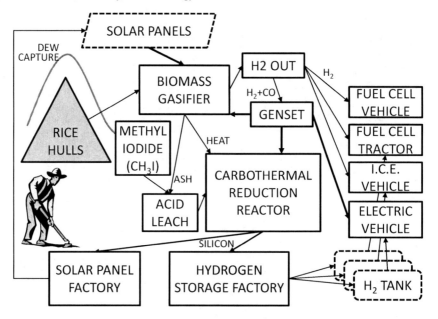

Figure 1. Self-perpetuating system for farm fuel and energy.

The process is based on rice hulls, the tough fibrous bran removed from rice grains in preparation for human consumption. Worldwide rice hull waste generation is enormous, some 660 million metric tons per year. Most of this biomass is left in heaps that decompose anaerobically, generating methane, which is a greenhouse gas 23 times more potent than carbon dioxide. Converting this waste to fuel not only displaces currently-used fossil fuels, but eliminates a potent source of greenhouse gases, making the system "carbon negative" – a sustainable outcome for the environment.

Rice hulls and other non-food agricultural residues can be processed thermophysically by gasification. A newly-invented method of gasification, developed by several of the authors[4], provides new advantages which make possible the system of Fig. 1. Gasification of biomass generates producer gas (hydrogen and CO plus some methane and some inert gases), heat, and a dry ash. All three output streams are put to use in multiple ways.

The producer gas can be fed directly into a generator set (a.k.a. "genset") to produce electricity. The genset may be an internal combustion engine modified for gaseous fuels, it may be a microturbine designed for high hydrogen and low methane levels, or it may be a solid oxide fuel cell (SOFC) which is tolerant of the CO that can poison other types of fuel cells. Hydrogen can be selectively removed from the producer gas via a suitable membrane to provide a pure fuel suitable to the types of fuel cells likely to be installed on vehicles, such as a polymer electrolyte membrane (PEM) fuel cell.

Rice hulls have a high mineral ash content[5] and this ash is rich in silica (SiO_2). Post-gasification biomass ash also includes soil nutrients such as potassium and phosphorous. The ash also may contain carbon in amounts which depend on the moisture level of the feedstock, and on the particulars of the gasifier operation. This mix can be processed to produce photovoltaic-grade silicon[6].

Rice hull ash must be acid-leached prior to processing to remove the phosphorus, which behaves as a dopant in silicon, and would render it unsuitable for photovoltaics if present in too large an amount. Note that it may be possible to recover phosphorus, which is a limited resource[7], and one likely to experience shortages in the future. Leaching can be performed with 1.25 M HCl plus acetic acid[8]. These reagents can be purchased, but it is also possible to produce acetic acid on the farm. This can be accomplished by first generating methanol from producer gas using a Zn-Cr or Cu catalyst at high pressure. Converting methanol to acetic acid can be done over an iron catalyst at moderate pressures with some addition of steam[17]. This reaction is facilitated by methyl iodide (CH3I), which is emitted from rice plantations, and can be harvested by gathering and separating the morning dew[9]. Leached ash is now ready for processing into silicon.

Carbothermal reduction is the means by which carbon and silica can be reacted to produce pure liquid silicon plus volatile gases such as CO, $SiO(g)$, and $SiC(g)$[10]. This process requires a suitable ratio of C to Si in the starting material, a reducing environment (readily available with hydrogen), and sufficient temperature (1500 to 2100 C) to favor Si over CO (see Figure 2)[11]. Adjusting the C:Si ratio is now possible with a new invention[12], such that the novel gasifier discussed above can automatically produce the proper feedstock. If CO from the producer gas is included in the cover gas, the partial pressure will suppress CO generation during carbothermal reduction, thereby permitting lower temperature operation, possibly down to 1500 C, thereby ameliorating materials selection challenges.

Silicon metal produced by the farmer can be bartered for finished products with two types of factories, both of which need the silicon as a raw material. One factory builds solar panels from the silicon[13]. The other factory produces hydrogen storage tanks based on silicon[14,18]. To achieve fuel independence early, the farmer gives preference to hydrogen storage tanks for installation on fuel cell tractors, vehicles, and converted internal combustion engine (ICE) vehicles. With enough storage, the farmer will next obtain solar panels (and an electric energy storage system) to help power the system of Fig. 1. The carbothermal reactor, in particular, operates on electric heating, so the more this can be supplied by solar panels, the more electricity is available to other farm operations.

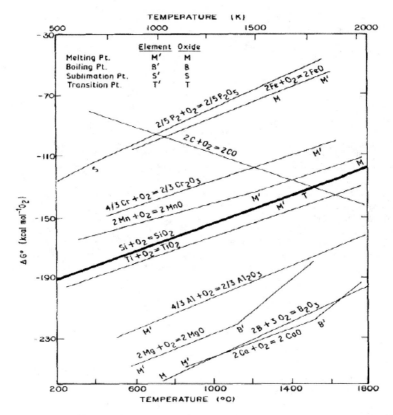

Figure 2. Free energy of formation of oxides of elements.

RESULTS

The novel indirectly-heated pyrolytic gasifier indicated in Fig. 1 is shown in Figure 3. This is the first of two full-scale prototypes capable of handling 1 to 3 tons per day of lignocellulosic material. Outputs are a dry producer gas of 300 BTU/cu.ft., mineral ash, and 250,000 BTU/hr. of waste heat from heat exchangers.

Figure 3. Novel, farm-scale indirectly-heated biomass gasifier.

Tests performed on the gasifier and on preliminary bench-top prototypes show the ability to produce ash ranging from a composition rich in char (fine, pure carbon charcoal, now called "biochar") to a white-gray ash with little or no char. To a first order the amount of char produced depends on feedstock moisture content. Assuming polymeric cellulose as the dominant plant material, a moisture content of approximately 20% or more will result in nearly-complete consumption of carbon into fuel gases. Dry feedstock, below 20% moisture, will result in a char-rich ash. By a combination of feedstock drying and steam injection, the carbon content can be subject to closed-loop feedback control using producer gas moisture content as the dependent variable. Once calibrated with char fraction for a given source material (wood will be different from corn stover, which will be different from rice hulls), this system has the ability to produce a C:Si ratio suitable for carbothermal reduction from a wide range of initial feedstock moisture content.

Electricity generation from the producer gas using the system of Fig. 3 has been demonstrated using a modified Lister-Petter 2.0 liter 4-cylinder spark-ignited ICE generating up to 15 kW of 3-phase 240 V AC power. Modifications include a custom-designed and controlled fuel injection system which mixes fuel and air prior to the intake manifold. A commercially-available "3-way" catalytic converter from a junked sport utility vehicle was used to reduce CO emissions in the exhaust stream. A separate project resulted in the development of a customized microturbine capable of producing 3-phase 480 V AC power in a high-reliability prime mover. Finally, although not yet tested, the producer gas generated by this novel thermophysical conversion system should be an ideal fuel for SOFCs[15]. SOFCs have high electrical conversion efficiencies relative to internal combustion engines, and once commercially available at competitive rates, will be the power generation technology of choice.

Carbothermal reduction of silicon is a commercially-available process, and was developed as early as the 1960s by companies such as Vulcan and ICI[11]. A key challenge is materials compatibility, a topic addressed by two researchers at the last meeting of the MCARE conference[8,16]. The heat required represents a significant parasitic draw on electric power generation. This can be reduced to some extent by using waste heat from the gasifier and the genset to pre-heat materials or the vessel. Final heating can come either from the combustion of hydrogen (2210 C in air) drawn as a slipstream off the producer gas, from electric arc heaters (large transformer and consumable graphite electrodes needed), or from conductive ceramic heater elements (e.g. SiC) if the process can be performed at the minimum temperatures of around 1500 C.

An economic model of how a farm operation can attain fuel self-sufficiency has been created to explain how such a scenario might evolve. The first step is for the farmer to purchase the minimum set of non-factory capital equipment shown in Fig. 1. Loan financing is assumed, with a 20 percent lump sum down payment on a $295,000 capital purchase price, followed by debt servicing through the 5 year period studied, assuming a zero interest rate. Three phases ensue.

1. Hydrogen is used for heat in the carbothermal reactor, producer gas is used to displace natural gas or propane use (e.g. for grain drying or space heating) and to generate acetic acid for leaching, with silicon bartered for hydrogen storage tanks.
2. Hydrogen is now used as a fuel for fuel cell vehicles, and possibly for portable electronics, silicon is now bartered for solar panels.
3. At full self-sufficiency, solar panels power much of the gasifier and carbothermal reactor, freeing more syngas for electricity, and for heat. Silicon is now sold on the commodity market, providing an additional "cash crop" for the farm operations.

Figures 4 through 7 show the economic results for this scenario. Figure 4 shows expenditures on capital, debt, and labor, the latter assumed at $30,000 per year. Figure 5 shows external purchases of diesel/gas, natural gas/propane, and electricity by the operation over time, with values approximating those of a 600 acre farm. Figure 6 shows the dedication of on-site generated hydrogen and silicon. Figure 7 displays the overall economic return on investment, with the solid line showing the payback. Break-even occurs within Year 3 in this scenario, corresponding to the end of the second phase listed above. Savings accumulate even faster starting in the third phase, when silicon is sold as a commodity. From this point onward, the farm operations are self-sufficient for fuel, and have a steady stream of a valuable, fungible commodity needed worldwide.

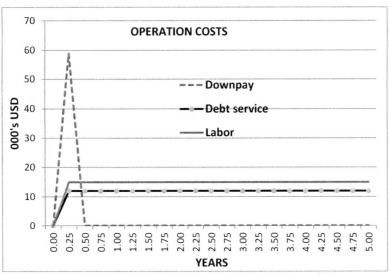

Figure 4. Non-fuel costs by year when using the system of Fig. 1.

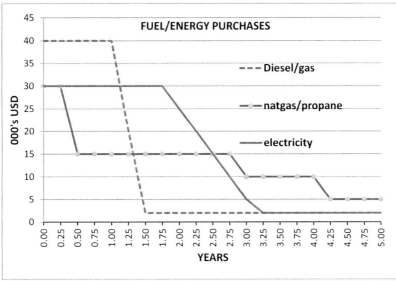

Figure 5. Fuel purchases by year when using the system of Fig. 1.

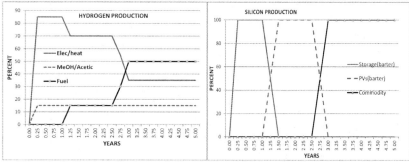

Figure 6. Dedication of hydrogen (left) and silicon (right) produced on-site from biomass.

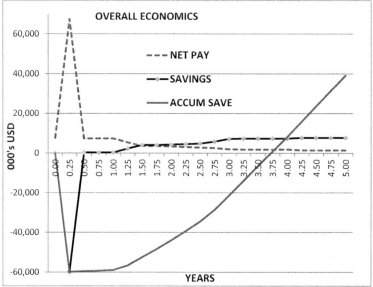

Figure 7. Overall economics for farm operations to reach fuel self-sufficiency within 4 years.

DISCUSSION AND SUMMARY

Many commercial operations generate biomass waste, notably the agricultural residues from food production. These wastes are a valuable source of energy and chemicals. It is now possible to convert and extract those resources with newly-developed technologies. The synergistic system described herein provides a novel means for energy self-sufficiency for farms and enterprises which generate lignocellulosic waste. This new system lies midway between the large-scale infrastructure-dependent US model, and the remote, rural, subsistence operations prevalent in much of the developing world. Local, modest-scale operations may not have the financial resources to purchase all the equipment needed for the system of Fig.1. This study describes a means for such operations to bootstrap up to fuel self-sufficiency, which relies on the existence of regional factories which produce finished goods and accept their own raw material as barter for those goods. By reducing the need for full financing, this sequence can be initiated with more reasonable levels of up-front capital investment. In this way, it may be possible to start the bootstrap sequence in villages or co-ops with the help of global financial institutions or charitable foundations.

A number of societally-important concerns can be lessened by such an approach. Biomass is used for cooking fuel in many remote settlements, resulting in desertification and in respiratory health problems associated with indoor combustion. Gasification captures more of the inherent chemical energy trapped in the biomass, so less biomass is needed for the same amount of energy delivered. Waste heat from the processes of Fig. 1 can also be used to provide smoke-free indoor cooking through a circulating loop of hot water fed through heat exchanger cook surfaces inside each dwelling. Atmospheric carbon emissions are reduced compared to current practices of allowing agricultural residues to decompose. It is even possible to sequester carbon via mixing biochar into topsoil. This soil augmentation practice helps boost yields by increasing water retention and providing frameworks for beneficial soil microbes. Valuable minerals extracted from the soil are concentrated in the ash, and used to help reduce the need for extractive mining and the associated environmental consequences.

Future developments related to the concept presented herein show even greater promise. Phosphorus costs are expected to rise considerably in the coming decades. This could become yet another fungible product from farm and biomass operations. Advances in silicon processing technology being developed for use on the moon may soon become viable for small-scale operations on earth. By adding a solar cell factory module and a hydrogen storage factory module to Fig. 1, an operation can become entirely self-sufficient. These are greater challenges, since many of the reagents needed are not readily available on farms (e.g. HF acid). With energy and commodity costs likely to increase dramatically, and soon, systems like this will be of increasing importance to maintaining and increasing the standard of living of humans everywhere.

ACKNOWLEDGEMENTS

This work was supported in part by USDA Rural Development Biomass R&D contract 68-3A75-5-607, Growth Dimensions of Belvidere County (Illinois) Biomass Commercialization Award Program, Illinois Department of Commerce and Economic Opportunity grant 07-0190, and the US Department of Energy grant DE-EE0001732. The authors are grateful to the engineering and technician staff who worked at Packer Engineering during the conduct of much of this work.

CORRESPONDING AUTHOR
Peter J. Schubert, Ph.D., P.E.
Director of the Richard G. Lugar Center for Renewable Energy
Professor of Electrical and Computer Engineering
IUPUI
pjschube@iupui.edu

REFERENCES
[1] U.S. Energy Information Administration, International Energy Outlook 2009, Report DOE/EIA-0484, 27 May 2009.
[2] Hirsch, R.L., Bezdek, R.H., Wendling, R.M., The Impending World Energy Mess, Apogee Prime 2010.
[3] http://www.dsireusa.org/ Dept. of Energy, EERE, North Carolina Solar Center.
[4] Pagnessi, J.E., Schubert, P.J., Wilks, A.D., "Biomass Gasification/Pyrolysis System and Process," US 20100275514, 4 Nov., 2010.
[5] http://www.ecn.nl/phyllis/, Energy Research Center of the Netherlands (ECN).
[6] Bose, D. N. , Govindacharyulu, P. A., and Banerjee, H. D., "Large Grain Polycrystalline Silicon from Rice Husk," Solar Energy Materials, vol. 7, (1982) pp. 319-321.
[7] Vaccari, D.A., "Phosphorus Famine: The Threat to Our Food Supply," Sci. Am. June 2009.
[8] Larbi, K.K., "Synthesis of Solar grade Silicon from Rice Husk Ash – An Integrated Process," Matls Challenges in Alt. and Renewable Energy Conf., Cocoa Beach, FL, 2010, Feb 21-25.
[9] K. R. Redeker, N.-Y. Wang, J. C. Low, A. McMillan, S. C. Tyler, and R. J. Cicerone (2000). "Emissions of Methyl Halides and Methane from Rice Paddies". Science 290 (5493): 966–969.
[10] Ceccaroli, B. and Lohne, O., "Solar-grade Silicon Feedstock," in Handbook of Photovoltaic Science and Engineering, Edited by A. Luque and S. Hegedus. Hoboken, NJ: Wiley, 2003, pp. 154-160.
[11] Bathey, B. R. and Cretella, M. C., "Review Solar-Grade Silicon," Journal of Materials Science, vol. 17, (1982) pp. 3077-3096.
[12] Schubert, P.J., "SYSTEM AND METHOD FOR CONTROLLING CHAR IN BIOMASS REACTORS," 3 Nov. 2011, US 20110266500.
[13] Schubert, P.J., "Solar Panels from Lunar Regolith," Intl. Space Dev. Conf., 28-31 May, 2010, Chicago, IL.
[14] Yen, D.W., Zhang, J, Christenson, J.C., Schubert, P.J., "APPARATUS AND PROCESS FOR FORMING AND HANDLING POROUS MATERIALS," 26 Jun. 2008, US 2008149594.
[15] Carter, B.C., Ceramic Materials, Springer, 2007.
[16] Mede, M., Italiano, P., Cooke, S. "Rice Hulls to Solar Grade Silicon," Matls. Challenges in Alt. and Renewable Energy Conf., Cocoa Beach, FL, 2010, Feb 21-25.
[17] Hokanson, A.E., and Rowell, R.M., "Methanol from Wood Waste: A Technical and Economic Study," USDA Forest Service General Tech. Report FPL 12, June 1977.
[18] Christenson, J.C., and Schubert, P.J., "PROCESSES AND APPARATUSES FOR PRODUCING POROUS MATERIALS," 13 Nov. 2008, US 20080277380.

Electric Grid

GALLIUM NITRIDE FOR GRID APPLICATIONS

Mike Soboroff
Program Manager – Power Electronics
Office of Electricity Delivery and Energy Reliability
U.S. Department of Energy

ABSTRACT

The U.S. Department of Energy's Office of Electricity Delivery and Energy Reliability (OE) is investing in the development of gallium nitride on silicon (GaN-Si) semiconductors to create advanced power electronics (PEs) as a part of an overall strategy to transition to the next-generation electric grid. GaN-Si PEs are expected to operate at significantly higher power levels than those fabricated on silicon and have the potential for cost effective, high volume manufacturing. The long range goal is devices operating at 20 kilovolts (kV) and 50 amps (A). Work to date has resulted in normally-off lateral GaN-Si transistors with operating voltages in excess of 1.5 kV. A variety of strategies are available to improve future performance of these devices and these will be explored as the program proceeds.

INTRODUCTION

As the United States transitions to a digital economy, the need to upgrade the nation's aging electric grid is becoming increasingly evident. Electricity demand is projected to increase by 30% between 2008 and 2035,[1] and the U.S. electricity delivery system must be able to meet this demand and ensure the continued supply of reliable, secure electricity.

Power electronics (PEs) will play a critical role in transforming the current electric grid into the next-generation grid. Today PEs operating at high power levels largely use a type of a silicon based semiconductor control device called a *thyristor*. They are usually stacked together into banks in order to sustain and manipulate the power levels required. However, silicon (Si)-based semiconductor technology cannot handle the power levels and switching frequencies required by next-generation utility applications—hundreds of kilovolts (kV) blocking voltages at tens of kilohertz (kHz) without the use of a significant amount of auxiliary support equipment.

PE devices based on wide bandgap (WBG) semiconductor materials, such as silicon carbide (SiC) and gallium nitride (GaN) could increase the reliability and efficiency of the next-generation electric grid without the need for supporting equipment. These materials are capable of supporting higher switching frequencies (kHz) and blocking voltages (upward of tens to hundreds of kV), while providing for lower switching losses, better thermal conductivities, and the ability to withstand higher operating temperatures with significantly reduced PEs footprints and weights. A number of challenges exist in using WBG PE devices to their full potential, including identifying new device topologies for high-power grid applications; developing the ability to consistently deliver robust devices; and creating a cost-effective, high-volume manufacturing process.

SiC vs. GaN

Given the current limitations of high-voltage silicon-based PEs, two material systems receive the most attention as potential replacements: silicon carbide and gallium nitride. SiC has been under investigated for the past 20 years, with great progress made towards high-voltage applications. SiC PEs currently operate at over 1,000 V; however, issues still remain regarding device reliability. One reliability concern is due to the fact that SiC can exist in multiple crystalline orientations. This material issue can adversely affect the device's reliability. A second limitation of SiC devices is the high cost associated with SiC material growth and production.

GaN material is also relatively expensive, but the cost of GaN has fallen much faster and in a shorter time period of study compared to SiC due to the multi-billion dollar markets enabled by solid-state lighting and low-voltage power electronics. However, GaN development has not progressed far enough along to allow for bulk material growth for fabricating cost-effective, high power electronics. While there are efforts advancing the development of bulk GaN, another avenue being pursued for low-cost medium and high power GaN PEs is the development of GaN epitaxial films on inexpensive, large diameter silicon substrates (GaN-Si). The use of GaN epitaxy allows for unique design architectures and shows promise for creating devices for high-power applications.[2] Furthermore, the technology is scalable and compatible with existing semiconductor fabrication technology. Figure 1 shows an example of a GaN-Si processed wafer.

Figure 1 GaN-on-Si wafer (Courtesy of Yole Développement and Imec)[3]

DOE-OE's FOCUS AND PATHWAY

The U.S. Department of Energy's (DOE's), Office of Electricity Delivery and Energy Reliability's (OE's) Power Electronics Research and Development Program[4] is investing in the development of gallium nitride-on-silicon (GaN-Si) based PEs to help realize the next generation electric grid. The program's stated goal is "To develop cost competitive, reliable power electronic (PE) devices/ components to improve functionality, reliability, and efficiency of next generation grid components and systems." The short-term goal (within five years) of the program is to demonstrate a device operating at greater than 5 kV and 15 A and longer term (5–15 years) to develop PE devices that can operate at greater than 20 kV and 50 A. Table I provides the specific approach for the GaN-Si developments.

Table I Program approach in OE's development of GaN-Si power electronics

Short Term (0–5 years)	• Conduct device and prototype component level R&D with commercially available GaN-Si substrates • Develop fabrication methods for GaN-Si devices through established wafer manufacturing methods • Design packaging for GaN-Si devices able to withstand heat loads and operating conditions Demonstrate a device operating at greater than 5 kV and 15 A
Long Term (5–15 years)	• Integrate GaN-Si components into power electronics modules and systems that can be incorporated into the grid Develop power electronic devices that can operate at greater than 20 kV and 50 A

The team designing and developing the GaN-Si PEs to meet short- and long-term goals outlined above includes Massachusetts Institute of Technology (MIT), MIT- Lincoln Laboratory (MIT-LL), and M/A-COM Technology Solutions (M/A-COM). Other collaborators include Oak

Ridge National Laboratory (ORNL) and Sandia National Laboratory (SNL). MIT will be predominantly responsible for device design and fabrication; MIT-LL will coordinate the development of the GaN-Si; and M/A-COM will perform advanced device fabrication and testing. ORNL and SNL will lend their expertise in the areas of device packaging and reliability testing. Establishing this collaboration and interaction from the outset enhances the chances for success in the development and commercialization of these next-generation power electronics.

CONSIDERATIONS FOR GaN-Si DEVICE DEVELOPMENT

There are some significant challenges to achieving the end goal of GaN-Si high power electronic devices. Some of the chief issues center on the design of the actual semiconductor device. The architecture currently under consideration is a dual-gate structure that is "normally-off." Normally-off ensures safety if and when the device is used in the electric grid, since the device would need to be intentionally activated for power to flow through it. The dual-gate structure—enhancement mode (E-mode) and depletion mode (D-mode) design—enables good control of the device as well as reduced on-resistance from the E-mode, while the D-mode allows for achieving very high breakdown voltages needed. Figure 2 shows the design of the dual-gate device.[5]

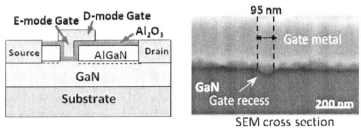

Figure 2 GaN-Si dual-gate enhancement mode transistor schematic and scanning electron microscope (SEM) image.

The other consideration is the geometry: vertical or lateral. This can impact current densities as well as power dissipation. Vertical devices can achieve higher current densities compared to lateral ones at higher breakdown voltages; however, when theoretical comparisons of vertical and lateral GaN-Si devices are made, the actual power densities of lateral devices are better. Lateral designs also require less GaN epitaxy, resulting in reduced device cost.

GaN-Si DEVICE RESULTS TO DATE

Figure 3 shows the current capability compared previous work for GaN-Si, GaN on silicon carbide, and GaN- Si devices where the silicon was removed. The stars inside the shaded oval reveal that the GaN-Si devices are beginning to achieve the results originally only seen on SiC substrates—approaching 2 kV breakdown voltages at comparable on-resistances. Figure 4 shows that silicon substrate removal technology is probably a pathway that could be used to improve the breakdown voltages for such GaN-Si devices.[6] Analysis shows that the silicon is what limits the breakdown voltage and after removal of the silicon and replacement with a non-conductive glass wafer, there appears to be a three-fold improvement in breakdown voltage.

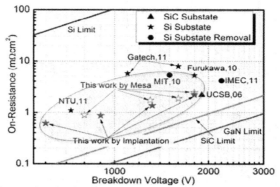

Figure 3 Breakdown voltages with GaN on various substrates. These results reveal GaN-Si breakdown voltage capability approaching the capability of GaN-Sapphire and GaN-SiC. Stars within the shaded oval represent the work supported by OE.

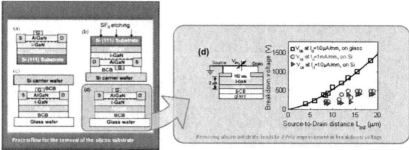

Figure 4 Substrate transfer technology (removing and replacing the silicon substrate) leads to significant improvements in breakdown voltages for GaN-based devices.

CURRENT DEVICES AND FUTURE PATHWAY

Based on the technology developments to date, this program has been able to reproducibly fabricate GaN on Si two-terminal devices with breakdown voltages >1,300V. Furthermore, on the question of manufacturability, the ability to produce 1.6 kV GaN-Si High electron mobility transistors (HEMT) in production environments with gold-free processing suggests that this technology could be transitioned to high-volume operations. There are a number of strategies currently being explored that could lead to higher power capabilities. These include the further exploitation of the substrate removal process, doping, and ion implantations to help increase the conductivities of the GaN films. To further help with thermal energy removal, the use of passive cooling materials will be exploited.

All these development options present a clear pathway towards the realization of high-power WBG PEs based on GaN-Si technology. Figure 5 shows the current status of the OE program to date along with the modeled development pathway for this program.

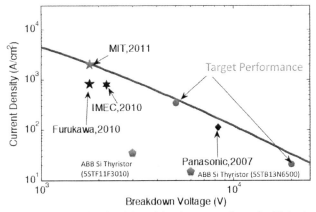

Figure 5 The modeled and anticipated development pathway for fabricating and realizing GaN-Si high power electronics.

CONCLUSIONS

The OE program's investment in developing GaN-Si power electronics has shown significant progress since its inception. Researchers are beginning to meet the power levels that are of value and interest to high power commercial and utility applications, that is devices with >1.5 kV breakdown voltages. In addition, there are a number of strategies that have been identified individually or in some combination that reveal a pathway forward to achieving higher breakdown voltages in these GaN-Si devices. Given development time, it is clear that meeting the intermediate and long-term goals of the OE program are entirely possible.

REFERENCES

[1] U.S. Energy Information Administration. *Annual Energy Outlook 2010.* Washington, DC: U.S. Energy Information Administration, 2010, 65–69.
http://www.eia.doe.gov/oiaf/aeo/pdf/trend_3.pdf
[2] Briere, Michael A. "GaN-based power devices offer game-changing potential in power-conversion electronics." *EE Times,* December 30, 2008.
http://www.eetimes.com/design/automotive-design/4010344/GaN-based-power-devices-offer-game-changing-potential-in-power-conversion-electronics
[3] "Dow Corning joins the imec GaN affiliation program." *I-Micronews.* January 2011.
http://www.i-micronews.com/news/Dow-Corning-joins-imec-GaN-affiliation-program,6305.html (accessed 31 January 2011).
[4] Office of Electricity Delivery and Energy. *Power Electronics Multi-Year Plan.* Washington, D.C.: U.S. Department of Energy, April 2011.

http://energy.gov/sites/prod/files/oeprod/DocumentsandMedia/OE_Power_Electronics_Progra m_Plan_April_2011.pdf

[5] Lu, B., O.I. Saadat, and T. Palacios. "High-Performance Integrated Dual-Gate AlGaN/GaN Enhancement-Mode Transistor." *Electron Device Letters, IEEE*31, no. 9 (September 2010): 990-992. http://www.mendeley.com/research/highperformance-integrated-dualgate-algangan-enhancementmode-transistor/

[6] Lu, B.; Palacios ,T. "High Breakdown (> 1500 V) AlGaN/GaN HEMTs by Substrate-Trasnfer Technology." *Electron Device Letters, IEEE*, 31, no. 9 (September 2010): 951-992. http://ieeexplore.ieee.org/xpl/freeabs_all.jsp?arnumber=5518357

Geothermal

HIGH-TEMPERATURE CIRCUIT BOARDS FOR USE IN GEOTHERMAL WELL MONITORING APPLICATIONS

Jennifer K. Walsh, Matthew W. Hooker, Kaushik Mallick, and Mark J. Lizotte
Composite Technology Development, Inc.
Lafayette, CO 80026

ABSTRACT

The mapping of geothermal resources involves the design and fabrication of sensor packages, including the electronic control modules, to quantify downhole conditions (temperature, pressure, seismic activity, etc.). Because of the extreme depths at which these measurements are performed, it is most desirable to perform the sensor signal processing downhole, and then transmit the information to the surface. The temperatures at these depths present a challenge for data logging because the multilayer circuit boards onto which the electronic modules are built were not specifically designed for use under these conditions. To address this issue, high-temperature circuit board materials are being developed for operation at temperatures up to 300°C. This work describes the development of multilayer circuit materials based on fiber-reinforced, high-temperature polymers. These materials have been found to exhibit suitable mechanical, electrical, and thermal properties for use in this application, while also being compatible with conventional manufacturing processes. The properties of these materials, as well as progress towards the fabrication of test circuits, are described in this report.

INTRODUCTION

The development of renewable and alternative energy sources will ensure the long-term energy independence of our nation. One of the key renewable resources currently being advanced is geothermal energy. Unlike wind and solar power, geothermal energy is not affected by changing weather and is therefore always available to meet peak power demands.

Geothermal power in the United States is currently produced from relatively shallow wells located primarily in California and Nevada. In these locations, energy is produced under nearly ideal circumstances that include porous rock and an ample supply of subsurface water. To tap into the large potential offered by generating power from the heat of the earth, and for geothermal energy to be more widely used, it will be necessary to drill deeper wells to reach the hot, dry rock located up to 10 km beneath the earth's surface. In this instance, water will be introduced into the well to create a geothermal reservoir. A geothermal well produced in this manner is referred to as an enhanced eothermal system (EGS) [1,2].

EGS reservoirs (Figure 1) are typically at depths of 3 to 10 km, and the temperatures at these depths have become a limiting factor in the application of existing downhole technologies. These high temperatures are especially problematic for electronic systems such as downhole data-logging tools, which are used to map and characterize the fractures and high-permeability regions in underground formations. Information provided by these tools is assessed so

Figure 1. Sensors are placed downhole in EGS reservoirs.

that underground formations capable of providing geothermal energy can be identified, and the subsequent drilling operations can be accurately directed to those locations.

Downhole signal processing requires the development of circuit boards that can withstand the elevated temperatures found at these depths. At present, the highest-temperature commercially available circuit boards are based on polyimide materials, and those have maximum use temperatures of 200 to 250°C [3].

In addition to thermal stability, downhole electronics must also be fabricated into high-aspect-ratio packages. For example, recent multilayer assemblies produced at Sandia National Laboratories (SNL) (Figure 2) were approximately 2.5 cm wide and 50 cm long [4]. Because of this very high form factor, glass-fiber-reinforced polymers are much more desirable than multilayer ceramic modules (MCM). MCMs have many advantages for some applications, but are susceptible to damage induced by the mechanical and vibrational loads

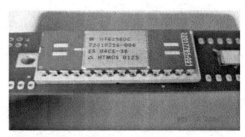

Figure 2. Multilayer data-logging circuit and surface-mount component previously produced by Sandia National Laboratories.

commonly experienced by data-logging tools. Thus, as EGS technology continues to advance, there is a strong need for multilayer electronics that can provide the necessary thermal performance while also being compatible with high-form-factor circuit designs.

EXPERIMENTAL APPROACH

In this work, both organic and inorganic polymer systems were evaluated for use in multilayer circuit production. The CTD-400-series materials described below are based on cyanate ester chemistries, whereas the CTD-1200-series materials are based on inorganic resin systems. For high-temperature circuit applications, the candidate materials must possess good thermal stability, mechanical strength, and high electrical resistivity, while also being compatible with conventional manufacturing processes.

In addition to the properties described above, these materials must also adhere strongly to copper. Copper is commonly used as the conductor in multilayer assemblies, and poor adhesion can lead to electromechanical failures at the copper/composite interfaces. In previous work [5], CTD developed screening procedure for assessing the adhesion composite materials to copper. A schematic illustration, as well as typical examples of the failure modes at the copper/composite interface, is shown in Figure 3. As shown below, a thin copper sheet is positioned at the center of a composite laminate and a short-beam-shear testing [6] is performed to assess the adhesion between the materials. For this study, 3.2-mm thick laminates were produced with a 0.17-mm thick copper sheet located at the mid-plane. The laminates were fabricated using a wet-layup process that approximates the fabrication of circuit boards.

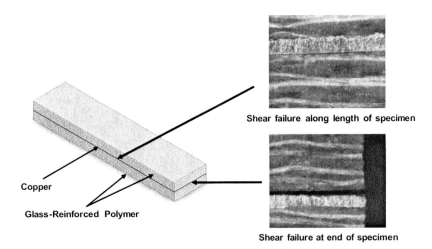

Shear failure along length of specimen

Shear failure at end of specimen

Figure 3. Specimen design and typical failure modes for copper/composite shear specimens.

Lap-shear testing was used to evaluate copper adhesion of two candidate materials (CTD-1212 and CTD-1280). That testing was conducted because the CTD-1280 material did not exhibit a short-beam-shear failure, and the second material was tested in this mode to provide a basis for comparative analysis. The copper/composite lap-shear specimens produced in this work were 19 cm long and 2.5 cm wide. The copper adherends were 0.24 cm thick, and the composite laminate was 0.5 mm thick.

Next, flat-plate composite laminates with a nominal thickness of 0.5 mm were produced for electrical testing. Characterization of the electrical properties included 30 kV withstand voltage and resistivity testing at 20 and 250°C, as well as measurement of the dielectric constant at 20°C.

Finally, the thermal characteristics of the best performing materials were evaluated. This included measurement of the thermal expansion in both the through-thickness and in-plane directions from 20 to 300°C, as well as Thermogravimetric Analysis (TGA) from ambient temperature to 800°C.

THEORETICAL APPROACH

In addition to the material testing and analysis, a finite element analysis (FEA) model was developed to predict the thermal stresses in high-temperature circuit boards based on the materials developed herein. The analysis focused on the interaction between the thermal expansion of the composite and copper in a typical via (vertical electrical connection between conductive layers of in the circuit board). The axi-symmetric model (Figure 4) consisted of a single representative via in a circuit board of dimensions significantly greater than the pin diameter [7].

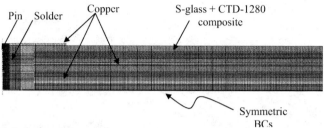

Figure 4. Axisymmetric finite element model of the pin, solder and via embedded in a circuit board with multiple copper layers

Symmetric boundary conditions (BC's) were imposed at the center of the board and ideal bond interfaces were assumed between the different components in the model. Orthotropic elastic properties for the composite material and thermal expansion in both in-plane and through-thickness directions were used to account for material non-linearity. The mechanical properties of the solder (assumed to be 100% lead) and the copper were defined by full elastic-plastic stress-strain data. The finite analysis consisted of multiple non-linear temperature increments ranging from 25°C to 300°C.

RESULTS AND DISCUSSION

Mechanical Performance and Copper Adhesion

As seen in Table I, two cyanate ester–based formulations (CTD-414 and CTD-430) exhibited good adhesive shear strengths to the copper at 20°C. Two of the inorganic systems, CTD-1202 and CTD-1205, did not bond to the copper, whereas CTD-1212 and CTD-1280 adhered to the metal layer. It should be noted that CTD-1212 and CTD-1280 are relatively low-modulus materials, so flexural testing does not necessarily provide a good estimate of shear strength. In the case of CTD-1280, the specimens deformed under load and returned to the original shape when the load was removed. This particular material did not exhibit a shear failure during this testing, and strong adhesion to the copper was evident throughout the fabrication, machining, and mechanical loading of these specimens. Figure 5 shows a photograph of a CTD-1280 short-beam-shear specimen after the part was subjected to mechanical deformation of 3 mm. As seen in this photograph, no adhesive failure between the copper and composite material was observed.

Table I. Shear strengths of candidate materials at 20°C.

Candidate Material	Short-Beam-Shear Strength at 20°C (68°F) (MPa)
CTD-414	46
CTD-430	50
CTD-1202	Did not adhere
CTD-1205	Did not adhere
CTD-1212	7.8
CTD-1280	Good adhesion; specimens did not fail in short-beam-shear test

Figure 5. Photograph of CTD-1280 short-beam-shear test specimen after mechanical loading. The internal copper layer remains adhered to the composite after 3-mm (0.12-in.) deformation.

In addition to room temperature testing, the short-beam-shear strengths of the copper/composite interfaces of the best-performing systems were also tested at 100 and 250°C. As seen in Figure 6, CTD-1212 and CTD-414 both retained approximately 80% and 65%, respectively, of their room-temperature strengths up to 250°C. Alternatively, the strength of the copper/composite interface in CTD-430 decreased significantly at 250°C. The higher mechanical strength of CTD-414 at 250°C (482°F), as compared with CTD-430, is attributable to the higher glass transition temperature of this cyanate ester–based formulation.

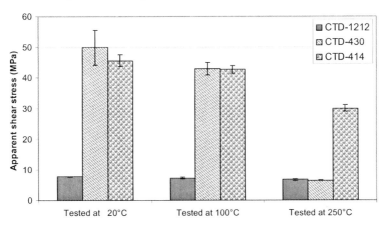

Figure 6. Short-beam-shear strength of CTD-1212, CTD-414, and CTD-430 composites with internal copper layers.

Because the CTD-1280 specimens did not fail in short-beam-shear testing, notched lap-shear specimens were fabricated and tested to assess the adhesion between the composite insulation and the copper. CTD-1212 lap-shear specimens were also produced to provide a basis for comparing data from the two test configurations. Once produced, the lap-shear specimens were loaded in tension at 20 and 250°C. As seen in Figure 7, the CTD-1280 composites exhibited strain-to-failures in excess of 2% at each test condition. Moreover, post-test evaluation (Figure 8) of the specimens indicated that the failure was within the composite itself (i.e., cohesive failure) and indicates that the material strongly adheres to the copper. Alternatively, the

CTD-1212 specimens possessed lower strain-to-failures at stress levels similar to those experienced by the CTD-1280 specimens. Furthermore, the CTD-1212 specimens failed at the copper/composite interface, indicating that the adhesive bond between the copper and composite is not as strong as that of CTD-1280. These results show that the CTD-1280 system exhibits both good strain tolerance and excellent adhesion to copper, both of which are needed for use in the production of multilayer circuits.

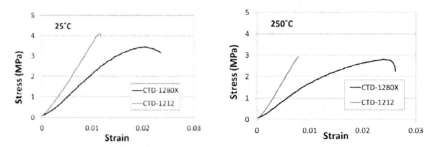

Figure 7. Typical stress/strain behavior of CTD-1212 and CTD-1280 at 25 and 250°C.

Figure 8. A) Adhesive failure at the CTD-1212 composite/copper interface and B) cohesive failure in the CTD-1280 composite.

Electrical Properties

Based on the results of the mechanical testing, CTD-1280 was selected for additional testing and evaluation. Characterization of the dielectric properties included the measurement of electrical resistivity and DC withstand voltage of composite laminates at 25 and 250°C, as well as measurement of the dielectric constant at room temperature. As seen in Table II, the resistivity of this material was greater than 120 $G\Omega$-cm (the limit of the test instrument) at 25°C, and 4.1 GΩ-cm at 250°C. These values are very high and indicate that the composite material is a very good electrical insulator at both ambient and elevated temperatures.

In addition, DC withstand voltage tests to assess the dielectric strength of the material. The material withstood applied voltages of 30 kVDC at both 25 and 250°C. These voltages are significantly higher than the anticipated operating voltages of the circuits and provide a good indication of both the high-temperature stability and quality of the insulator materials. The performance of the materials between 250 and 300°C is expected to be similar and consistent with the needs of circuit board design and operation.

Finally, the capacitance of CTD-1280 was measured at room temperature at 1 MHz, and the dielectric constant was estimated to be 4.2 under these conditions. This value is typical for other fiberglass-reinforced materials currently used in the manufacture of printed circuit boards.

Table II. Electrical properties of CTD-1280 at 25 and 250°C.

	25°C	250°C
Resistivity at 6 kVDC	>120 GΩ-cm	4.1 GΩ-cm
30 kVDC withstand voltage	Pass	Pass
Dielectric constant (1 MHz)	4.2	—

Thermal Properties

The thermal stability and thermal expansion properties of CTD-1280 were characterized to determine the upper use temperatures and expansion behavior of the composite materials. Thermogravimetric Analysis (TGA) was used to assess the thermal stability of of CTD-1280. As shown in Figure 9, CTD-1280 has a multistep decomposition, with the first step occuring at approximately 340°C and the second step at 440°C. The material retained nearly 80% of its mass, indicating that the material will withstand operating temperatures on the order of 300°C and that no significant degradation would occur below 400°C, which is well above the anticipated conditions for EGS applications.

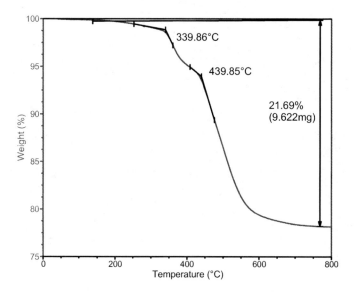

Figure 9. Decomposition behavior of CTD-1280.

The thermal expansion of CTD-1280 was measured from 25° to 300°C in both the through-thickness and in-plane directions (Figure 10). The thermal expansion in fiber-reinforced composites is highly dependent upon the direction in which the measurement is performed. In this instance, the through-thickness thermal expansion is approximately ten times higher than the in-plane expansion. The data in Figure 10 also includes copper, since this material is commonly used as the conductive material in multilayer circuits. The thermal expansion of the CTD-1280/fiberglass composite is similar to that of copper in the in-plane direction, and this is important because the differential stresses at the copper/composite interface will be minimal.

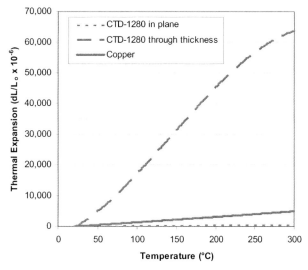

Figure 10. Thermal expansion of CTD-1280 in the through-thickness and in-plane directions as compared with copper.

FEA Modeling of Multilayer Assemblies

A finite element analysis (FEA) model was developed to estimate the magnitude of the stresses in multilayer circuits caused by the thermal expansion. The data collected in this work was used as the basis for this model, and a via (or through-thickness electrical connection) was modeled because this element of the circuit contains the most dissimilar materials. The through-thickness deformation of the model assembly is shown in Figure 11. The thermal expansion of the composite in the through-thickness direction is significantly greater than in the in-plane direction, which induces tensile stresses in the copper. Figure 12 shows a contour plot of the von Mises stress in the copper portion of the model. The maximum von Mises stress in the copper is estimated to be 94 MPa which is significantly below the ultimate failure stress of 210 MPa for copper at 300°C [8]. Thus, the stresses induced by the thermal expansion are not large enough to damage the circuit during operation at elevated temperatures.

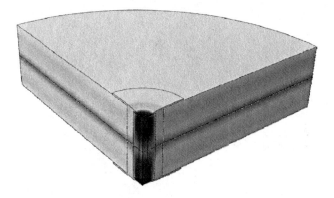

Figure 11. Contour plot of the through-thickness deflection in a multilayer circuit at 300°C.

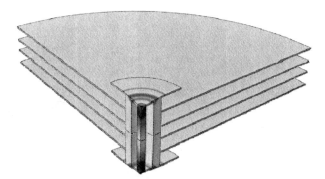

Figure 12. Contour plot of the von Mises stress in a via at 300°C.

FUTURE WORK

Testing of CTD-1280/fiberglass composites has shown that the material has suitable properties for use in high-temperature, multilayer circuit board applications. Based on these findings, CTD plans to fabricate and test prototype circuits fabricated with these new materials. The test articles will be fabricated and tested in collaboration with Calumet and Sandia National Laboratory. Circuit board testing at Sandia National Laboratory will include exposing the boards to temperatures up to 300°C and evaluating trace adhesion, pad adhesion, and material loss.

CONCLUSIONS

A new material for high-temperature circuit board applications was developed and tested. Testing included measurement of the mechanical, electrical thermal properties of the material at elevated temperatures. This new material, designated CTD-1280, was shown to exhibit suitable performance for use in high-temperature, downhole applications, and finite element analyses indicate that the thermal stresses induced during high-temperature operation are well within the capabilities of this material. Future work will include the fabrication and testing of multilayer devices to demonstrate their performance at the system level.

ACKNOWLEDGEMENTS

The work described herein was funded, in part, by the U.S. Department of Energy through grant number DE-EE0002751. The authors gratefully acknowledge this support, as well as the support of our collaborators, Sandia National Laboratory and Calumet.

REFERENCES

1. "The Future of Geothermal Energy: Impact of Enhanced Geothermal Systems (EGS) on the United States in the 21st Century." Prepared by the Massachusetts Institute of Technology under DOE Contract DE-AC07-05ID14517, 2006.

2. "Geothermal Technologies Program–Multi-Year Research, Development, and Demonstration Plan." U.S. Department of Energy, 2008.

3. "Specifications for Base Materials for Rigid and Multilayer Printed Boards." IPC-4101B, Association for Connecting Electronics Industries, June 2006.

5. M.W. Hooker, P.E. Fabian, S.D. Grandlienard, D.E. Codell, and M.J. Lizotte. "Shear Strengths of Copper/Insulation Interfaces for Fusion Magnet Applications." *Advances in Cryogenic Engineering* 52B:306–313 (2006).

6. ASTM Test Standard D2344/D2344M-00e1, "Standard Test Method for Short-Beam Shear Strength of Polymer Matrix Composite Materials and Their Laminates." ASTM International, West Conshohocken, PA, 2000.

8. N. van Steenberge, B. Vandevelde, I. Schildermans, and G. Willems, "Analytical and finite element models of the thermal behavior for lead-free soldering processes in electronic assembly", Proceedings of the 6th International Conference on Thermal, Mechanical and Multi-Physics Simulation and Experiments in Micro-Electronics and Micro-Systems, 2005.

9. Atlas of Stress-strain curves, ASM international, 2002.

SELF-DEGRADABLE CEMENTITIOUS SEALING MATERIALS IN ENHANCED GEOTHERMAL SYSTEM

Toshifumi Sugama, Tatiana Pyatina, and Thomas Butcher
Brookhaven National Laboratory
Upton, NY 11973

ABSTRACT

Objective is to develop temporary cementitious fracture sealing materials possessing self-degradable properties in Enhanced Geothermal System (EGS). The ideal sealer's self-degradation occurs when the sealer heated at temperatures of $\geq 200°C$ comes in contact with water. The study centers on formulating sodium silicate-activated slag/Class C or F fly ash blend cements and investigating the ability of sodium carboxymethyl cellulose (CMC) as thermal degradable additive to promote the disintegration of the sealer. We developed self-degradable slag /C or F fly ash blend cements to meet the following criteria including an initial setting time ≥ 60min at $85°C$, compressive strength >2000 psi for $200°C$ autoclaved specimen, and a self-disintegration. The calcium silicate hydrate and tobermorite crystal phases, and geopolymer as amorphous phase were identified as hydrothermal reaction products responsible for developing compressive strength. Sodium hydroxide derived from hydrolysis of sodium silicate activator not only initiated the pozzolanic reaction of slag and fly ash, but also played a vital role in generating *in-situ* exothermic heat that significantly contributed to promoting self-degradation of cementitious sealers. The source of this exothermic heat was interactions between sodium hydroxide and gaseous reactants generated from thermal decomposition of CMC additive in aqueous medium. In addition, we evaluated the usefulness of magnesium oxide, MgO, in expanding the volume of sealer. The MgO was converted into crystalline brucite, $Mg(OH)_2$ in hydrothermal environment. The *in-situ* growth of brucite crystals was responsible for the sealer's expansion, which improved their plugging performance by enhancing their adherence to the inner surface of fracture.

INTRODUCTION

A critical operation in assembling and constructing Enhanced Geothermal Systems (EGSs) is the creation of a hydrothermal reservoir in a hot rock stratum at temperatures $\geq 200°C$, located at ~ 3-10 km below the ground surface. In this operation, water is pumped down from injection well to stimulate the hot rock stratum. This hydraulic stimulation in terms of hydro-shearing, known as hydraulic fracturing, initiates the opening of existing fractures. Also, multi-injection wells are required to create a desirable reservoir of a permeable fracture flow network. After forming the reservoir, a production well is installed within the fracture's network.

During such construction of wellbores, operators pay considerable attention to the pre-existing fractures and the pressure-generated ones made in the underground foundation during drilling. These fractures in terms of lost circulation zones often cause the wastage of a substantial amount of the circulated water-based drilling fluid or mud. To deal with this problem, operators must seal or plug the lost circulation zones with the proper materials to avoid depleting the drilling fluid or mud. Once this problem is resolved, the drilling operation can resume and continue until the wellbore structure is completed. Thereafter, the hydraulic –stimulation process begins. Next, the inevitable concern is the fact that all sealing materials used to plug the fractures in hydrothermal reservoir's rock stratum during drilling operations to avoid lost circulation must be disrupted by water under a certain pressure to reopen the sealed fractures during fracturing operations.

In response to this need, the objective of this project is to develop temporary cementitious fracture sealing materials with self-degradable properties. The ideal sealer's self-degradation must take place when sealers at temperatures of ≥200°C come in contact with water. If successful, outcome of our research will reduce the total costs of the sealing and multi-fracture drilling operations, and lower the overall expense of raw materials, while eliminating the following three critical problems: 1) lost of circulation; 2) the need for additional isolation liners; and, 3) the managed pressure drilling. Unlike a conventional R&D project designing tough, durable cementitious materials, our current project aimed at developing a temporary cementitious sealer is required that sealer exhibits good mechanical-and sealing behaviors, while self-degrading under a combination of high temperature and water injection. Thus, the challenge we faced was to find ways how to disintegrate a hard hydraulic cementitious material in hot EGS wells by injecting water without any added chemical reagents.

In our previous DOE project aimed at developing cost-effective acid-resistant cements for the conventional geothermal wells, researchers at Brookhaven National Laboratory (BNL) formulated alkali-activated cementitious materials (AACMs) that resisted acid much better than did the OPCs [1,2]. The AACMs were prepared hydrothermally using two major components: One was inexpensive industrial by-products with pozzolanic properties, such as the slag from steel industries and the fly ashes from coal-combustion power plants; the other was sodium silicates along with various mol. ratios of Na_2O/SiO_2 as the alkali activators to initiate the pozzolanic reactions. Our assessment of these by-product resources strongly highlighted the beneficial environmental effects of their usage. According to the U.S. Geological Survey, in 2000, U.S steel industries generated 8.9 million tons of slag as wastes and by-products [3], and in 2009, 5.7 million tons of fly ash by-products were yielded from U.S. coal combustion power plants [4]. Of these amounts, 5.1 million tons of slag and 2.2 million tons of fly ash were recycled into industrial products. Thus, this abundant resource of these by-products as the starting materials of our cementitious sealer assures their local availability across the U.S.A.

From the standpoint of assuring the self-degradation of cementitious materials, the biodegradable biopolymers including the starch, cellulose acetate, gelatin, and poly($_L$-lactic acid) in the form of powder, microsphere, and fiber, are currently used as additives, which promote the partial biological degradation of biocompatible bone cements [5-8]. When these additives in bone cements come in contact with body fluids, they disintegrate, so creating high porosity therein, and assuring the development of inter-connective channels for the bone tissue's growth and facilitating the resorption of such cements. Our particular interest in biopolymers was their intriguing mechanism of thermal degradation. The cellulose and cellulose-related compounds are degraded thermally in air and water at temperatures around 200°C; yielding acetic acid as by-products [9-11]. This information inspired us to investigate the potential of cellulose compounds as thermal degradable additives to promote the self-degradation of temporary cementitious sealing materials at temperatures ≥ 200°C after the water has penetrated through the sealer.

Based upon this concept, our focus in this study centered on formulating AACMs, in particular, slag/Class C and F fly ashes combined system, as the binder in temporary sealing material, and investigating the ability of sodium carboxymethyl cellulose (CMC) as thermal degradable additive to promote the disintegration of AACM-based sealer. CMC and cellulose-related compounds were not new materials in the drilling industries. These compounds frequently are used as additives in water-based drilling mud to reduce fluid loss, to confer water-holding properties on mud, and to assure its appropriate rheology at elevated temperature [12-14]. In addition, the AACM systems do not incorporate any Ordinary Portland Cement (OPC) that emits 0.9 metric ton CO_2 gas [15] and a great deal of mercury [16] during the manufacture of every ton of OPC; this greatly underscores the value of our cementitious sealer in eliminating carbon footprints.

Another important property in designing sealing materials is to assure an expansion in their volume upon exposure to hydrothermal environments at ≥200°C. Such a volume expansion of sealers emplaced in the fractures improves their plugging performance by enhancing their adherence to the inner surface of fracture: This ensures that the sealers have adequate stability and reliability without their being washed out and removed by a locally disbanded sealer during further drilling operations. The commonest ways to increase the volume of cementitious materials is to utilize the *in-situ* growth of the crystalline hydration products formed in the cementitious bodies. Among those products, the sulfoaluminate (ettringite)[17,18]-, and Ca and Mg oxides[19-23]-based hydration products are well recognized as satisfactorily expandable ones. The representative crystalline phase of first hydration product is ettringite, $3CaO.Al_2O_3.CaSO_4.32H_2O$, containing a large volume of water, and it is formed by mixing five starting materials; Portland cement, anhydrous hauyne ($3CaO.3Al_2O_3.CaSO_4$), gypsum ($CaSO_4$), quick lime (CaO), and water [24]. However, ettringite displays one major drawback in a high temperature environment; namely, its formation decomposes at a heating temperature > 70°C [25-29]. Thus, this volume expansion technology was inapplicable to our sealer.

When the CaO and MgO expansive additives embedded in OPC neat paste are autoclaved, the additive's hydration initiates promptly in an alkaline environment created by the hydrolysis of OPC, resulting in the conversion of the Ca- and Mg-oxides into crystalline Ca and Mg hydroxides. Afterward, the pressure generated by their *in-situ* growth extends the expansion and swelling of hydrothermally cured OPC. However, in our previous study on the phase identification in autoclaved sodium silicate-activated slag/Class C fly ash cementitious material [30], we found that all free lime, CaO, present in Class C fly ash hydrothermally reacted with the slag to form crystalline calcium silicate hydrates, such as calcium silicate hydrate (I) and 1.1 nm tobermorite. We did not detect $Ca(OH)_2$ –related crystal phase, which aids in expanding the autoclaved cementitious material. This finding seemingly suggested that the CaO expansive additive might not be useable in the autoclaved sodium silicate-activated slag/Class C fly ash system because of its reactivity with silicate species, rather than forming $Ca(OH)_2$. Thus, our interest focused on investigating the usefulness of dead-burned MgO as an expansive additive for AACMs.

The sealers to be developed were required to meet all the following nine material criteria: 1) One dry mix component product; 2) plastic viscosity, 20 to 70 cp at 300 rpm; 3) be suitable for conventional cement squeezing technology; 4) maintenance of pumpability for at least 1 hour at 85°C; 5) compressive strength >2000 psi; 6) be self-degradable by injection with water under a certain pressure; 7) expandable and swelling properties ≥0.5 % of total volume of sealer; 8) excellent penetration through fractures of ~0.04 in. wide spacing; and 9) anti-filtration properties.

EXPERIMENTAL PROCEDURE
Two industrial by-products with pozzolanic properties, granulated blast-furnace slag under trade name "New Cem," and Class C and F fly ashes were used as the hydraulic pozzolana cement. The slag was supplied by Lafarge North America, and fly ashes were obtained from Boral Material Technologies, Inc. Table 1 lists their chemical constituents. A sodium silicate granular powder under trade name "Metos Bead 2048," supplied by The PQ Corporation was used as the alkali activator of these pozzolana cements; its chemical composition was 50.5 mol. wt % Na_2O and 46.6 mol. wt % SiO_2. The formula of the dry pozzolana cements employed in this test had slag/Class C and /F fly ash ratios of 100/0, 80/20, 60/40, 40/60, and 20/80 by weight. Sodium silicate powder was added as alkali

activator at 2.8, 4.1, and 6.8 % by the total weight of pozzolana cement to prepare the dry mix cementitious reactant. Commercial Class G well cement was used as a control.

Table 1. Chemical composition of Ground Granulated Blastfurnace Slag, and Class C and F fly ashes.

	CaO, wt%	SiO₂, wt%	Al₂O₃, wt%	MgO, wt%	Fe₂O₃, wt%	TiO₂, wt%	Na₂O, wt%	K₂O, wt%	SO₃, wt%	Loss in ignition for C, wt%
Slag	38.5	35.2	12.6	10.6	1.1	0.4	-	-	0.1	1.5
Class C fly ash	25.6	36.2	20.1	5.4	6.7	-	1.7	0.5	1.7	2.1
Class F fly ash	0.47	39.3	38.6	1.4	12.9	-	1.5	1.9	1.2	2.7

Sodium carboxymethyl cellulose (CMC) under the production name "Walocel CRT 30 PA," supplied by Dow Chemical Corp., was used as the thermal degradable additive to promote the disintegration of the slag/fly ash-based sealers; it dissolves in water forming an anionic polymer. The CMC was produced by the etherification of cellulose from renewable resources, like wood, and had the chemical structure shown in Figure 1. We used it at 0.7 % by total weight of dry slag/fly ash mixture. Hard burned magnesium oxide (98.3 wt% MgO) under the trade name "MagChem 10 CR" from Martin Marietta Magnesia Specialties, LLC was used as expansive additive. Its physical properties included 97 % passing 200 mesh (0.074 mm) and surface area ranging from 0.2 to 0.3 m²/g. The contents of MgO additive were 2.0, and 3.0 % by the total weight of cement components.

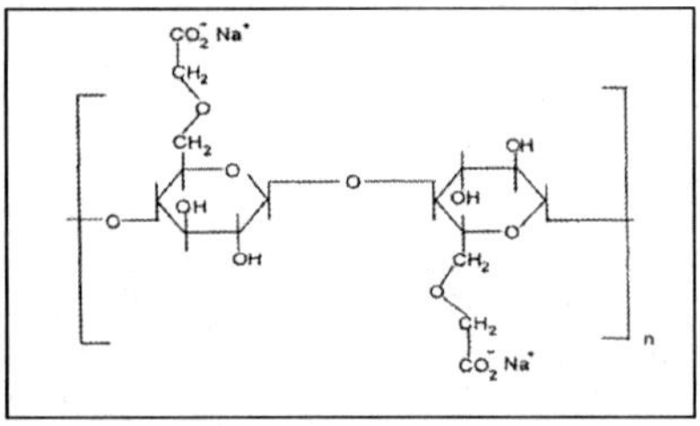

Figure 1. Chemical structure of CMC.

The sodium silicate alkali-activated slag/Class C and /F fly ash (AASC and AASF) slurries with CMC were prepared by adding an appropriate amount of water to the dry mix cementitious component, and then left them in an atmospheric environment until the cement set. Afterward, these room temperature-set AASC and AASF samples were placed into an autoclave at 200°C for 5 hours under the hydrothermal pressure ranging from 300 to 1000 psi. Some autoclaved samples then were heated for 24 hours in oven at 200° and 250°C. The commercial Class G well cement modified with 4.1 % sodium silicate and 0.7 % CMC also was exposed to the same conditions as those of AASC and AASF samples.

Measurements

The initial- and final-setting times of the CMC-containing AASC and AASF sealing slurries at 85°C were determined in accordance with modified ASTM C 191-92 [31]. In it, a slurry-filled conical mold was placed in screw-topped round glass jar to avoid any evaporation of water from the slurry during heating. The slurry was examined every 30 minutes to determine its setting time. The specimen with size of 35 mm diameter x 70 mm length was used for determining compressive strength. Thermogravimetric analyzer (TGA) gives us information on the thermal decomposition and decomposition kinetic of CMC; further, we adapted Fourier transform infrared spectroscopy (FT-IR) to reveal the chemical behavior of the decomposed CMC in AACM. X-ray powder diffraction (XRD) and FT-IR were used to identify the crystalline and amorphous hydrothermal reaction products of the autoclaved samples. The high-resolution scanning electron microscopy (HR-SEM) was employed to survey any alterations in the microstructure of sealers before and after its self-degradation, and also to explore the morphologies of the crystalline and amorphous hydrothermal reaction products formed in sealers. To identify these reaction products, we used energy-dispersive x-ray (EDX) concomitant with HR-SEM.

With a Type K thermometer, in conjunction with temperature data logger, we monitored the heat energy generated in a self-degrading process of the 200°C-heated AASC and AASF specimens after their contact with water.

RESULTS AND DISCUSSION

Properties of Slag/Class C and F fly ash Systems

Our first experiment was directed toward gaining data on the setting time and compressive strength of CMC-containing AASC and AASF samples made with slag (S)/Class C (C) and /F (F) ratios of 100/0, 80/20, 60/40, 40/60, and 20/80. Table 2 shows the changes in water content, expressed as the ratio of water (w) to dry cement (c), as a function of the S/C and /F ratio used in preparing the AASC and AASF slurries; it includes their initial- and final-setting times at 85°C. The water content in these AASC and AASF systems was adjusted to attain a consistency similar to that of Class G well cement slurries made with the w/c ratio of 0.54. As is evident, the w/c ratio decreased with the replacement of some portion of slag by Class C and F fly ashes. For commercial Class G well cement, the initial setting time of this cement without CMC was ~ 90 min (not shown). With 0.7 % CMC, there was no evidence of initial setting time within 360 min. Undoubtedly, CMC acted as a set retarder of Class G cement. In contrast, CMC was not as effective in retarding the setting of all AASC and AASF slurries, except for 20/80 slag/fly ash F ratio slurry; these slurries were set in curing period of time up to ~ 210 min. However, the slurries made with three formulas related to 100/0, 80/20, and 60/40 S/C ratios failed to meet the criterion ≥60 min. Thus, no further test was conducted for these formulas.

Table 2. Water/cement ratio, and initial and final setting times at 85°C for slurries made with various slag/Class C and /F fly ash ratios and Class G well cement.

Formula	Water/cement, w/c ratio	Initial setting time at 85°C, min	Final setting time at 85°C, min
Class G well cement	0.54	>360	Unknown
100/0 S/C	0.47	<30	<60
80/20 S/C	0.46	<30	<60
60/40 S/C	0.45	~ 30	~ 60
40/60 S/C	0.45	~60	~90
20/80 S/C	0.44	~120	~180
80/20 S/F	0.45	~60	~120
60/40 S/F	0.42	~150	~270
40/60 S/F	0.40	~210	~360
20/80 S/F	0.40	>210	Unknown

Table 3 shows the compressive strength and hydrothermal reaction product for specimens made with the screened formulas including 40/60 and 20/80 S/C ratios, and 80/20, 60/40, 40/60, and 20/80 S/F ratios, after autoclaving for 5 hours at 200°C. The reaction products were identified using XRD, FT-IR, and SEM-EDX. As is seen, the development of compressive strength depended on the S/C and S/F ratios. For the S/C blend system, the specimens with a 40/60 ratio has a compressive strength of 3836 psi; those made with the 20/80 ratio displayed a ~ 46 % decline of strength to 2187 psi. Two crystalline hydrated reaction products, 1.1 nm tobermorite [$Ca_5(OH)_2Si_6O_{16}.5H_2O$] and calcium silicate hydrate (I) [$CaO.SiO_2.H_2O$, C-S-H (I)] phases, were responsible for strengthening autoclaved 40/60 and 20/80 ratio specimens. In contrast, the effect of autoclaving on the development of strength in commercial Class G well cement was poor. Its compressive strength was ~ 3.6- and ~ 2.1-fold lower than that of 40/60 and 20/80 S/C ratios, respectively. For the S/F blend system, the aluminum (Al)-substituted 1.1 nm tobermorite, $Ca_5AlSi_5O_{16}.5H_2O$, played the major role in improving compressive strength.

Table 3. Hydrothermal reaction products and compressive strength for 200°C-autoclaved specimens made with various slag/Class C and /F fly ash ratios.

Formula	Hydrothermal reaction product		Compressive strength, psi
	Major phase	Miner phase	
Class G well cement	Unknown	-	1060
40/60 S/C	tobermorite, C-S-H (I)	-	3836
20/80 S/C	tobermorite, C-S-H (I)	-	2187
80/20 S/F	Al-substituted tobermorite	-	5983
60/40 S/F	Al-substituted tobermorite	geopolymer	5271
40/60 S/F	geopolymer	Al-substituted tobermorite	4790
20/80 S/F	geopolymer	Al-substituted tobermorite	789

Figure 2 shows the HR-SEM image of well-formed tobermorite, which was identified by the interlocking conformation of thin plate-like crystals. Correspondingly, the highest compressive strength of 5983 psi in this test series was determined from 80/20 S/F ratio specimens predominated with tobermorite. The specimens made with 60/40 S/F ratio had two hydrated reaction products: One was tobermorite as the major crystal hydration product; the other was geopolymer as the minor amorphous hydration product. More replacement of slag by the Class F fly ash led to the formation of geopolymer; in fact, the 40/60 and 20/80 S/F ratio specimens were comprised of geopolymer as the major reaction product and tobermorite as the minor one. To visualize the morphology of amorphous geopolymer as cementitious binder, we explored the fracture surface of autoclaved 40/60 ratio specimen using HR-SEM coupled with EDX (Figure 3). The HR-SEM image revealed the cluster of fine fibrous-like reaction products growing on the surfaces of Class F fly ash spheres, wherein, the cohesively agglomerated masses of these cluster acted as cementitious structure. The geopolymer reaction products formed on the surfaces of fly ash spheres, accounting for the interaction between sodium silicate and fly ash. The image also revealed that this agglomerated microstructure contained numerous voids, denoting that the geopolymer-based cementitious structure was poorly formed in the 40/60 ratio

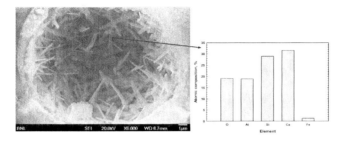

Figure 2. SEM image coupled with EDX atomic compositions for aluminum-substituted 1.1 nm tobermorite crystal formed in 200°C-autoclaved 60/40 slag/Class F fly ash ratio specimen.

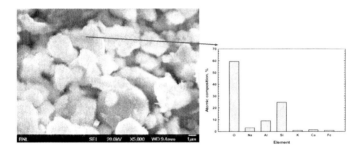

Figure 3. Microstructure and atomic composition of amorphous geopolymer formed in 200°C-autoclaved 40/60 slag/Class F fly ash ratio specimen.

specimen. Thus, the 40/60 and 20/80 ratio specimens caused the reduction of compressive strength; in fact, the values of 4790 and 789 psi compressive strength, respectively, corresponded to ~ 20 % and ~ 87 % lower than that of 80/20 ratio specimens, which did not form any geopolymer.

CMC Self-degradation promoting additive

TGA was employed to obtain information on the quantitative analysis and kinetic study of CMC's thermal decomposition (Figure 4). The TGA curve revealed that its decomposition began at 168°C, and was completed at 303°C. The total loss in weight of CMC occurring between 168° and 303°C was 42.6 %, implying that a large volume of gaseous species was emitted during the decomposition of CMC. The thermal decomposition kinetic computed by integrating the closed areas of curves with the baseline revealed that a half of CMC's total weight loss took place at temperatures from 168° to ~ 260°C.

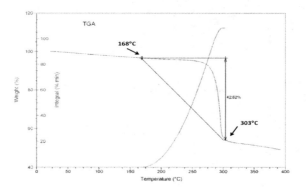

Figure 4. TGA curve of CMC.

Our study next shifted to clarifying the chemistry of CMC's thermal decomposition. To do so, FT-IR analyses in the wavenumber region, 3500-700 cm⁻¹ were carried out for non-heated and heated CMC at 200°, 250°, and 300°C for 24 hours (Figure 5). Based upon the chemical structure of CMC in Figure 1, in conjunction with literature survey [32-34], the spectrum (a) of the 80°C-made CMC film included several prominent absorption bands at 2925 and 2864 cm⁻¹ associated with C-H anti-symmetric and symmetric stretching vibrations, at 1603 cm⁻¹ belonged to aromatic C=C stretching mode, at 1576 cm⁻¹ shoulder attributed to C-O anti-symmetric stretching in the carboxylate ion group, -COO⁻, at 1400 cm⁻¹ due to -CH₂ scissoring overlapped with C-O symmetric stretching in –COO⁻, at 1329 cm⁻¹ related to both –CH₂ scissoring and C-OH bending, at 1157 cm⁻¹ referred to C-O-C stretching in the ether group, and both at 1057 and 1020 cm⁻¹ assignable to C-O stretching in >CH-O-CH₂- linkage. Compared with this, FT-IR spectrum (b) of the 200°C-heated CMC film was characterized by declining the band intensity at 1329, 1057, and 1020 cm⁻¹, and disappearing the band at 1157 cm⁻¹, demonstrating that heating the CMC at 200°C entailed the breakage and rupture of several linkages, such as C-OH, C-O-C, and >CH-O-CH₂- within CMC's molecular structure. Further increasing temperature to 250°C incorporated five new bands at 1685, 1457, 879, 835, and 780 cm⁻¹ into the spectrum (c). The possible contributors to these new bands at 1685 and 1457 cm⁻¹ were the carbonyl, C=O, in ketone species formed by intra-ring dehydration [35, 36] and a carbonated compound like sodium carbonate, Na₂CO₃ [37], respectively. The other new bands emerging in low wavenumber region < 900 cm⁻¹ might be associated with the organic fragments yielded by thermal degradation of CMC. This spectrum also strongly represented that the band at 1576 cm⁻¹ attributed to –COO⁻ ionic group had become one of the major peaks, while the aromatic- and >CH-O-CH₂- linkage-related bands had vanished. Thus, the thermal decomposition of CMC at 250°C seemingly generates many –COO⁻ groups that are converted into organic acid in an aqueous media. Several investigators [9, 11] reported that this thermal decomposition generated volatile decomposition products, such as CO, CO₂, CH₄, water, and acetic acid. Among those, the major decomposition product at 250°C was acetic acid, which occupies more than 40 % of the total decomposition products. This information strongly supported the likelihood of –COO⁻ → organic acid transformation as well as the formation of sodium carbonate from the interactions between two decomposition by-products of CMC, sodium oxide and

CO_2. At 300°C, the spectrum (d) revealed the incorporation of more C=O groups and Na carbonate into the thermally degraded CMC, as reflected in the pronounced growth of bands at 1685 and 1457 cm^{-1} coexisting with the new band at 849 cm^{-1} attributed to carbonate.

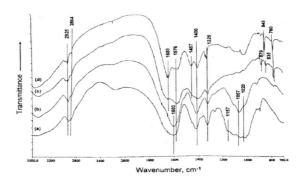

Figure 5. FT-IR spectra for 80°C-made CMC film (a), and 200°C- (b), 250°C-(c), and 300°C-heated CMCs (d).

Self-degradation

We questioned whether CMC-modified AACM really degrades after contacting with water. In response to this question, three different tests were conducted to visualize its self-degradation: The first test was to directly immerse hot specimens with 60/40 and 20/80 S/C ratios, and 80/60, 60/40, 40/60, and 20/80 S/F ratios, and also Class G cement modified with CMC into water at 25°C; second test was to impregnate water into the air-free AACM and Class G cement specimens; and third one was to expose in an autoclave for a short- and long-term of 5 and 24 hours. The detailed test procedures were as follows; for the first test, 1) the specimens were autoclaved at 200°C for 5 hours, and then heated at 200°C for 24 hours, 2) shortly after removing hot specimens form an oven, they were immersed in the water at 25°C, and 3) visual observation allowed us to evaluate the magnitude of its self-degradation. For the second test, 1) the specimens were autoclaved at 200°C for 5 hours, and then heated at 200°C for 24 hours, 2) the specimens were cooled for 24 hours at room temperature, 3) thermocouple-embedded specimens were placed in a vacuum chamber at -1000 kPa for 10 min to eliminate any air present in the specimens, 4) air-free voids in the vacuumed specimens were filled with water (water/specimens volume ratio= 17/1), while determining the generation of *in-situ* exothermic heat by temperature data logger, and 5) visual observation was made to evaluate the magnitude of self-degradation. For the third test, we observed the changes in appearance of specimens after exposure for 5 and 24 hours in an autoclave at 200°C.

For the first test, all 200°C-heated specimens, except for 40/60 and 20/80 S/F ratios, had developed numerous cracks in the specimen shortly after immersing them in water 25°C. Furthermore, the propagation of cracks with elapsed immersion time led to their disintegration. In contrast, CMC-free specimen showed no signs of degradation. Contrarily, incorporating CMC into the Class G cement didn't affect degradation and did not develop any cracks. This information strongly demonstrated that CMC conferred potential self-degrading properties on AACM, but not on commercial Class G well cement. A possible interpretation is that when water was permeated into AACM containing thermally decomposed CMC, it served in disintegrating the hard cementitious structure possessing a compressive strength of more than 2000 psi. Conceivably, such disintegration by water may be due to the generation of water-catalyzed energy in AACM. If this assumption is valid, two resources were involved, one from AACM, and the other from the decomposed CMC.

To verify this assumption, we determined the exothermic heat energy generated in AACM by water penetration. Plotting the relation curve between the elapsed time and exothermic temperature for AACM (20/80 S/C ratio specimen) in Figure 6 shows that ~ 130 sec after impregnating the vacuumed specimen with water, its internal temperature rapidly rose from 22.2°C to a peak of 41.5°C. The total elapsed time from the onset to the maximum temperature was only 110 sec. We observed the development of numerous cracks in the water-impregnated specimens along with the increase in their exothermic temperature.

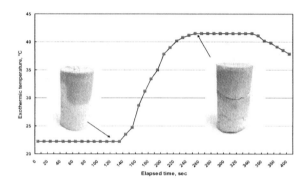

Figure 6. *In-situ* exothermic temperature vs. elapsed time for CMC-containing AACM after beginning impregnation with water.

Figure 7 plots the *in-situ* exothermic temperature generated in the vacuumed 80/20, 60/40, 40/60, and 20/80 S/F ratio specimens as a function of elapsed time after they were impregnated with water. For 80/20 ratio, it generated the *in-situ* exothermic heat at 40°C promoting its self-degradation. Although the similar feature of curve was recorded in the 60/40 ratio specimen, its peak temperature was 9°C lower than that of 80/20 ratio, suggesting that more displacement of slag portion by Class F fly ash caused the reduction of *in-situ* exothermic heat evolved in the water-impregnated specimen. We also observed that this 60/40 ratio specimen was degraded by water penetration. In contrast, there was no generation of the exothermic peak for 40/60 and 20/80 ratios. Their curves revealed a slight increase in temperature after ~ 130 sec from the beginning of water impregnation, but beyond this time, they leveled off. Furthermore, these specimens did not show any signs of the self-degradation and - breakage, and nor were visual cracks and fissures observed in the specimens.

Based upon information described above, Figure 8 delineated the self-degradation mechanism for AACM. In this mechanism, there were two important factors: One factor was the presence of free sodium hydroxide as reactant derived from the hydrolysis of sodium silicate activator in the AACM slurries; the other was related to the formation of ionic carbonic acid from wet carbonation of CO_2 emitted from thermal degradation of CMC. The sodium hydroxide directly reacted with ionic carbonic acid to form sodium bicarbonate. Then, the chemical affinity of sodium bicarbonate with CH_3COOH from the decomposed CMC led to the evolution of CO_2 as reaction product. *In-situ* exothermic energy was generated throughout this process, including dissolution, interactions, and evolution. Nevertheless, we considered that this *in-situ* CO_2 gas

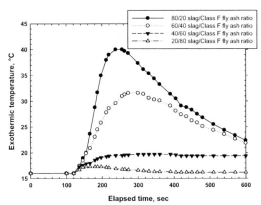

Figure 7. *In-situ* exothermic temperature vs. elapsed time for autoclaved 80/20, 60/40, 40/60, and 20/80 slag/Class F fly ash ratio specimens after beginning impregnation with water.

evolution and sodium acetate as exothermic reaction product led to volume expansion of AACM, thereby resulting in the breakage and disintegration of AACM. As is evident from this mechanism, the sodium hydroxide hydrolysate played a pivotal role in promoting the self-degradation of sealers. As described in Table 3, in the 40/60 and 20/80 S/F ratio specimens comprising geopolymer as the major reaction product, the amount of free sodium hydroxide dissociated in the water-impregnated specimens may be very little, if any, because the polymeric network structure of geopolymer was constituted of sodium aluminum silicate compounds, which were formed by the interactions between mullite in Class F fly ash and sodium hydroxide hydrolysate (Figure 9). Thus, it is reasonable to assume that the intercalation of sodium in its polymeric networks impaired the production of sodium hydroxide, thereby declining the evolution of *in-situ* exothermic heat promoting the self-degradation. Correspondingly, the 20/80 ratio attributed to more consumption of sodium hydroxide to form geopolymer had the lowest heat evolution. The lack of free sodium hydroxide was the major reason why these geopolymer-based 40/60 and 20/80 ratio specimens remained intact in terms of non-self degradation. Thus, although the geopolymer-based sealers had a porous structure, its structure did not help in promoting the sealer's self-degradation. The magnitude of self-degradation depended on the exothermic temperature generated in the water-impregnated sealers; a higher exothermic temperature increased the breakages of the sealer.

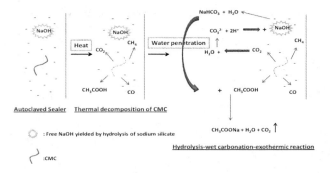

Figure 8. Self-degradation scheme of heated CMC-containing sodium silicate-activated pozzolan cementitious materials after contact with water.

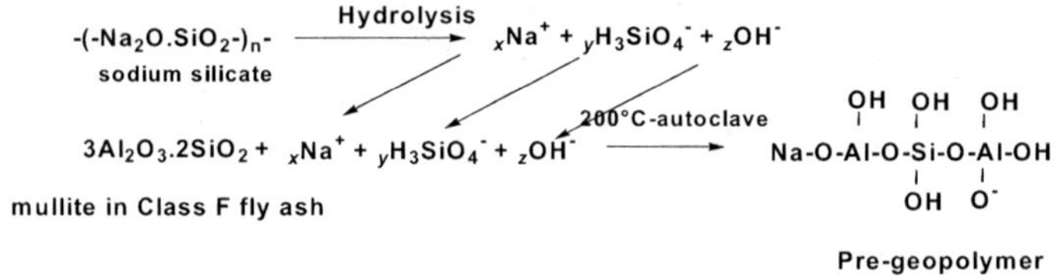

Figure 9. Hydrothermal reactions between sodium silicate and Class F fly ash.

Also, we found that hydrothermal pressure controls the reactivity of sodium silicate; a high pressure enhanced its reactivity, reflecting the dearth of the free sodium hydroxide reactant. In fact, under pressure of 1000 psi, the peak exothermic temperature declined with a decreasing amount of sodium silicate, from 38.5°C for 6.8% sodium silicate to 27.4°C for 2.8% sodium silicate (Figure 10). Thus, we believe that a 2.8 % content of sodium silicate was insufficient to generate an appropriate amount of free sodium hydroxide reactant for 20/80 S/C ratio specimens under 1000 psi pressure.

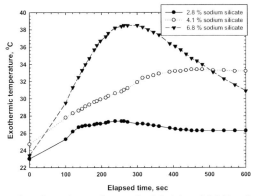

Figure 10. Comparison of exothermal curve features for 2.8, 4.1, and 6.8 % sodium silicate-containing 20/80 S/C ration specimens made under 1000 psi pressure.

Expansion

Our focus centered on monitoring the changes in the expansion rate of 2.0 and 3.0 % MgO-incorporated AASC cements by varying pressure from 300 to 1500 psi (Figure 11). For both the 2.0 and 3.0 % MgO specimens, the resulting data revealed that the expansion rate declined with a

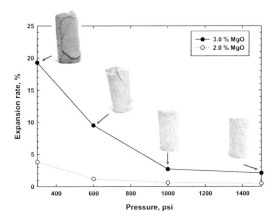

Figure 11. Changes in expansion rate of 2.0 % and 3.0 % MgO-incorporated cements as a function of pressure, ranging from 300 psi to 1500 psi.

rising pressure. With 3.0 % MgO, increasing the pressure to 600 psi from 300 psi conspicuously reduced its expansion rate to 9.5 % from 19.3 %. A further increase in pressure to 1000 psi gave a minimum expansion of 2.7 %. A similar expansion rate was recorded from specimens made under

1500 psi pressure, suggesting that the pressures between 1000 and 1500 psi had no significant effect on suppressing the expansion of cements. As is evident from the photos in Figure 12, although the rate of expansion was minimized by these high pressures, we visually observed the development of cracks in the specimens. In contrast, no cracks were visible in the 2.0 % MgO-incorporated specimens made under pressures of ≥600 psi. Thus, 2.0 % seems to be the effective amount of MgO needed in creating crack-free cement along with a moderate rate of expansion.

The XRD studies were undertaken to identify the crystalline hydration reaction products contributing to the extension of volumetric expansion for MgO-incorporated AASC cements after autoclaving at 200°C. XRD tracings were conducted for cements containing 0, 2, and 3 % MgO, respectively. For specimens without MgO, the XRD pattern (not shown) revealed the formation of two hydrothermal-catalyzed pozzolanic reaction products, 1.1 nm tobermorite [$Ca_5(OH)_2Si_6O_{16}.5H_2O$] phase and the calcium silicate hydrate (I) [$CaO.SiO_2.H_2O$, C-S-H (I)] phase. The other phases detected, such as quartz and tricalcium aluminate ($3CaO.Al_2O_3$, C_3A), were associated with the non-reacted Class C fly ash and slag as the starting materials. The specimens containing 2.0 % MgO showed similar features in their XRD pattern (not shown), except for the presence of brucite [$Mg(OH)_2$] yielded by the hydrothermal reaction of MgO. The XRD line intensity of the brucite-related d spacings rose with an increasing content of MgO, while the quantity of the other reaction products and the non-reaction ones remained unchanged. Thus, the *in-situ* phase transformation of MgO to $Mg(OH)_2$ during autoclaving at 200°C appears to serve in expanding and swelling the cementitious specimens.

To visualize the alteration of microstructure caused by *in-situ* MgO→$Mg(OH)_2$ phase transition, we explored the fracture surfaces of 3.0 % MgO-containing cements after autoclaving at 200°C, by SEM-EDX (Figure 12). This image was characterized by a radical growth of sunflower-like crystals. As is seen in the EDX atomic composition for these crystals, there were two major elements, O and Mg, strongly suggesting that this morphology can account for being a well-formed crystal structure of brucite, $Mg(OH)_2$ contributing to the volumetric expansion of sealer.

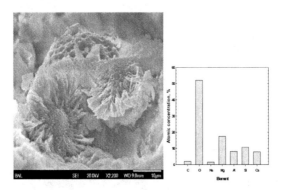

Figure 12. SEM images coupled with EDX atomic composition for the fracture surfaces of 200°C-autoclaved cements without (top) and with MgO (bottom).

CONCLUSIONS

Self-degradable slag (S)/Class C (C) or F (F) fly ash blend pozzolana cements were formulated, assuming that they might serve well as alternative temporary fracture sealers to cure lost circulation during well drilling in Enhanced Geothermal System (EGS) wells operating at temperatures of $\geq$ 200°C. The candidate formulas were screened based upon material criteria including an initial setting time $\geq$ 60 min at 85°C, compressive strength $\geq$ 2000 psi for a 200°C autoclaved specimen, and the extent of self-degradation of cement heated at $\geq$200°C after it was contacted with water. After mixing with water and being autoclaved at 200°C, the calcium silicate hydrate (I) [C-S-H (I)] and aluminum-substituted 1.1 nm tobermorite, and non-substituted 1.1 nm tobermorite crystal phases, and geopolymer as amorphous phase were identified as hydrothermal reaction products responsible for the development of a compressive strength > 2000 psi. All the S/C systems displayed the self-degradation. In the S/F systems, the 200°C-autoclaved cements had the combined phases of tobermorite as its major reaction product and amorphous geopolymer as its minor providing a compressive strength of 5270 psi. Sodium hydroxide derived from the hydrolysis of sodium silicate activator not only initiated the pozzolanic reaction of slag and fly ash, but also played an important role in generating *in-situ* exothermic heat that significantly contributed to promoting self-degradation of cementitious sealers. The source of this exothermic heat was the interactions between sodium hydroxide, and gaseous CO_2 and CH_3COOH by-products generated from thermal decomposition of sodium carboxymethyl cellulose (CMC) additive at $\geq$200°C in an aqueous medium. Thus, the magnitude of this self-degradation depended on the exothermic temperature evolved in the sealer; a higher temperature led to a severe disintegration of sealer. In contrast, the excessive formation of geopolymer phase due to more incorporation of Class F fly ash into this cementitious system affected its ability to self-degrade, reflecting that there was no self-degradation. The geopolymer was formed by hydrothermal reactions between sodium hydroxide from sodium silicate and mullite in Class F fly ash. Thus, the major reason why geopolymer-based cementitious sealers did not degrade after heated sealers came in contact with water was their lack of free sodium hydroxide.

We identified hard-burned magnesium oxide (MgO) as a suitable expansive additive for self-degradable sealers. MgO extended the volumetric expansion of sealers during their exposure in a hydrothermal environment at 200°C. The expansion rate of sealers depended on two factors: One was the content of MgO; the other was hydrothermal pressures, ranging from 300 to 1500 psi. With 3.0 wt% MgO, an impressive expansion rate of 19.3 % was observed at the lowest pressure of 300 psi. Increasing the pressure to 1500 psi resulted in the expansion rate of only 2.1 %. The *in-situ* growth of brucite crystal formed by the hydrothermal hydration of MgO was responsible for the sealer's expansion improving their plugging performance by enhancing their adherence to the inner surface of fracture.

The two final formulations were 19.3% slag-76.7% Class C fly ash-2.8% MgO-1.2% CMC for slag/Class C fly ash system and 38.3% slag-57.7% Class F fly ash-2.8% MgO-1.2% CMC for slag/Class F fly ash system, with 10 to 4% sodium silicate by total weight of cement. These formulations provided self-degradable materials with more than 2000 psi compressive strength and volume expansion of more than 0.5%. Studies on the material flow behavior and fracture penetration involving API slot-tests for lost circulation materials are currently ongoing.

REFERENCES
1. T. Sugama and L. Brothers, 2004. "Sodium-silicate-activated Slag for Acid-resistant Geothermal Well Cements." J. Adv. in Cem. Res. 16, p.77-87.
2. T. Sugama, L. Brothers, and T. Van de Putte, 2005. "Acid-resistant Cements for Geothermal Wells: Sodium Silicate Activated Slag/Fly Ash Blends." J. Adv. in Cem. Res. 17, p.65-75.
3. R.S. Kalyoncu "Slag-iron and steel," U.S. Geological Survey Minerals Yearbook, 2000. http://minerals.usgs.gov/minerals/pubs/commodity/iron_&_steel_slag/790400.pdf.
4. T.D. Kelly, D.E. Sullivan, and H. van Oss "Coal combustion products statistics," U.S. Geological Survey, Last modification: December 1, 2010. http://minerals.usgs.gov/ds/2005/140/coalcombustionproducts.pdf.
5. I. Espigares, C. Elvira, J.F. Mano, B. Vazquez, J. San Roman, and R.L. Reis, 2002. "New partially degradable and bioactive acrylic bone cements based on starch blends and ceramic fillers, Biomaterials." 23, p.1883-1895.
6. L.F. Boesel, S.C.P. Cachinho, M.H.V. Fernands, and R.L. Reis, 2007. "The in vitro bioactivity of two novel hydrophilic, partially degradable bone cements." Acta Biomaterialia, 3, p.175-182.
7. Y. Zuo, F.Yang, J.G.C. Wolka, Y. Li, and J.A. Jansen, 2010. "Incorporation of biodegradable electrospun fibers into calcium phosphate cement for bone regeneration." Acta Biomaterialia, 6, p.1238-1247.
8. W.J.E.M. Habraken, H.B. Liao, Z. Zhang, J.G.C. Wolke, D.W. Grijpma, A.G. Mikos, J. Feijen, and J.A. Jansen, 2010. "In vivo degradation of calcium phosphate cement incorporated into biodegradable microspheres." Acta Biomaterialia, 6, p.2200-2211.
9. R.M. Aseyeva, T.N. Kolosova, S.M. Lomakin, Y.Y. Libonas, G.Y. Zaikov, and V.V. Korshak, 1985. "Thermal degradation of cellulose diacetate, Polym." Sci. U.S.S.R., 27, p.1917-1926.
10. P. Jandura, B. Riedl, and B.V. Kokta, 2000. "Thermal degradation behavior of cellulose fibers partially esterified with some long chain organic acids." Polym. Degrad. Stab. 70, p.387-394.
11. M. da Conceical, C. Lucena, A.E.V. de Alencar, S.E. Mazzeto, S. de A. Soares, 2003. "The effect of additives on the thermal degradation of cellulose acetate." Polym. Degrad. Stab. 80, p.149-155.
12. F. Farbbri and M. Vidali, 1970. "Drilling mud in geothermal wells." Geothermic, 2, p.735-741.
13. A.L. Alldredge, M. Elias, and C.C. Gotschalk, 1986. "Effects of drilling muds and mud additives on the primary production of natural assemblages of marine phytoplankton." Marine Envi. Res. 19, p.157-176.
14. I.S. Dairanieh and S.M. Lahalih, 1988. "Novel polymeric drilling and viscosifiers." Eur. Polym. J. 24, p.831-835.
15. S. Anand, P. Vrat, and R.P. Dahiya "Application of a system dynamics approach for assessment and mitigation of CO_2 emission from the cement industry," J. Environmental Management, 79 (2006) 383-398.
16. T. L. Mlakar, M. Horvat, T. Vuk, A. Stergarsek, J. Kotnik, J. Tratnik, and V. Fajon "Mercury species, mass flows and processes in a cement plant," Fuel, 89 (2010) 1936-1945.
17. I. Teoreanu and C. Dumitrescu, "Mechanisms of controlled expansion cements hardening," Cem. Concre. Res., 12 (1982) 141-155.
18. Y. Fu, P. Gu, P. Xie, and J.J. Beaudoin, "Effect of chemical admixtures on the expansion of shrinkage-compensating cement containing a pre-hydrated high alumina cement-based expansive additive," Cem. Concre. Res., 25 (1995) 29-38.
19. S. Chatterji, "Mechanism of expansion of concrete due to the presence of dead-burnt CaO and MgO," Cem. Concre. Res., 25 (1995) 51-56.
20. T. Liang and T. Mingshu, "Correlation between reaction and expansion of alkali-carbonate reaction," Cem. Concre. Res., 25 (1995) 470-476.

21. Y. Qing, C. Huxing, W. Yuqing, W. Shangxian, and L. Zonghan, "Effect of MgO and gypsum content on long-term expansion of low heat portland slag cement with slight expansion," Cem. Concre. Comp., 26 (2004) 331-337.
22. C. Maltese, C. Pistolesi, A. Lolli, A. Bravo, T. Cerulli, and D. Salvioni, "Combined effect of expansive and shrinkage reducing admixtures to obtain stable and durable mortars," Cem. Concre. Res., 35 (2005) 2244-2251.
23. L. Mo, M. Deng, and M. Tang, "Effects of calcination condition on expansion property of MgO-type expansive agent used in cement-based materials," Cem. Concre. Res., 40 (2010) 437-446.
24. S. Nagataki and H. Gomi, "Expansive admixtures (mainly ettringite)," Cem. Concre. Res., 20 (1998) 163-170.
25. R. Yang, C.D. Lawrence, C.J. Lynsdale, and J.H. Sharp, "Delayed ettringite formation in heat-cured Portland cement mortars," Cem. Concre. Res., 29 (1999) 17-25.
26. P. Yan and X. Qin, "The effect of expansive agent and possibility of delayed ettringite formation in shrinkage-compensating massive concrete," Cem. Concre. Res., 31 (2001) 335-337.
27. T. Ramlochan, M.D.A. Thomas, and R.D. Hooton, "The effect of pozzolans and slag on the expansion of mortars cured at elevated temperature Part II: Microstructural and microchimical investigations," Cem. Concre. Res., 34 (2004) 1341-1356.
28. R. Barbarulo, H. Peycelon, S. Prene, and J. Marchand, "Delayed ettringite formation symptoms on mortars induced by high temperature due to cement heat of hydration or later thermal cycle," Cem. Concre. Res., 35 (2005) 125-131.
29. N. Meller, K. Kyritsis, and C. Hall, "The hydrothermal decomposition of calcium monosulfoaluminate 14-hydrate to katoite hydrogarnet and β-anhydrite: An in-situ synchrotron X-ray diffraction study," J. Solid State Chem., 182 (2009) 2743-2747.
30. T. Sugama, T. Butcher, L. Brothers, and D. Bour "Self-degradable cementitious sealing materials," BNL-94308-2010-IR, October 2010.
31. Designation: ASTM C-191-92, "Standard test method for time of setting of hydraulic cement by vicat needle."
32. S. Soares, N.M.P.S. Ricardo, S. Jones, and F. Heatley, High temperature thermal degradation of cellulose in air studied using FT-IR and ^{1}H and ^{13}C solid-state NMR, Euro. Polym. J., 37 (2001) 737-745.
33. V. Pushpamalar, S.J. Langford, M. Ahmad, and Y.Y. Lim, Optimization of reaction conditions for preparing carboxymethy cellulose from sago waste, Carbohydrate Polym., 64 (2006) 312-318.
34. H. Jin, Q. An, Q. Zhao, J. Qian, and M. Zhu, Pervaporation dehydration of ethanol by using polyelectrolyte complex membranes based on poly(N-ethyl-4-vinylpyridinium bromide) and sodium carboxymethyl cellulose, J. Membr. Sci., 347 (2010) 183-192.
35. J. Scheirs, G. Camino, and W. Tumiatti, Overview of water evaluation during the thermal degradation of cellulose, Eur. Polym. J., 37 (2001) 933-942.
36. R.M. Aseyeva, T.N. Kolosova, S.M. Lomakin, Y.Y. Libonas, G.Y. Zaikov, and V.V. Korshak, Thermal degradation of cellulose diacetate, Polym. Sci. U.S.S.R., 27 (1985) 1917-1926.
37. L.J. Bellamy, The infra-red spectra of complex molecules, pp. 385-387, Third editions, 1975, published by Chapman and Hall Ltd.

Hydrogen

THERMODYNAMICS AND KINETICS OF COMPLEX BOROHYDRIDE AND AMIDE
HYDROGEN STORAGE MATERIALS

Andrew Goudy, Adeola Ibikunle, Saidi Sabitu and Tolulope Durojaiye
Department of Chemistry, Delaware State University, Dover, Delaware

ABSTRACT
    In this study the hydrogen storage characteristics of several new destabilized borohydride systems were compared to the hydrogen storage behavior of MgH_2. The mixtures included: $Mg(BH_4)_2/Ca(BH_4)_2$; $Mg(BH_4)_2/CaH_2/3NaH$; and $Mg(BH_4)_2/CaH_2$; systems. Temperature programmed desorption, TPD, analyses showed that the desorption temperature of $Mg(BH_4)_2$ can be lowered by ball milling it with these additives. The PCT isotherm of the resulting mixtures displayed well-defined plateau regions. The desorption kinetics of $Mg(BH_4)_2$, $Ca(BH_4)_2$ and their 5:1 mixture were also compared in the two-phase region at the same temperature and thermodynamic driving force. The rate of hydrogen desorption from the $Mg(BH_4)_2/Ca(BH_4)_2$ mixture was faster than that from either of the constituents. Modeling studies showed hydrogen release from $Mg(BH_4)_2$, during the first 80% of the reaction, is diffusion controlled while in $Ca(BH_4)_2$ it is phase boundary controlled. In the mixture the rate appears to be under the mixed control of both processes. Lithium amide / magnesium hydride mixtures with an initial molar composition of $2LiNH_2 + MgH_2$ were also studied with and without the presence of 3.3 mol% potassium hydride dopant. TPD analyses showed that the KH doped samples had lower onset temperatures than their corresponding pristine samples. The de-hydriding kinetics of the doped and pristine mixtures was compared at a constant pressure driving force. The addition of KH dopant was found to significantly increase hydrogen desorption rate from the ($2LiNH_2 + MgH_2$) mixture.

1. INTRODUCTION
    Borohydrides of alkali and alkali-earth metals are being studied due to their high hydrogen content.[1-4] However, their application as solid state hydrogen storage materials has been hindered by unfavorable thermodynamics and kinetics. A number of approaches including addition of catalyst/additives have been used to improve the hydrogen storage properties of these borohydrides.[5-10] Addition of MgH_2 was found to lower the enthalpy of $LiBH_4$ by 25kJ/mol in the presence of $TiCl_3$ with 8-10% hydrogen capacity.[11] Metal oxides and chlorides lowered the onset desorption temperature of $LiBH_4$ from 400 °C to 200 °C.[10] Ibikunle et al [6] studied the $CaH_2/LiBH_4$ system with $TiCl_3$, TiF_3, V_2O_5 and TiO_2 additives and found that the mixtures released hydrogen between 400-450 °C. Ball-milling of small amounts of metal chlorides with α-$Mg(BH_4)_2$ was found to result in the formation of transition metal doped nanocomposites.[5] Its onset temperature was lowered by more than 100 °C with the addition Nb- or Ti chloride while the mixture of the Nb-and Ti chloride in the nanocomposite lowered the desorption temperature by 125 °C. The mixture of $Ca(BH_4)_2$ and MgH_2 was reported to be reversible up to 60 % after rehydogenation for 24 h under 90 bar hydrogen pressure at 350 °C.[12] It was also found that the dehydrogenated products for both pure $Ca(BH_4)_2$ and mixture contains CaB_6.
    A mixture of $LiBH_4/Mg(BH_4)_2$ in ratio 1:1 showed a lower onset dehydrogenation temperature due to in-situ formation of dual-cation (Li, Mg) bororohydrides.[13] It was shown that the dehydrogenation involves different mechanistic pathway from that of the constituents. MgB_2 was reversibly hydrogenated and rehydrogenated to $Mg(BH_4)_2$, at 900 bar H_2 and 400 °C with hydrogen capacity greater than 11wt% by Severa et al.[14] Fichtner et al.[15] also reported the reduction in the activation energy of bulk $Mg(BH_4)_2$ by a factor of two when infiltrated into activated carbon.

Theoretical calculations have also been used to predict the decomposition equation of several borohydrides systems.[16, 17] Some of these reactions are:

$$5Mg(BH_4)_2 + Ca(BH_4)_2 \rightarrow CaB_{12}H_{12} + 5MgH_2 + 13\ H_2 \qquad (1)$$
$$3Mg(BH_4)_2 + CaH_2 + 3NaH \rightarrow 3NaMgH_3 + CaB_6 + 10H_2 \qquad (2)$$
$$3Mg(BH_4)_2 + CaH_2 \rightarrow 3MgH_2 + CaB_6 + 10H_2 \qquad (3)$$

The hydrogen storage characteristics of these systems were reported in a previous study done in our laboratory.[7] The TPD analysis of a mixture containing $Mg(BH_4)_2$ and $Ca(BH_4)_2$ show a desorption temperature that was considerably lower than either of the two constituents. One of the goals of the present study is to examine the kinetics of $Mg(BH_4)_2$ and $Ca(BH_4)_2$ and a mixture of the two using constant pressure thermodynamic forces. Modeling studies will also be done to determine the process controlling their dehydrogenation rates.

2. EXPERIMENTAL

The unsovalted $Mg(BH_4)_2$ was prepared from MgH_2 and triethylamine borane as described by Chłopek et al.[18] The $Ca(BH_4)_2$ hydrogen storage grade and other hydrogen storage materials were obtained from Sigma-Aldrich. All sample preparations were carried out in an argon-filled Vacuum Atmospheres glove box with oxygen and moisture levels below 1ppm. The sample was ball-milled for 10 h using a SPEX 8000D mixer/mill. Prior to carrying out the kinetics experiments on the samples, it was necessary to determine the pressure at the middle of the plateau, P_m, from the pressure composition isotherms (PCI) for each of the samples. For easy comparison, the kinetics was run at the same temperature and thermodynamic driving force. About 2.0 g of sample was placed in the reactor, and glass wool was placed on it before closing it with quick connects connection. A thermocouple was allowed to run to the bed of the sample in the reactor to monitor the temperature of the sample. The initial pressure, P_i was set using the calibrated pressure transducer on the apparatus. An opposing pressure which corresponds to the thermodynamic driving force used was then set with the aid of a back pressure regulator. Hydrogen was then allowed to flow into the reservoir from the sample across the back pressure regulator while monitoring its pressure as a function of time. The reaction was allowed to continue until no further change in pressure was noticed.

3. RESULTS AND DISCUSSION

3.1 Temperature Programmed Desorption, Kinetics and Modeling Studies on MgH₂

Several mixtures were made in which MgH_2 was ball milled with 4 mol% of TiH_2, Nb_2O_5 or Mg_2Ni. TPD measurements were done in order to determine the effect of each catalyst on the hydrogen desorption properties of MgH_2. The profiles in Figure 1 show that pure MgH_2 has the highest onset temperature of about 310 °C. The onset temperatures for all the catalyzed mixtures are in the order: Pure $MgH_2 > TiH_2 > Nb_2O_5 \geq Mg_2Ni$. The plots also show that all of the mixtures released greater than 6 wt% hydrogen except the Nb_2O_5 catalyzed mixture, which released about 5 wt% hydrogen. This lower weight percentage could possibly result from partial oxidation of the Mg in the alloy caused by the presence of oxide in Nb_2O_5.

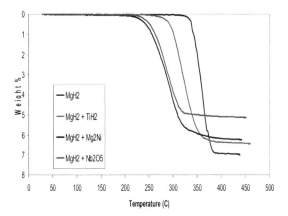

Figure 1, TPD Profiles for MgH$_2$ mechanically alloyed with TiH$_2$, Mg$_2$Ni or Nb$_2$O$_5$

Desorption kinetics experiments were done on MgH$_2$ at 400 °C in order to determine the catalytic effect of several additives on the hydrogen desorption rates from MgH$_2$. The additives used were TiH$_2$, Nb$_2$O$_5$ and Mg$_2$Ni. In these measurements constant pressure thermodynamic driving forces were used on all samples. This was done by adjusting the hydrogen pressure in the reactor to a value that is a little higher than that of the mid-plateau pressure (P$_m$), to make sure that only the hydrogen rich phase was present initially and sealing off the reactor. The pressure in the remaining system (P$_{op}$) was then adjusted to a value such that the ratio between the mid-plateau pressure and the opposing pressure (P$_m$/P$_{op}$) was a small whole number. This ratio is defined as the N-value. In these experiments, the N-value was set at 5 for all the sample mixtures. Figure 2 contains an isotherm for MgH$_2$ catalyzed by TiH$_2$. The plateau pressure at 400 °C is 20 atm. In this case an opposing pressure of 4 atm was applied so that the pressure ratio was 20/4 and the N-value was 5. Further details about the technique of constant pressure thermodynamic forces are published elsewhere.[19-21] Figure 3 contains plots of the reacted fraction against the time for hydrogen desorption from the MgH$_2$ mixtures. It can be seen that the mixture of MgH$_2$ + 4 mol% Nb$_2$O$_5$ has the fastest desorption kinetics while the pure MgH$_2$ sample has the slowest desorption kinetics rate.

An attempt was also made to determine the rate-controlling process in these samples by doing kinetic modeling. The method used was the one that was used by Smith and Goudy[21] to determine the process that controlled the rates of hydrogen desorption from a series of LaNi$_{5-x}$Co$_x$ alloys. The equations that were used are shown below:

$$\frac{t}{\tau} = 1 - \left(1 - X_B\right)^{1/3} \tag{4}$$

Where $\tau = \dfrac{\rho_B R}{b k_s C_{Ag}}$

$$\frac{t}{\tau} = 1 - 3(1 - X_B)^{2/3} + 2(1 - X_B) \qquad (5)$$

Where $\tau = \dfrac{\rho_B R^2}{6bD_e C_{Ag}}$

Where t is the time at a specific point in the reaction, X_B is the fraction of the metal reacted. R is the initial radius of the hydride particles, 'b' is a stoichiometric coefficient of the metal, C_{Ag} is the gas phase concentration of reactant, D_e is the effective diffusivity of hydrogen atoms in the hydride, ρ_B is the density of the metal hydride and k_s is a rate constant.

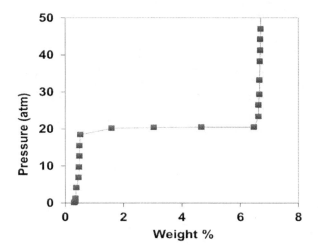

Figure 2. Desorption PCI for MgH$_2$+TiH$_2$ at 400 °C

    The model based on Eq. (4) will have chemical reaction at the phase boundary controlling the reaction rate whereas a model based on Eq. (5) is one in which diffusion controls the overall reaction rate. When this modeling technique was also applied to the present study the curves in Figure 4 were generated. The figure contains three plots for MgH$_2$. In the graph, one curve is an experimental curve taken from Figure 3, a second curve was calculated from equation 5 with the overall rate being controlled by diffusion and a third curve was calculated from equation 4 with chemical reaction controlling the rate. In order to determine the theoretical curves, it is important to determine a value for τ. This was achieved through a series of statistical data analyses. The results show that the data generated from equation 4, with chemical reaction controlling the overall rate, fits the experimental data better than the data generated from the diffusion controlled model. The phase boundary controlled model also worked well for all the catalyzed samples.

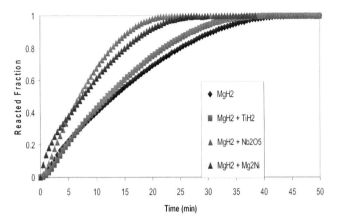

Figure 3. Kinetics Plots for Some MgH₂-Based Systems

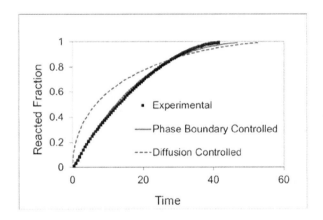

Figure 4. Modeling curves for pure MgH₂

3.2 Temperature Programmed Desorption Measurements, Kinetics, and Modeling Studies on a $Mg(BH_4)_2/Ca(BH_4)_2$ Destabilized System

Since first principles calculations have predicted that a system consisting of a mixture of $Mg(BH_4)_2$ and $Ca(BH_4)_2$ should be thermodynamically suitable for hydrogen storage, a TPD curve was constructed for the $Mg(BH_4)_2/Ca(BH_4)_2$ mixture in the stoichiometric ratio of 5:1. This curve is shown in Figure 5 along with those for pure $Mg(BH_4)_2$ and pure $Ca(BH_4)_2$. The curves show that the mixture releases hydrogen at a lower temperature than $Mg(BH_4)_2$ or $Ca(BH_4)_2$. In addition, all of these materials have lower onset temperatures than MgH_2.

Figure 6 shows the PCIs for the samples whose reactions are shown in equations 1-3. Pressure ratios (N-values) could be established based on the plateau pressures in these isotherms and kinetics could be done at constant pressure thermodynamic driving forces. The kinetics of the $Mg(BH_4)_2/Ca(BH_4)_2$ mixture as well as those of the pure $Mg(BH_4)_2$ and $Ca(BH_4)_2$ components were performed. Figure 7 contains plots in which the kinetics of the borohydrides are compared to that for MgH_2. It can be seen that the $Mg(BH_4)_2/Ca(BH_4)_2$ mixture has faster kinetics than either $Mg(BH_4)_2$ or $Ca(BH_4)_2$. All of the borohydrides desorb hydrogen faster than pure MgH_2 under the conditions used.

Equations (4) and (5) were fitted to the kinetic data for $Mg(BH_4)_2$, $Ca(BH_4)_2$ and $Mg(BH_4)_2+Ca(BH_4)_2$. Figures 8-10 contain modeling plots for the each of the borohydrides based on the kinetics data in Figure 7. In each graph, one curve is an experimental curve, a second curve is based on the overall rate being controlled by diffusion, and a third curve is calculated based on chemical reaction controlling the rate. The plots in Figure 8 show a good fit for a diffusion controlled model up to 80% of the reaction but not beyond, while those in Figure 9 show a good fit for phase boundary controlled up to 80%. This means that the process controlling hydrogen desorption of this reaction at the later stage of the reaction is different from that at the beginning. The plots in Figure 10 do not fit with either of the process, not even in early part of the reaction. It seems that the desorption of hydrogen by the $Mg(BH_4)_2/Ca(BH_4)_2$ mixture may be under the mixed control of both processes.

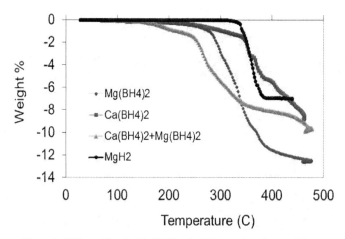

Figure 5. TPD profiles for $Mg(BH_4)_2$, $Ca(BH_4)_2$ and a mixture of the two compounds

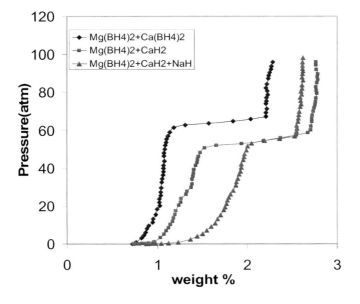

Figure 6. The isotherms were done at 450 °C.

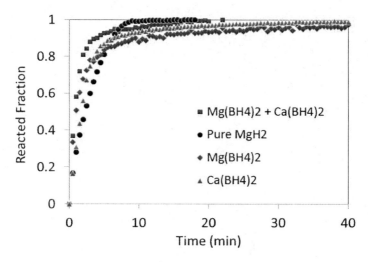

Figure 7. Kinetics of Ca(BH$_4$)$_2$, Mg(BH$_4$)$_2$ and the Mixture

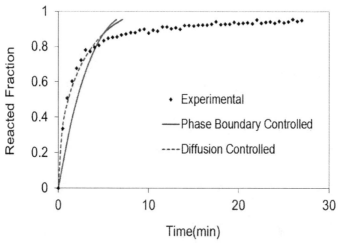

Figure 8. Modeling of Mg(BH$_4$)$_2$

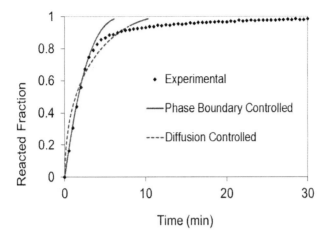

Figure 9. Modeling of Ca(BH$_4$)$_2$

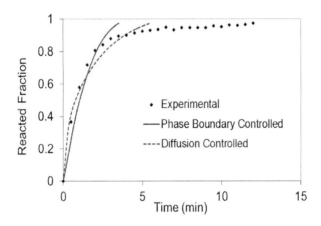

Figure 10. Modeling of Mg(BH$_4$)$_2$+Ca(BH$_4$)$_2$

3.3 Temperature Programmed Desorption Kinetic Measurements on the $LiNH_2+MgH_2$ System

Temperature Programmed Desorption (TPD) measurements were done to determine the temperatures at which the sample mixtures released hydrogen. Figure 11 represents the TPD analyses of the doped and pristine samples of ($LiNH_2+MgH_2$) mixtures. Each sample mixture was heated to approximately 350°C at a rate of 4°C/min. It can be seen that the doped samples have lower desorption temperatures than their pristine samples. The mixture ($1.9LiNH_2+MgH_2+0.1KH$) has an onset desorption temperature of 75.3°C whereas its corresponding pristine sample has an onset desorption temperature of 109°C. These results confirm that KH is an effective dopant for reducing the desorption temperature of the ($2LiNH_2/MgH_2$) mixture.

Desorption kinetics measurements were done on each sample at 210°C and N = 10. A temperature of 210 °C was chosen because temperatures in excess of 400 °C, as used for the borohydrides, were too high for this system. From Figure 12, it can be seen that the KH catalyzed sample has much faster desorption kinetics than its corresponding pristine sample. The addition of potassium hydride dopant was found to have a 25-fold increase on the desorption rates of the $2LiNH_2$ + MgH_2 mixture.

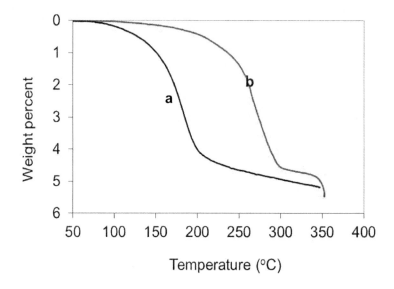

Figure 11. TPD curves for (a) $1.9LiNH_2+MgH_2+0.1KH$ and (b) $2LiNH_2+MgH_2$ samples.

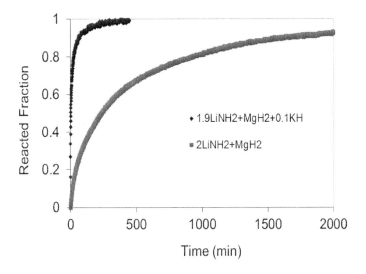

Figure 12. Kinetics of the LiNH₂/MgH₂ System

4. CONCLUSION

The results of this study show that a mixture of $Mg(BH_4)_2$ and $Ca(BH_4)_2$ desorbs hydrogen at a faster rate than its components. Modeling studies indicate that the rate determining process for hydrogen desorption from $Mg(BH_4)_2$ is diffusion controlled whereas reaction at the phase boundary controls hydrogen desorption rates from $Ca(BH_4)_2$. The rate of hydrogen evolution from a $Mg(BH_4)_2/Ca(BH_4)_2$ mixture is likely under the mixed control of both processes. The results of this study have also shown that KH is a very effective catalyst for improving hydrogen desorption rates from a $2LiNH_2/MgH_2$ mixture.

ACKNOWLEDGMENT
This work was financially supported by grants from the Department of Energy and the U.S. Department of Transportation

REFERENCES

[1]   Mauron P, Buchter F, Friedrichs O, Remhof A, Bielmann M, Zwicky CN, Zuttel A. Stability and reversibility of LiBH4. J. Phys. Chem. B 2008; 112: 906-10.
[2]   Kojima Y. Hydrogen storage and generation using sodium borohydride. R&D Review of Toyota CRDL; 2005: 40 (2):31-5.

[3]   Siegel DJ, Wolverton C, Ozoliņš V. Thermodynamic guidelines for the prediction of hydrogen storage reactions and their application to destabilized hydride mixtures. Physical Review B 2007; 76: 134102-6.

[4]   Rönnebro E. Development of group II borohydrides as hydrogen storage materials. Current Opinion in Solid State and Materials Science 2011; 15(2): 44-51.

[5]   Badaji EG, Hanada O, Fichtner M. Effects of several metal chloride on the thermal decomposition behaviour of α–Mg(BH$_4$)$_2$. International journal of hydrogen energy 2011; 36: 12313-8.

[6.]   Ibikunle A, Goudy AJ, Yang H. Hydrogen storage in CaH$_2$/LiBH$_4$ destabilized metal hydride system. J. Alloys and compounds 2009; 475:110-15.

[7]   Durojaiye T, Ibikunle A, Goudy AJ. Hydrogen storage in destabilized borohydride materials, International journal of hydrogen energy 2010; 35: 10329-33.

[8]   Yu XB, Yang ZX, Guo YH, Li SG. Thermal decomposition performance of Ca(BH$_4$)$_2$/LiNH$_2$ mixtures. J. Alloys and Compounds 2011;509: S724– 7.

[9]   Lee JY, Ravnsbæk D, Lee Y, Kim Y, Cerenius Y, Shim J, Jensen TR, Hur NH, Cho YW. Decomposition reactions and reversibility of the LiBH$_4$-Ca(BH$_4$)$_2$ Composite. J. Phys. Chem. C 2009; 113: 15080–6.

[10]   Au M, Jurgensen A. Modified lithium borohydrides for reversible hydrogen storage. J. Phys. Chem. B 2006; 110: 7062-7.

[11]   Vajo JJ, Skeith SL. Reversible storage of hydrogen in destabilized LiBH$_4$, J. Phys. Chem. B2005; 109(9): 3719-22.

[12]   Kim Y, Reed D, Lee Y-S, Lee JY, Shim J-H, Book D, Cho YW. Identification of the dehydrogenated product of Ca(BH$_4$)$_2$. J. Phys. Chem. C 2009; 113: 5865–71.

[13]   Fanga Z-Z, Kanga X-D, Wanga, Li H-W, Orimo S-I. Unexpected dehydrogenation behavior of LiBH$_4$/Mg(BH$_4$)$_2$ mixture associated with the in situ formation of dual-cation borohydride. J. of Alloys and Compounds 2010; 491: L1–4.

[14]   Severa G, Ronnebro E, Jensen C. Direct hydrogenation of magnesium boride to magnesium borohydride: demonstration of >11 weight percent reversible hydrogen storage. Chem Commun 2010; 46: 421-3.

[15]   Fichtner M, Zhao-Karger Z, Hu J, Roth A, Weidler P. The kinetic properties of Mg(BH$_4$)$_2$ infiltrated in activated carbon. Nanotechnology 2009; 20: 204029-33.

[16]   Ozolins V, Majzoub EH, Wolverton C. First-principles prediction of thermodynamically reversible hydrogen storage reactions in the Li-Mg-Ca-B-H system. J. Am. Chem. Soc 2009; 131:230-7.

[17]   Alapati SV, Johnson JK, Scholl DS. Large-scale screening of metal hydride mixtures for high-capacity hydrogen storage from first-principles calculations. J Phys Chem C 2008; 112: 5258-62.

[18]   Chłopek K, Frommen C, Léon A, Zabara O, Fichtner M. Synthesis and properties of magnesium tetrahydroborate, Mg(BH$_4$)$_2$ .J. of Materials Chemistry 2007; 17: 3496-503.

[19]   Koh JT, Goudy AJ,Huang P, Zhou G. A comparison of the hydriding and dehydriding kinetics of lanthanum nickel compound (LaNi5) hydride. J. Less Common Me t1989; 153: 89-100.

[20]   Donohue RM, Goudy AJ. Desorption kinetics of 1:3 erbium cobalt intermetallic hydride. Inorg. Chem 1991; 30: 175-9.

[21]   Smith G, Goudy AJ. Thermodynamics, kinetics and modeling studies of LaNi$_{5-x}$Co$_x$ hydride systems. J. of Alloys and Compounds2001; 316: 93-8.

MICROSTRUCTURE AND CORROSION BEHAVIOR OF THE Cu-Pd-M TERNARY ALLOYS FOR HYDROGEN SEPARATION MEMBRANES

Ö.N. Doğan[1], M.C. Gao[1,2], R. Hu[1,2]
[1]DOE National Energy Technology Laboratory, 1450 Queen Avenue, SW, Albany, OR 97321, USA
[2]URS Corporation, P.O. Box 1959, Albany, OR 97321, USA

ABSTRACT

Surface poisoning and corrosion are the most significant degradation mechanisms acting on the membrane materials at elevated temperatures in syngas derived from coal containing impurities such as H_2S. Cu-Pd alloys demonstrated significant potential for being resistant against these degradation mechanisms. It was also shown that Cu-Pd compositions containing ordered bcc (B2) crystal structure exhibit hydrogen permeability comparable to pure palladium. The B2 structure of the membrane alloys was successfully stabilized at higher temperatures by ternary element additions. Furthermore, the ternary alloy coupons were exposed to simulated syngas environments at 500°C. While additions of Al and Mg to the Cu-Pd alloys did not change the mass gain, additions of Y, La, Ti, and Hf increased the mass gain during the exposures.

INTRODUCTION

To produce high purity hydrogen fuel from coal, hydrogen must be separated from synthesis gas (syngas), a product of coal gasification. Membrane technology can be used to achieve this goal. The U.S. Department of Energy established a set of performance targets for hydrogen separation membranes for the syngas applications in its Hydrogen from Coal program[1]. Although a variety of hydrogen separation membrane materials exist today, none of them are shown to be suitable for use in contaminant laden syngas at elevated temperatures. Surface poisoning and corrosion are the most significant degradation mechanisms acting on the membrane materials at elevated temperatures in gases containing carbon dioxide, carbon monoxide, water vapor, and impurities (hydrogen sulfide, arsenic, selenium, and mercury).

Cu-Pd alloys demonstrated significant potential for being resistant to these degradation mechanisms[2-4]. In addition, Cu-Pd compositions containing ordered bcc (B2) crystal structure exhibit hydrogen permeability comparable to pure palladium[5]. The superior hydrogen permeability of Cu-Pd membranes with primarily B2 structure exists through temperatures at which the alloy transforms into an fcc structure, triggering a sharp drop in permeance.

An integrated approach that combines first-principles density functional theory (DFT) calculations and key experiments, was used in this study to accelerate new CuPdM alloy design that expand the bcc phase field for hydrogen separation from syngas derived from coal.

DFT CALCULATIONS

The first principles package of Vienna *ab initio* simulation (VASP)[6-7] was used to calculate the total energies using electronic density functional theory (DFT). Projector augmented-wave (PAW) potentials[8] were employed as supplied with VASP. The Perdew-Burke-Ernzerhof[9] gradient approximation to the exchange-correlation functional was utilized. The Brillouin zone integrations were performed using the Monkhorst–Pack k-point meshes[10], and a smearing parameter of 0.2 eV was chosen for the Methfessel–Paxton[11] technique. All structures were fully relaxed (both lattice parameters and atomic coordinates) until energies converged to a precision of 1 meV/at. A high precision setting was used. The plane-wave energy cutoff was held constant at 500 eV. The semi-core 3p, 4p and 5p electrons of transition metal elements were explicitly treated as valence. To obtain enthalpy of formation values ΔH_f, a composition-weighted average of the pure elemental cohesive energies was subtracted from the cohesive energy of a given composition. The resulting energy was an "enthalpy" because its volume was relaxed at zero pressure. A 2x2x2 supercell was built and an individual Pd atom was substituted for the alloying elements. Fig. 1 shows the enthalpy of formation for hypothetical B2 $Cu_8Pd_{8-x}M_x$ (x=0-8) ternary alloys for transition metals and non-transition metals. The results indicated that elements Sc, Ti, Zn, Y, Zr, Hf, La, Al and Mg were strong stabilizers under the assumption that the competing phases in the individual ternaries, configurational entropy effect and lattice vibration at finite temperatures were ignored. Phase stability study for the complete Cu-Pd-M ternaries at finite temperatures are necessary to draw comprehensive conclusions, but are beyond the scope of the present work.

EXPERIMENTAL PROCEDURE

Based on the present DFT calculations, seven alloying elements (Ti, Zr, Hf, Y, La, Al and Mg) were chosen for experimental verification. Accordingly, fourteen ternary alloys and two benchmark binary alloys were selected in order to locate the B2 phase field in the Cu-Pd-M ternary system. The alloy compositions in atomic percent were nominally $Cu_{66}Pd_{27.75}M_{6.25}$ and $Cu_{50}Pd_{43.75}M_{6.25}$ where M=Pd, Ti, Zr, Hf, Y, La, Al, Mg and designated as Cu-28Pd-6M and Cu-44Pd-6M, respectively.

The starting materials were high purity Cu, Pd, Ti, Zr, Hf, Y, La, Al and Mg elements in pellet or sponge form. The sixteen alloys were prepared by melting in a vacuum arc furnace back-filled with high purity argon. Each sample weighed ~40g. After a homogenization at 900°C for 72 hours, the alloys were subjected to an equilibration annealing at 400°C for 21 days.

Coupon corrosion tests were performed in a severe environment exposure laboratory. A simulated syngas mixture (representative of a composition downstream from water-gas shift reactors) containing (in volume %) $50H_2$, $30CO_2$, $1CO$, $19H_2O$, and varying amounts of H_2S (0, 100, and 1000 ppm) was prepared using mass flow controllers. Using a syringe pump, a predetermined amount of distilled water was introduced directly into the hot zone of the furnace at a constant rate.

Test coupons cut from the buttons mentioned above with approximate dimensions of 20 mm x 10 mm x 1 mm were surface ground to a 600 grit finish, then cleaned and weighed on a scale to five decimal points. The samples were placed individually in high-density alumina crucibles on their edges so that more than 95% of the surface area was exposed to the gas mixture. The crucibles were situated in the hot zone of the tube furnace on an alumina tray. After the ends of the tube were sealed, helium was flowed for two hours to flush the air out at room temperature. Then the furnace was turned on and heated at a rate of 200°C per hour to 500°C. During heating, the He flow was maintained. Once the furnace reached 500°C, the He flow was stopped; the gas mixture was allowed to flow through the tube and the water pump was turned on. The samples were exposed to these conditions for 24 hours before the gas mixture and water flow stopped and the He flow restarted. After one hour of He flow, the furnace cooled to room temperature. The samples were then taken out of the furnace for weighing. This procedure was repeated five times to provide a 120 hour exposure for the samples at each H_2S level.

Wavelength dispersive x-ray fluorescence spectroscopy (WDXRF) (Rigaku, ZSX Primus II) was employed for the chemical analysis. Cu-Zn alloys were used for calibrating the WDXRF. The phase identification was done using x-ray diffraction (XRD) with a high-temperature stage (Rigaku, Ultima III with Jade analysis software). Scanning electron microscopy (SEM) (FEI, Inspect F50 scanning electron microscope with Oxford INCA Microanalysis) was employed for microstructural characterization.

RESULTS AND DISCUSSIONS

The content of major elements, as determined by WDXRF, is shown in Table 1. The table indicates that the impurity content in each sample is less than 1 wt.%. The relatively large deviation of the Mg content in one Cu-Pd-Mg sample from the nominal value may be due to element evaporation since Mg has a lower melting point and high vapor pressure.

Microstructure:

For Cu-44Pd-6M alloys, XRD results indicated the presence of the B2 phase for M= Ti, Y, La, Al and Mg (Fig. 2). All Cu-28Pd-6M alloys showed either Cu_3Pd (Tetragonal, P4/mmm, $L1_0$) or Cu_4Pd (Tetragonal, $P4_2/m$) as the major phase, and two alloys (containing Hf and Mg) indicated the B2 phase as the minor phase, listed in Table 2. The results demonstrated that the B2 phase field does not shift to the Cu-rich side in the Cu-Pd-M ternary (M=Ti, Zr, Y, La, Al). This is due to the presence of the stable compounds in the Cu-rich Cu-M system (e.g. Al_4Cu_9) or the extension of stable Pd-rich Pd-M compounds (e.g. YPd_3, $HfPd_3$, $LaPd_5$, $ZrPd_3$) in the ternary system. No observation of the binary Cu_3Pd $L2_1$ phase in the samples was made.

Using ImageJ software based on the compositional contrast, the volume percentage of B2 phase in each sample was estimated from the back-scattered SEM images (e.g. in Fig. 3). Listed in Table 3 are the compositions of all phases identified using XRD as well as the estimated volume percentage of B2 phase where identified. The estimated volume percentage of the B2 phase in Cu-44Pd-6M (M=Mg, Al, and Y) is 100%, 75% and 70% at 400°C respectively. The present experiments indicate that Mg, followed by Al and Y, is the strongest B2 stabilizer.

High-temperature XRD experiments done on the B2-containing Cu-44Pd-6M alloys indicated that the temperature at which the B2→FCC transformation occurs was significantly raised by addition of Mg, Al, La, Y or Ti (listed in Table 4). For example, addition of 6 at% Mg to the Cu-50Pd alloy raised the T_f from 480°C to above 860°C.

Corrosion:

Mass change of the coupons with time during the exposure to the simulated syngas at 500°C is shown in Fig. 4. In general, addition of 6 at.% alloying element (Ti, Zr, Hf, Y, La, Al, Mg) to Cu-34Pd by replacing Pd resulted in an increase in mass during the exposure to the syngas containing no H_2S (Fig. 4a). While the mass gain in most of the Cu-28Pd-6M alloys was comparable, the Cu-28Pd-6La alloy's mass gain was significantly higher than the rest. Addition of 100 ppm H_2S to the syngas stream did not significantly change the mass gain results of the Cu-28Pd-6M alloys compared to the results of the test without H_2S except for the Hf and La containing alloys (Fig. 4c). While the mass gain of the Hf containing alloy increased with the introduction of H_2S, the mass gain of the La containing alloy decreased. The Y, Al, Ti, and Mg containing alloys (Cu-28Pd-6M) performed better than the pure Pd in the 100 ppm H_2S test. In the 1000 ppm H_2S exposure, the Cu-34Pd and Cu-28Pd-6Mg alloys performed the best with negligible mass gain. Most of the other Cu-28Pd-6M alloys showed increased mass gain except for the La containing alloy whose mass gain continued to decrease with increasing H_2S content in syngas. The mass gain of the pure Pd coupon also increased significantly with increasing H_2S. The results of the syngas exposure tests done on the Cu-28Pd-6M alloys demonstrated that the Cu-34Pd and Cu-28Pd-6Mg alloys consistently performed well regardless of the H_2S content in the syngas.

In general, the corrosion rates for the Cu-44Pd-6M alloys were lower and did not exceed 1.5 mg/cm² during the 120 hour exposure for any alloy. In addition, none of the alloys in this group showed significantly higher rates than the rest of the group. Furthermore, it was possible to group the behavior of the alloys based on the alloying elements. When the Cu-50Pd alloy was alloyed with a transition metal from group IIIA (Y or La), the ternary alloys gained mass at the fastest rate. Addition of IVA group elements (Ti or Zr or Hf) to the binary Cu-50Pd caused an intermediate level of corrosion during the exposure to syngas regardless of the H_2S content. The lowest corrosion rates were recorded when Cu-50Pd was used unalloyed or when alloyed with Mg or Al.

This study utilized predictive DFT calculations to screen for potential alloying elements that can stabilize the CuPd B2 phase field in the Cu-Pd-M ternaries. The screening process was carried out for all transition metals and many non-transition elements without any experimental input. The present DFT calculations were crude in the sense that the competing phases against B2 in individual Cu-Pd-M ternary system, the configurational entropy effect and phonon vibration at finite temperatures were ignored. Nonetheless, out of seven elements recommended for experiments based on the DFT calculations, five elements (Mg, Al, Y, Ti, and La) were experimentally verified to expand the phase field of B2 CuPd into the ternary system at 400°C. Therefore, the present study suggests that utilizing predictive DFT calculations can significantly cut down the number of trial-and-error experiments, and thus accelerate new materials design for hydrogen separation. Based on the microstructural and corrosion investigation, Mg and Al were the most promising alloying elements in the Cu-50Pd alloys for use in syngas membrane applications.

CONCLUSIONS

1. The present study suggests that utilizing predictive DFT calculations can significantly cut down the number of trial-and-error experiments, and thus accelerate new materials design for hydrogen separation.
2. Al, Mg, Y, La, and Ti, when added to Cu-Pd, were shown to extend the stability of the B2 structure to higher temperatures than that in the binary Cu-Pd alloy.
3. Alloying with 6 at.% Mg or Al did not degrade the corrosion resistance of the Cu-34Pd and Cu-50Pd alloys.

ACKNOWLEDGEMENTS

This research was performed in support of the Fuels Program of Strategic Center for Coal at DOE National Energy Technology Laboratory.

This report was prepared as an account of work sponsored by an agency of the United States Government. Neither the United States Government nor any agency thereof, nor any of their employees, makes any warranty, express or implied, or assumes any legal liability or responsibility for the accuracy, completeness, or usefulness of any information, apparatus, product, or process disclosed, or represents that its use would not infringe privately owned rights. Reference herein to any specific commercial product, process, or service by trade name, trademark, manufacturer, or otherwise does not necessarily constitute or imply its endorsement, recommendation, or favoring by the United States Government or any agency thereof. The views and opinions of authors expressed herein do not necessarily state or reflect those of the United States Government or any agency thereof.

REFERENCES

1. Office of Fossil Energy, *Hydrogen from Coal Program, RD&D Plan, External Draft*, U.S. Department of Energy, September 2010.
2. B.D. Morreale, M.V. Ciocco, B.H. Howard, R.P. Killmeyer, A.V. Cugini, R.M. Enick, Journal of Membrane Science 241 (2004) 219-224.
3. O. Iyoha, R. Enick, R. Killmeyer, B. Morreale, Journal of Membrane Science 305 (2007) 77-92.
4. A. Kulprathipanja, G.O. Alptekin, J.L. Falconer, J.D. Way, Journal of Membrane Science 254 (2005) 49-62.
5. B.H. Howard, R.P. Killmeyer, K.S. Rothenberger, A.V. Cugini, B.D. Morreale, R.M. Enick, F. Bustamante, Journal of Membrane Science 241 (2004) 207-218.
6. Kresse, G. and J. Hafner, Phys. Rev. B, 1993. **47**: p. 558.
7. Kresse, G. and J. Furthmueller, Phys. Rev. B, 1996. **54**: p. 11169.
8. Blochl, P.E., Phys. Rev. B, 1994. **50**(24): p. 17953.
9. Perdew, J.P., K. Burke, and M. Ernzerhof, Phys. Rev. Lett., 1996. **77**: p. 3865.
10. Monkhorst, H.J. and J.D. Pack, Phys. Rev. B, 1976. **13**(12): p. 5188.
11. Methfessel, M. and A.T. Paxton, Phys. Rev. B, 1989. **40**(6): p. 3616.
12. P.R. Subramanian, D.E. Laughlin, Journal of Phase Equilibria, 12 (1991) 231-243.

Table 1. Chemical composition (wt.%) of the experimental alloys as determined by XRF.

Alloy	Cu	Pd	Ti	Mg	Zr	Hf	Y	Al	La
Cu-50Pd	38.38	61.52							
Cu-34Pd	54.82	45.08							
Cu-44Pd-6Ti	40.41	56.49	3.10						
Cu-28Pd-6Ti	58.51	38.22	3.26						
Cu-44Pd-6Zr	37.53	56.04			6.42				
Cu-28Pd-6Zr	56.50	36.34			7.15				
Cu-44Pd-6Hf	34.98	51.52				12.81			
Cu-28Pd-6Hf	52.81	33.69				13.14			
Cu-44Pd-6Y	37.41	56.57					5.61		
Cu-28Pd-6Y	55.64	37.70					6.30		
Cu-44Pd-6La	37.88	53.26							8.85
Cu-28Pd-6La	55.80	36.93							7.23
Cu-44Pd-6Al	40.94	56.26						1.81	
Cu-28Pd-6Al	57.89	39.07						2.11	
Cu-44Pd-6Mg	40.94	57.39		1.67					
Cu-28Pd-6Mg	57.92	40.98		0.87					

Table 2. Phases observed in the Cu-28Pd-6M alloys by XRD.

Alloys	Phase 1	Phase 2	Phase 3	Phase 4
Cu-34Pd	(CuPd)_FCC			
Cu-28Pd-6Ti	Cu_4Pd_tetra	Cu_3Pd_tetra	Pd_5Ti_3	$TiPd_3$
Cu-28Pd-6Zr	Cu_3Pd_tetra	$ZrPd_3$		
Cu-28Pd-6Hf	Cu_4Pd_tetra	$HfPd_3$	PdCu_B2	
Cu-28Pd-6Y	Cu_3Pd_tetra	YPd_3		
Cu-28Pd-6La	Cu_4Pd_tetra	$LaPd_5$		
Cu-28Pd-6Al	Cu_3Pd_tetra	$Cu_{0.78}Al_{0.22}$	Al_4Cu_9	$Cu_{5.75}Al_{4.5}$
Cu-28Pd-6Mg	Cu_3Pd_tetra	$Cu_{0.7}Pd_{0.3}$	PdCu_B2	

Table 3. Phases observed in the Cu-44Pd-6M alloys by XRD.

Alloys	Phase 1	Phase 2	Phase 3	Volume fraction of B2
Cu-50Pd	(CuPd) FCC			0
Cu-44Pd-6Ti	FCC	B2	Pd$_3$Ti	0.05-0.1
Cu-44Pd-6Zr	FCC	Pd$_{0.845}$Zr$_{0.155}$		0
Cu-44Pd-6Hf	Cu$_3$Pd_L1$_2$	HfPd$_3$		0
Cu-44Pd-6Y	B2	YPd$_3$	FCC	0.7
Cu-44Pd-6La	B2	LaPd$_5$		0.05-0.1
Cu-44Pd-6Al	B2	FCC		0.75
Cu-44Pd-6Mg	B2 #1	B2 #2		1

Table 4. Start and finish temperatures for the B2→FCC

Alloy	T$_s$ (°C)	T$_f$ (°C)
Cu-50Pd*	<300	480
Cu-44Pd-6Mg	~640	>860
Cu-44Pd-6Y	575-600	675-700
Cu-44Pd-6Al	650-675	825-850
Cu-44Pd-6Ti	<400	775-800
Cu-44Pd-6La	<400	625-650

*Reference 12

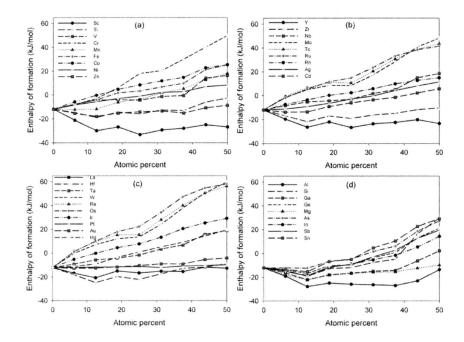

Fig. 1. Enthalpy of formation of hypothetical B2 $Cu_8Pd_{8-x}M_x$ (x=0-8) alloys predicted from the present DFT calculations

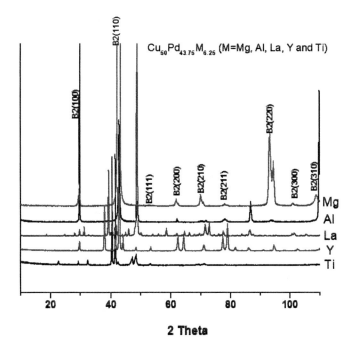

Fig. 2. XRD plots for B2-containing Cu-44Pd-6M alloys

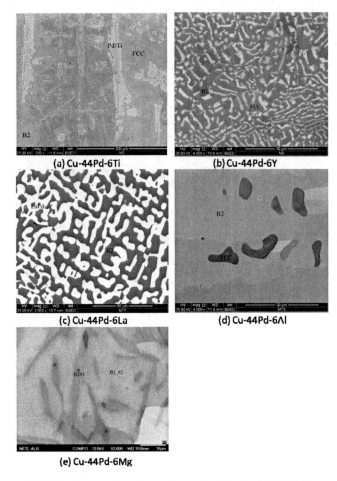

Fig. 3. Back-scattered electron micrographs of the Cu-44Pd-6M alloys

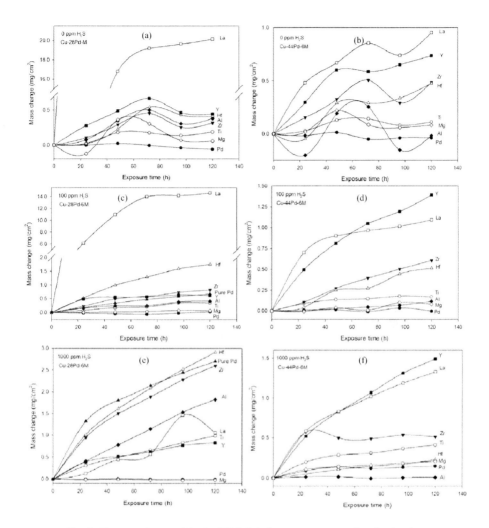

Fig. 4. The mass change of the Cu-28Pd-6M alloy coupons exposed to simulated syngas composition containing 0 ppm H_2S (a), 100 ppm H_2S (c), and 1000 ppm H_2S (e). The mass change of the Cu-44Pd-6M alloy coupons exposed to simulated syngas composition containing 0 ppm H_2S (b), 100 ppm H_2S (d), and 1000 ppm H_2S (f). The exposures were carried out at 500°C.

METAL-HYDROGEN SYSTEMS: WHAT CHANGES WHEN SYSTEMS GO TO THE NANO-SCALE?

A.Pundt, Institut für Materialphysik, Friedrich-Hund-Platz 1, 37075 Göttingen, Germany

ABSTRACT:

For hydrogen storage applications, nano-scale metal-hydrogen systems are suggested to reduce loading and unloading times. But, nano-scaling the system does also change other physical properties. Many changes can be related to micro-structure and mechanical stress, but also to new structures only evolving in the nanometer-range. This paper shortly summarizes changes of metal-hydrogen systems turned to nano-scale and discusses the findings with regard to storage applications.

INTRODUCTION

With regard to a successful use of metal-hydrogen storages for many mobile applications, light weight storage with short loading times are required. Gas storage in high pressure vessels is one option. But, by materials failure due to external forces or by hydrogen embrittlement especially of fittings, the large quantities of escaping hydrogen gas can lead to severe safety problems. A more safe way to store hydrogen is, therefore, the storage inside of a metal.[1] In a metal, hydrogen is solved on interstitial lattice sites.[2] Hydrogen release is, for such a case, very mild. It is limited by hydrogen permeation which is the product of hydrogen diffusion and solubility, in the metal. For some metals, like the light weight Mg, low permeation rates even hinder the materials application for storage.[3,4]

To yield reasonable hydrogen loading and unloading times is one important issue. After successful absorption, loading time, on one hand, is determined by interstitial hydrogen diffusion and hydrogen tunneling in the metal.[5] On the other hand it is determined by the diffusion length. Thus, one way to shorten hydrogen loading times is to shorten the hydrogen travel paths, by reducing the system size. Thereby, nano-scale systems, with sizes between 1000 nm and 1 nm come into play. However, in this range of sizes it is expected that materials change their physical properties. This is well-known from molecular chemistry and computer simulations very small size systems below 5 nm diameter.[6] [7]However, even for larger scales of about 100 nm, changes are reported.[8]

Physical properties such as hydrogen solubility and M-H thermodynamics are important for hydrogen storage in a metal. These properties are provided by the PCT-isotherm (Pressure-Concentration dependency at constant Temperature). A typical single-stage isotherm of a bulk metal, here Pd-H, is shown in Fig.1. The isotherm gives the relation between the environmental hydrogen gas pressure p_H or the chemical potential μ_H and the internal hydrogen concentration c_H . The hydrogen concentration c_H is usually given in H/M, the number of hydrogen atoms per metal atom. After phase transition, it is c_H =0,6 H/Pd for Pd, and for Mg it is c_H =2 H/Mg, at 300K.

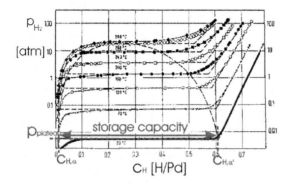

Fig.1:

PCT-curves of Pd-H, as measured by H. Frieske and E. Wicke.[2] Concentration limits, $c_{H,\alpha}$ and $c_{H,\alpha'}$ and the plateau pressure $p_{plateau}$ at 20°C, are highlighted.

(With permission from Springer Science + Business Media, from Ref[2])

For the scope of this paper, the isotherms information is divided into two different aspects:

1) the pressure range where the concentration changes strongly upon a small pressure change. Often materials undergo at least one phase transition in this range, yielding the formation of hydrides. The related equilibrium pressure is called "plateau pressure, $p_{plateau}$" because it stays constant for bulk metals or increases slightly for bulk alloys. The plateau pressure is given by the enthalpy and the entropy of hydride formation, meaning the change of enthalpy and entropy upon hydride formation from the solid solution phase. The plateau pressure determines the environmental conditions for storage. Of importance are the loading and the unloading plateau pressures, which are usually different and open a hysteresis.

2) the concentrations limits (solubility limits , in Fig. 1 marked with $c_{H,\alpha}$ and $c_{H,\alpha'}$), bordering the strong concentration change at the plateau pressure. They control the concentration increase for a metal upon hydride formation, or the real width of the miscibility gap(s). The limits mainly give the storage capacity $c_{sc} = c_{H,\alpha'} - c_{H,\alpha}$ in H/M.

Both aspects can change for nano-materials. This paper focuses on typical micro-structures and mechanical stress contributions of nano-scale systems and how they can influence the two described aspects.

MICROSTRUCTURAL CONTRIBUTIONS[9]

When the metal is prepared with nano-scale extension, the impact of micro-structural defects becomes important. These defects often act as hydrogen trapping sites that are occupied at very low pressures and that cannot be unloaded under normal conditions. Thereby, they increase the α-phase solubility limit. When the local hydrogen concentration in the traps is smaller than the hydride concentration, they also decrease the hydride phase solubility limit. In total, traps usually reduce the hydrogen storage capacity of the metal.

A chunk of metal is shown in Fig. 2, summarizing the important micro-structural contributions discussed here. Hydrogen trapping sites are marked with red dots. Beside a) the conventional interstitial lattice site, b) surfaces, c) subsurface sites as well as d) grain boundaries and interfaces, and e) dislocations become important.[10] For nano-metals the relative number of defect sites is much higher than in a conventional bulk material.

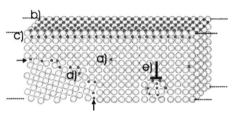

Fig.2:

Sketch on the micro-structure of nano-metals. Beside the conventional interstitial lattice site (a), surfaces (b), subsurface (c), grain boundaries and interfaces (d), as well as dislocations (e) become important. Red dots mark hydrogen locations. Most defects act as hydrogen traps.[9]

Republished with permission of **ANNUAL REVIEWS**, from [9]; permission conveyed through Copyright Clearance Center, Inc.

For different defect-rich bulk materials, these different micro-structural contributions have been isolated and studied. Studies on clean cut surfaces (Ni or Pd) show that surfaces can adsorb even more than 1 H/M-Atom.[11] Additionally, for several layers below the surface, i.e. in the sub-surface, the hydrogen content is often increased.[12,13,14,15,16] In the dilatation field of an edge dislocation further hydrogen trapping sites are present.[17,18] In heavily deformed bulk Palladium, Maxelon et al. detected hydrogen enriched regions of cylindrical shape with radii in 1-2 nm extension.[19,20] The lcoal hydogen concentration was suggested to be that of the hydride, even though the metal matrix was still in the α-phase.[21] Grain boundaries also act as traps for hydrogen, as determined by Mütschele and Kirchheim on nano-crystalline Pd-H.[22,23,24] They found a strongly increased solubility limit when the grain size was reduced and correlated this to low-energy sites available in the grain boundary region. Weissmüller and Lemier, however, showed that solubility changes can also result from lattice straining by grain boundaries.[25,26] Studies on 3-6 nm Pd-clusters (stabilized in tetraoctylammoniumbromide, TOAB) showed that, for clusters, surface and subsurface sites become the dominant hydrogen trapping sites. They increase the α-solubility limit and decrease the width of the miscibility gap.

For all these contributions an increase is expected for the total solubility in the solid solution, with regard to the bulk system.

For each nano-system, the impact of these micro-structural contributions can be calculated as function of dimension. This is exemplarily done in Fig.3 for Nb thin films. For the calculations, the main assumption was made that the local contributions are thickness-independent and just weighted by their volume content. Surface (S) contributions inversely scale on the film thickness and increase the solubility limit c_α by

$$c_{\alpha,S} = \frac{(d - d_{Surf})}{d} c_0 + \frac{d_{Surf}}{d} c_{surf} \tag{1}$$

The total surface and subsurface thickness is approximated with d_{surf} and the related mean hydrogen concentration with c_{surf}. With regard to the above mentioned results, $d_{surf} = 0.7$ nm with $c_{surf}=1$ H/M-atom was used, at room temperature.

Grain boundaries (GB) increase the solubility limit by

$$c_{\alpha,GB} = \frac{(d - d_{GB})^2}{d^2} c_0 + \frac{2d_{GB}d}{d^2} c_{GB} \tag{2}$$

when the grain size equals the film thickness. For approximation, the local grain boundary concentration c_{GB} can be averaged to be between that of the hydride and the α-phase. For Nb $c_{GB}=0.4$ H/Nb. A grain boundary width d_{GB} of 0.9 nm was used.

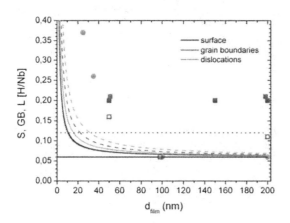

Fig.3:

Micro-structural impact on the solubility limit of Nb-H films of different thickness d_{film}. Calculation is done for the impact of surface (S), grain boundaries (GB) and dislocations (D), the sum of surface and grain boundaries (dashed red) as well as surface and dislocations (dashed green). Experimental data are also implemented. For details see text.

The sum of both contributions (marked with red dashed line in Fig.3) shows that, for nano-crystalline Nb films ($d_{film}=d_{grain}$) with thicknesses above 200 nm, micro-structural impacts on solubility limits are negligible. But the solubility limit of a nano-crystalline Nb-film increases strongly below 50 nm: It is more than doubled below 20 nm film thickness. For bulk Nb-H it is 0.06 H/Nb (marked with black bold line) while for the film, it increases to 0.12 H/Nb (marked with black dotted line).

For epitaxial films, misfit dislocations often occur between the film and the substrate. For their presence, also the influence of elevated temperatures has to be considered when the thermal expansion of the film and the substrate differs. These dislocations usually exist at the film/substrate interface, with dislocation distance a_x and a_y that might differ in x- and y-directions, respectively.[38] Their contribution shifts the solubility limit by

$$c_{\alpha,disl} = c_0 + \frac{1}{d} \cdot \left(\frac{1}{a_x} + \frac{1}{a_y} \right) \cdot \pi r^2 (c_D - c_0) \tag{3}$$

with the mean local concentration c_D and the cylindrical enrichment zone with radius r. The radius of the enrichment zone has been fitted here, to be about r = 0.3 nm. This estimates the influence of the local position of the dilatation field of the edge dislocation at the interface: The dilatation field only affects the α-phase solubility limit, when it is located inside the metal. The calculation is done applying the local concentration c_D=0.7 H/Nb of the hydride. Two different dislocation distances are determined for epitaxial Nb films on Sapphire substrates, with a_x=2.1 nm for dislocation lines in [$\bar{1}$12]-direction and a_y=14 nm for dislocations in [$\bar{1}1\bar{2}$]-direction.[27] As sum of surface and dislocation trapping the estimated change of the solubility limit is plotted in Fig.3, with the green dashed line.

Again, changed become important for film thickness below 50 nm. The solubility limit is more than doubled for films below 30 nm film thickness. It should be further considered, that films below a thickness of several 10 nm do not possess misfit dislocations and the film is just elastically strained.

Micro-structural impact on hydrogen storage properties of nano-metals

In total, from the above considerations, the α-solubility limit will be increased by the typical micro-structure of nano-materials, especially for sizes below 50 nm. Strong effects are expected below about 30 nm film thickness. Micro-structure, in this region, strongly reduces the storage capacity of the metal because of hydrogen trapping. However, for very thin epitaxial films where misfit dislocations are presumably not present, stronger changes of the storage capacity are expected for less than 10 nm film thickness. One aim should be, therefore, to minimize the amount of micro-structural defects in the nano-sized storage material only leaving surface trapping sites. These might be saturated by some surface cover layers.

STRAIN ENERGY CONTRIBUTIONS

In Fig.3, next to the calculated defect contributions, some experimental results on different Nb-films are also shown (marked in red for nano-crystalline films, and green for epitaxial films). As can be seen, the measured solubility limits do not match the theoretical curves, for many films. Measured solubility limits of nano-crystalline Nb-films (red signs) are shifted to much higher concentrations. For thick epitaxial films (green signs), the curves match experimental values, but for films of 35 nm and 25 nm, shifts are even stronger. This proves that there are further contributions shifting solubility limits not implemented yet. We relate this to mechanical stress.

Metal–Hydrogen Systems: What Changes When Systems go to the Nano-Scale?

Metals of nano-scale size are usually not stable because of their large surface to volume fraction. For many cases surfaces increase the total system's enthalpy. Therefore, any contact between nano-materials results in agglomeration to reduce the total surface. Consequently, to maintain nano-materials they need stabilizers. Thin films and multi-layers are stabilized by substrates, clusters by scaffold- or shell-forming materials.

These stabilizers often are strongly bond to the nano-metal, thereby also affecting the metals properties.[28] For M-H systems, stabilization leads to extreme mechanical stress since the lattice intends to expand upon hydrogen loading. For thin films, mechanical stress of several GPa has been measured. This strain energy shifts the plateau pressure of the nano-system to increased pressures.

The elastic stresses and strains can be calculated by using theory of linear elasticity. The model is demonstrated for the thin film condition in Fig.4.

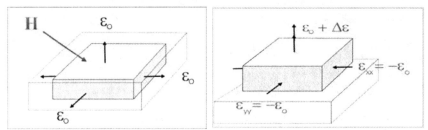

Fig.4: Model for hydrogen-induced expansion of (a) a free film and (b) a film that is fixed on a hard substrate. For the latter case, in-plane expansion is not possible. For calculations, the freely expanded film has to be back-strained to its original in-plane area.

While the free bulk sample can expand by ε_0 in all three room directions upon hydrogen loading (Fig.4 a)), this is not possible for the fixed thin film on an ideally hard substrate (Fig.4 b)). In this case, in-plane expansion is not possible. Only the vertical direction can expand freely. The resulting in-plane stress can be determined by back-straining the freely expanded film to the initial area of the fixed film. The biaxial modulus M relates this in-plane back-strain $\varepsilon = -\varepsilon_0$ with an in-plane mechanical stress σ.

$$\sigma = M \cdot \varepsilon \qquad (4)$$

For an isotropic cubic film the biaxial Modulus is given by $M = \dfrac{E}{(1-v)}$, with Poisson's number v. E is the modulus of elasticity. For bulk Nb it is E=103 GPa, and Poisson's number v =0.387.[29] The lattice expansion scales with hydrogen concentration. For Nb-H it is $\varepsilon_0 = 0.058 \cdot c_H$.[30] Thus, for Nb thin films, hydrogen induced in-plane stress is[31,32]

$$\sigma = -M \cdot 0.058 \cdot c_H = -9.67 \text{GPa} \cdot c_H \qquad (5)$$

This in-plane stress adds a strain $\Delta\varepsilon$ to the conventional out-of-plane expansion ε_0. The total vertical strain is given by[33,34]

$$\varepsilon_{zz,tot} = \varepsilon_0 + \frac{2\upsilon}{1-\upsilon}\varepsilon_0 = \left(1 + \frac{2\upsilon}{1-\upsilon}\right)\cdot 0.058\cdot c_H = 0.131\cdot c_H \tag{6}$$

For films with preferential in-plane orientation more detailed calculations have to be performed, considering the tensors of stress and strain as well as the elastic stiffness tensor of the lattice structure and material. For elastically anisotropical systems, in-plane stress depends on the in-plane directions.

Measurements verify the initial strain and stress to be linear elastic. This is exemplarily shown in Fig.5 for Nb-films, as well as for Pd and Y-films. The initial slope of the curves depends linear on the hydrogen concentration. For Fig. 5a) the vertical lattice expansion measured by X-ray diffraction on Nb-H films follows the calculated slope of 0.131 of Eq. 6 (marked with red line) up to 0.08 H/Nb. The stress is slightly smaller compared to the calculated value of Eq.5, marked with the red line in Fig5b). This might be due to grain boundaries that reduce the total stress in the film.

For larger concentrations strain and stress increments are smaller. In this region, the film escapes from high stress by plastic deformation or film detachment from the substrate. Other stress release processes are also possible. Plastic deformation results in surface roughening on the nano-scale. Each formed dislocation releases stress by bringing metal atoms to the film surface. This gives a line-pattern on the film surface that can be monitored by STM. A typical surface pattern is shown in Fig6 a), for a 12 nm Gd film on a W-substrate. Plastic deformation results in reduced stress increase upon hydrogen loading. Thereby, the maximum stress is reduced from -9.6 GPa·0.7=6.7 GPa at 0.7 H/Nb to about -3.4 GPa in nano-crystalline films, or to about -1.7 GPa for epitaxial films.

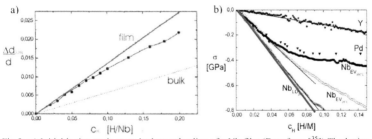

Fig.5.: a) Initial lattice strain upon hydrogen loading of a Nb-film.(Data from [35]) The lattice strains linearly by hydrogen concentration. The red lines mark the calculated values from theory of linear elasticity. b) Initial stress increase by hydrogen loading of Nb-films of different micro-structure, Pd- and Y-films.(Data from [9,34,38]) For all films, the initial stress change scales linear with the hydrogen concentration.

Film detachment results in buckled films that are only partly clamped to the substrate. For 200 nm Nb films, buckles are elongated by 100 μm and have widths of about 10 μm and elevation heights of 2 μm.[36] Thus, buckle pattern can be detected with an optical microscope. Buckled film volumes are still compact, but they easily break. Film detachment can result in complete stress release.

The impact of mechanical stress on the α-solubility limit can be seen in Fig.3. Next to results on Nb-film deposited on hard substrates, the figure also contains data points from Nb-films deposited on thin, elastically soft polymer substrates (open squares). These substrates stretch upon hydrogen loading and, thereby, the in-plane stress is strongly reduced compared to the film on a thick and hard substrate. But, the micro-structure is similar to the nano-crystalline film. As can be seen, the α-solubility limit is closer to the bulk value. One data point (at about 100 nm film thickness) is determined for a film that was completely detached from the substrate. This data point closest matches to the bulk value, even though the micro-structure is similar to the nano-crystalline film. These results confirm the strong impact of mechanical stress on the solubility limit.

Wagner et al. have, further, demonstrated an impact of mechanical stress also on the plateau pressure.[37] For about 200 nm thick Pd-films (2 % Fe) with different clamping conditions, as shown in Fig 6. By in-situ resistivity measurements, Wagner et al. determined an increased plateau pressure of 400 mbar for clamped Pd-H films (loaded with hydrogen for the first time), compared to that of Pd-H bulk of 18 mbar Fig.6 C1). In Fig. 6, the plateau pressure is visible by a strong increase in resistivity. Stress release by plastic deformation results in a reduced plateau pressure, but it is still larger than that of bulk Pd-H. For the second loading (C2), the plateau region is strongly sloped. Its mean value of about 70 mbar is closer to that of bulk. For a film strongly detached from the substrate (A), the plateau pressure resembles that of bulk. For a partly detached film (B), two different plateau regions where detected. Their contents are related to the volume fractions of detached and attached film.[37]

The strain energy increases with the square-dependency on the hydrogen-concentration.[38] This shifts the Free Energy of the hydride more strongly compared to the solid solution. The chemical potential at the phase transition, thereby, is increased compared to the stress-free state. Up to now, increases of about +3 kJ/mol have been achieved.[37] Because of the same argument, plateau regions for films stressed by hydrogen, are expected to be sloped. Measurements on hydrogen gas loaded Nb films have confirmed that, then, for constant pressures in the phase transition region, the phase transition freezes.[39]

High stress states are expected for very thin films. As Nörthemann et al. have recently shown, for Nb-films of less than 28 nm film thickness, it is energetically not favorable to form dislocations upon hydride formation.[40] Further, Wagner et al. demonstrated for 25 nm Pd films hydride formation with absence of dislocations at the metal-hydride interface (coherent metal/hydride interface).[41] But, for these films, stress releasing dislocations were found at the metal/substrate interface. This research is actually ongoing.

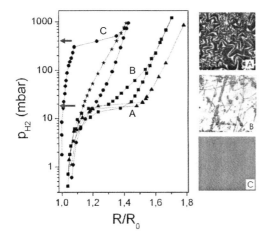

Fig.6:

Mechanical stress impact on plateau pressures for about 200 nm Pd (with up to 2%Fe) films.[37] For a highly detached film (C.) the plateau pressure is close to the bulk value of 18 mbar. For the strongly bond film (A), it increases to 400 mbar during first loading. The partly bond film (B) shows a mixture of plateaus. Light microscopy images of the films on the right side. For further details see text.

Stress impact on hydrogen storage properties of nano-metals

As can be seen from the above considerations, when destabilization of the hydride and an increase of the plateau pressure is desired, the high stress state is helpful. Thick and mechanically hard substrates with strong adhesion to the metal should be used for this case. Furthermore, stress release by dislocations should be avoided by using small storage metal sizes that prefer to strain elastically. For the high stress state, the collateral reduction of storage capacity has to be accepted.

When maximum storage capacity is desired, mechanical stress should be kept as small as possible. Mechanically soft and thin stabilizers like polymers should be considered. Metals should be on larger scales to support stress release by dislocation formation. Buckled or free-standing films are presumably not relevant for application because ease of film disruption.

INTERFACES BETWEEN METALS

Actually, nano-metals stacking is used to improve the hydrogen loading conditions, for example by using Mg/Ti films to avoid the blocking effect in Mg. But, when interfaces between metals are used, further aspects come into play that are demonstrated, here, for model on a Fe/V film package. For metal-metal interfaces Hjörvarsson et al. suggested changes in the local hydrogen solubility by electron transfer.[42,43] Others relate depletion zones to local lattice strains.[44] The change could result in a hydrogen depleted volume close to the interface in the hydrogen absorbing metal.

Recently, Gemma et al. confirmed the existence of such depletion zones at interfaces by tomographical atom probe (APT).[45] This method allows identifying metal and hydrogen (deuterium) atoms and their local position with 1 A depth resolution and 5 A lateral resolution. For Fe/V packages Gemma et al. found a depletion zone of about 1 nm in the V-layer. Further, he detected a chemical intermixing zone of about 1 nm width between Fe and V layers, when sputtered at 20°C. According to Gemma et al. this results in an additional hydrogen depletion zone because of the alloying effect. A typical APT measurement on a Fe/V package is shown in Fig. 7.

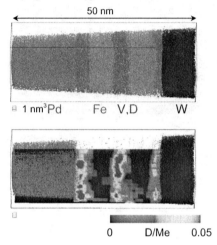

Fig.7:

APT results on a deuterium loaded Fe/V package. The package is deposited on a W-substrate and covered with Pd for protection. (positions of Fe: red dots, V: green dots, Pd: grey dots and W: blue dots) Deuterium solves in the V-layers. Depleted layers are found at the interfaces to Fe. In the color-scale plot the concentration of deuterium varies inside the V-layers.[45]

Impact of interfaces on hydrogen storage properties of nano-metals

Interfaces between different metals can result in hydrogen depletion zones, because of chemical intermixing, electron-transfer or lattice strains. They, thereby, reduce the storage capacity of the system. The contribution of the depletion zones scales with the amount of interfaces. From the viewpoint of maximum storage capacity, the number of interfaces should be kept as small as possible.

STRUCTURAL CHANGES IN NANO-METALS

When nano-systems, like Pd-clusters (TAOB-stabilized, single-size), have diameters of about 5 nm new crystal structures come into play that can strongly change the hydrogen storage properties of the metal.[46]28 In this size range, surface energy minimization leads to more spherical cluster shapes, possessing five-fold symmetry structures like icosahedrons or others. As Suleiman et al. reported on differently stabilized, but still uniform-sized Pd-clusters, the presence of five-fold symmetry structures depends on the choice of the stabilizer.[47,48] Mechanical stress might be one factor to influence the cluster structure and hydrogen storage. As P. Jena suggested at the MCARE 2012, another factor might be the electronic exchange between the stabilizer and the cluster.

Impact of new structures on hydrogen storage properties of nano-metals

For very small sizes of less than about 10 nm, surface and sub-surface trapping sites reduce the storage capacity of the metal. Additionally, structural transitions into new phases that do not exist for the bulk M-H system, have to be considered. They may differ in their hydrogen storage capacity. For icosahedral Pd-clusters, the storage capacity is reduced compared to their cubic counterparts. However, also an increase of the hydrogen storage capacity might result from the presence of new phases.

SUMMARY

Hydrogen loading and unloading times can be strongly shortened when nano-metals are used. But, also the hydrogen storage properties change when the system size is reduced. Results on hydrogen absorption in different nano-metals, like thin films, film packages and clusters are regarded with respect to hydrogen storage properties. From results on model systems, general suggestions are developed:

1) Micro-structural defects usually act as hydrogen trapping sites, their contents need to be minimized for maximum storage capacity.

2) Mechanical stress, evolving between the nano-metal and its required stabilizer, also reduces the storage capacity of the metal. Soft and thin stabilizers are suggested to be suitable for maximum storage capacity.

3) When hydride destabilization is desired, mechanical stress can be also used. For this, hard and thick substrates with strong adhesion are preferred as well as metals with small dimensions in the 20 nm range.

4) Metal/metal interfaces can lead to hydrogen depletion zones by chemical intermixing, electron transfer or elastic lattice strain. The number of metal/metal interfaces should be kept small.

5) For very small metals, like for example clusters, new (five-fold symmetry) structures have to be considered that do not show up in the bulk phase diagram. Their storage properties are expected to differ from conventional structures.

ACKNOWLEDGEMENT

Financial support by the Deutsche Forschungsgemeinschaft DFG via several projects (SFB602, PU131-7/2, PU131-9/1, PU131-10/1) is gratefully acknowledged.

[1] N. Stetson, Plenary talk on MCARE 2012, Clearwater, Florida.
[2] E. Wicke, H. Brodowsky, H. Züchner in Hydrogen in metals II, Eds. G. Alefeld, J. Völkl, Springer Verlag Berlin-Heidelberg, 1978.
[3] J. Renner, H.J. Grabke, Z. Metallkde 69 (91978) 639-642.
[4] A. Krozer, B. Kasemo, J. Vac. Sci. Technol. A5 (1986) 1003-1005.
[5] J. Völkl, G. Alefeld, Ch. 12 in „Hydrogen in Metals I", Eds. G. Alefeld, J. Völkl, Springer-Verlag Berlin 1978.
[6] P. de Jongh, P. Adelheim, Ch. 10 in "Handbook of Hydrogen Storage", Ed. M. Hirscher, Wiley-VCH 2010.
[7] H. Lüth, Solid surfaces, interfaces and thin films, Springer Verlag Berlin Heidelberg 2001.
[8] G. Song, M. Geitz, A. Abromeit, H. Zabel, Phys. Rev. B 54 (1996)14093-14101

[9] A. Pundt, R. Kirchheim, Ann. Rev,. Mat. Sci. 36 (2006) 555–608.
[10] R. Kirchheim, Sol. State Phys. 59 (2004) 203-305.
[11] H. Conrad, G. Ertl, E.E. Latta, Surf. Sci. 41 (1974.) 435–46.
[12] K.H. Rieder, M. Baumberger, W. Stocker, Phys. Rev. Lett. 51 (1983) 1799–802.
[13] K. Christmann, Mol. Phys. 66 (1989).
[14] R.J. Behm, V. Penka, M.G. Cattina, K. Christmann, G. Ertl, J. Chem. Phys. 78 (1983) 7486–90
[15] K. Christmann, Surf. Sci. Rep. 9 (1988) 1–163.
[16] U. Muschiol, P.K Schmidt, K. Christmann, Surf. Sci. 395 (1998) 182–204.
[17] R. Kirchheim, J. Met. 33 (1981) A28–29.
[18] D.K. Ross, K.L. Stefanopoulus, Z. Phys. Chem. 183 (1994) 29.
[19] M. Maxelon, A. P undt, W. Pyckhout-Hintzen, R. Kirchheim, Scripta Mat. 44 (2001) 817-822.
[20] B.J. Heuser, J.S. King, G.S. Sommerfield, F. Boué, J.E. Epperson, Acta metall.39 (1991) 2815.
[21] D. R. Trinkle, H. Ju, B. J. Heuser, T. J. Udovic, Phys. Rev.B 83, (2011) 174116 1-7.
[22] T. Mütschele, R. Kirchheim, Scr. Metall. 21 (1987) 1101–4.
[23] T. Mütschele, R. Kirchheim, . Scipta Metall. 21 (1987) 135–40
[24] H. Natter, B. Wettmann, B. Heisel, R. Hempelmann, J. Alloys Comp. 253–254 (1997)) 84–86.
[25] J. Weissmüller, C. Lemier, Philos. Mag. Lett. 80 (2000) 411–18.
[26] J. Weissmüller, in Nanocrystalline Metals and Oxides: Selected Properties and Applications, ed. P. Knauth, J. Schoonman, Boston: Kluwer Acad. Publ. (2001) pp. 1–39.
[27] E. J. Grier, M. L. Jenkins, A. K. Petford-Long, R. C. C. Ward, M. R. Wells, Thin Solid Films 358 (2000) 94-98.
[28] P. Jena, Invited talk on MCARE 2012, Clearwater, Florida.
[29] D.I. Bolef, J. Appl. Phys. 32 (1961) 100.
[30] H. Peisl, Ch.2 in „Hydrogen in Metals I", Eds. G. Alefeld, J. Völkl, Springer-Verlag Berlin 1978.
[31] Q.M. Yang, G. Schmitz, S. Fähler, H.U. Krebs, R. Kirchheim, Phys. Rev. B 54 (1996) 9131.
[32] P.M. Reimer, H. Zabel, C.P. Flynn, A. Matheny, K. Ritley, J. Steiger, S. Blasser, A. Weidinger, Z. Phys. Chem. 181 (1993) 367.
[33] H. Zabel, A. Weidinger, Comm. Condens. Mater. Phys. 17 (1995) 239–262.
[34] U. Laudahn, S. Fähler, H. U. Krebs, A. Pundt, M. Bicker, U. v. Hülsen, U. Geyer, R.Kirchheim, J. Appl. Phys. 74 (1999)647-649.
[35] U. Laudahn, A. Pundt, M. Bicker, U. v. Hülsen, U. Geyer, T. Wagner, R. Kirchheim, J. Alloys Comp. 293–295 (1999) 490–494.
[36] A. Pundt, P. Pekarsky, Scripta Materialia 48 (2003) 419–423.
[37] S.Wagner, A. Pundt, Appl. Phys. Lett. 92 (2008) 051914 1-3.
[38] M. Dornheim, dissertation thesis, Göttingen 2002.
[39] K. Nörthemann, A. Pundt, Phys. Rev. B 83 (2011) 155420 1-10.
[40] K. Nörthemann, A. Pundt, Phys. Rev. B. 78 (2008) 014105 1-9.
[41] S. Wagner, H. Uchida, V. Burlaka, M. Vlach, M. Vlcek, F. Lukac, J. Cizek, C. Baehtz, A.Bell, A. Pundt, Scripta Materialia 64 (2011) 978–981.
[42] B. Hjörvarsson, J. Ryden, E. Karlsson, J. Birch, and J.-E. Sundgren, Phys. Rev. B 43, (1991) 6440.
[43] G. Andersson, B. Hjörvarsson, and P. Isberg, Phys. Rev. B 55, (1997) 1774.
[44] V. Meded and S. Mirbt, Phys. Rev. B 71, (2005) 024207.
[45] R. Gemma, dissertation thesis, Göttingen 2011.
[46] A. Pundt, M. Suleiman, M..T. Reetz, C. Bähtz, N.M. Jisrawi, Mater. Sci. Eng. B 108 (2004) 19–23.
[47] M. Suleiman, C. Borchers, M. Guerdane, N. M. Jisrawi, D. Fritsch,R. Kirchheim1, A. Pundt, Z. Phys. Chem. 223 (2009) 169–181
[48] M. Suleiman, J. Faupel, C. Borchers, H.-U. Krebs, R. Kirchheim, A. Pundt, J. All.oys Comp. 404–406 (2005) 523–528.

Materials Availability
for Alternative Energy

PREPARATION OF ORGANIC-MODIFIED CERIA NANOCRYSTALS
WITH HYDROTHERMAL TREATMENT

Katsutoshi Kobayashi, Masaaki Haneda and Masakuni Ozawa
Advanced Ceramics Research Center, Nagoya Institute of Technology
Asahigaoka 10-6-29, Tajimi, Gifu 507-0071, Japan

ABSTRACT
    The effect of the organic stabilizing agents on the structure and the surface states of CeO_2 nanocrystals prepared by the hydrothermal synthesis has been investigated with infrared (IR) spectroscopy, X-ray diffraction (XRD) measurement and transmission electron microcscope (TEM) observation. Investigation on the temperature dependence of the nanocrystal size and shape in the hydrothermal treatment using oleate as a stabilizing agent clearly revealed the acceleration of the crystalline growth at high temperature. Hydrothermal syntheses of carboxylate-modified CeO_2 nanocrystals were performed using laureate and linoleate species in addition to oleate. XRD measurement indicated the formation of nanocrystalline CeO_2 in any case and no remarkable difference in crystalline structure or crystallite size. IR spectroscopy revealed the different surface states of CeO_2 nanocrystals, depending on the molecular structure of stabilizing agents, that laureate species having a short straight hydrocarbon chain made a simple bidentate coordination to CeO_2, while oleate made mixed coordination state of bidentate and unidentate. Linoleate species having a convoluted hydrocarbon chain indicated the bidentate coordination and the existence certain amount of free linoleic acid or linoleate species among CeO_2 nanocrystals. TEM observation indicated that laureate adsorbed on CeO_2 made the uniform interdigitation, while linoleate made more complicated connections with free linoleate species or linoleic acid.

INTRODUCTION
    Recently, morphological control of inorganic nanocrystals (NCs) has been one of the most important subjects in the field of the advanced materials. Inorganic NCs containing rare earth elements are well known to exhibit unique and superior electrical, optical and magnetic properties depending on their sizes and shapes. In particular, NCs containing cerium dioxide (CeO_2, ceria) have attracted strong attention in both the academic and the industrial circles due to their wide-ranging application, such as electrolyte materials of solid oxide fuel cells [1], solar cells [2], three-way catalysts in automobiles [3], ultraviolet-shielding materials [4], and polishing agents [5]. To date, a number of synthetic methods including solvo-hydrothermal treatments [6], sol-gel processes [7] and thermal evaporation methods [8], have generally been studied for the production of metal-oxide NCs. In 2006, Yang et al. achieved the production of uniform CeO_2

NCs which have cubic structures enclosed with {200} facets, applying a solvothermal process in toluene [9]. However, the solvothermal method is not suitable for the industrial application because of its high cost and environmental load. Comparing the above mentioned methods, the environmentally benign and relatively inexpensive hydrothermal method is most favorable route for CeO_2-NC production. From the above viewpoint, CeO_2 NC preparation under a mild condition has been challenged. In 2009, Taniguchi et al. found that CeO_2 NCs could also be prepared by the ordinal hydrothermal route using oleate-precipitation method [10,11], where oleic acid was used as a stabilizing agent for preventing nascent nuclei from re-dissolving and for suppressing agglomeration of generated primary nanoparticles in the NC growth process. Oleic acid was also used with various synthetic methods by several researchers' groups [9-13], while some researchers employed other carboxylic acid for the production of CeO_2 NCs or CeO_2-containing NCs [14,15]. Since these researches have been separately performed under the different conditions, the effect of the organic stabilizing agents on the morphology and surface state of CeO_2 NCs has not been discussed.

Recently, as the methodology of the NC synthesis develops rapidly, improvement of the NC handling technique has become strongly required. How to load morphology-controlled NCs on the target substrate, how to modify the alignment of NCs, and how to remove the organic stabilizing agents from the NC surface without contamination have been more and more important. However, researches concerning the NC handling are small in number to our knowledge. For these purposes, it is indispensable to understand the behavior of the organic stabilizing agents in the NC handling and also to search for the suitable stabilizing agents.

As an initial step for understanding the role and behavior of organic stabilizing agents in the NC handling, in this study, we have attempted to investigate the effect of the structure of carboxylic acid used as a stabilizing agent in a hydrothermal process on CeO_2-NC formation. First, for grasping the CeO_2-NC growth process in the hydrothermal treatment, the morphological difference of CeO_2 NCs obtained at different hydrothermal temperatures was investigated using the oleic acid. Then, CeO_2-NC formation was carried out using several types of carboxylic acid with different molecular structures. Here, lauric acid with carbon number of 12, which is a saturated fatty acid, and linoleic acid with carbon number of 18, which has two *cis* double bond in its hydrocarbon (HC) chain, were employed in addition to the oleic acid having a single *cis* double bond with carbon number of 18. Molecular shapes are strongly affected by the existence of *cis* double bond which bends the HC chain, and the carbon number represents the chain length. Through TEM morphological observation and surface analysis by IR spectroscopy, the effect of the length and shape of carboxylic acid on the produced CeO_2 NCs was discussed.

EXPERIMENTAL

Carboxylic Acid as Stabilizing Agents

Structural formulae, names and abbreviations of carboxylic acids used in this study were listed in Fig. 1. All the carboxylic acids were added into the reaction media in the form of potassium carboxylates as stabilizing agents. Potassium oleate ($C_{17}H_{33}COOK$, 19% solution, Wako Pure Chemical Industries, Ltd.) were used as received. Potassium laureate ($C_{11}H_{23}COOK$) and potassium linoleate ($C_{17}H_{31}COOK$) were prepared by mixing lauric acid ($C_{11}H_{23}COOH$, 98.0+%, Wako Pure Chemical Industries, Ltd.) and linoleic acid ($C_{17}H_{31}COOH$, 88.0+%, Wako Pure Chemical Industries, Ltd.) with potassium hydroxide aqueous solution (KOH, 1 mol/L solution, Wako Pure Chemical Industries, Ltd.) at equivalent amounts, respectively. All the chemicals were used as received without further purification.

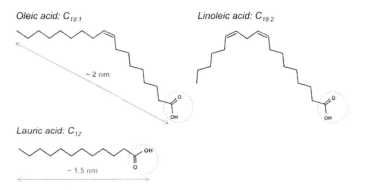

Figure 1. Structural formulae and abbreviations of carboxyl acids used as stabilizing agents.

Synthesis of CeO_2 Nanoparticles

Cerium solution and each carboxylate solution were prepared by dissolving 7 mmol of diammonium cerium(IV) nitrate (($NH_4)_2Ce(NO_3)_6$, 95+%, Wako Pure Chemical Industries, Ltd.) and 7 mmol of potassium carboxylate with 30-mL distilled water, respectively. Then, the carboxylate solution was added into the cerium solution at room temperature under vigorously-stirred condition, followed by addition of 10 mL of 25-wt% ammonia aqueous solution (Wako Pure Chemical Industries, Ltd.). A portion of CeO_2 nanoparticles generated by the addition of ammonia solution were collected from the solution before a hydrothermal treatment and were analyzed for comparison. CeO_2 nanoparticles obtained without a hydrothermal treatment were washed with distilled water and freeze-dried under vacuum.

Hydrothermal Synthesis of CeO_2 Nanocrystals

The solution mixture containing cerium complex and carboxylate salt and ammonia solution was then transferred to a Teflon-lined stainless steal autoclave. Subsequently, the sealed autoclave was heated at 150°C or 200°C for 6-48 hours under stirred condition at the rate of 500 rpm. Afterward, the system was cooled to room temperature under the ambient environment.

Nanocrystalline products were separated from the reaction solution by centrifugation at 3000 rpm for 30 min. The precipitates were washed with distilled water three times and dried at 90°C for 24 hours in air. A portion of the dried precipitates were then dispersed in a nonpolar solvent, namely toluene ($C_6H_5CH_3$, 99.5+%, Wako Pure Chemical Industries, Ltd.) in this study, without any purification process. CeO_2 NCs before or after toluene dispersion were used for further analyses.

Characterization

Synthesized nanoparticles were placed on a non-reflecting Si plate and characterized by X-ray diffraction (XRD; MiniFlex II, Rigaku) measurement with Cu-$K\alpha$ radiation ($\lambda = 1.5418$ Å) at 40 mA and 40 kV. Fourier-transform infrared (IR) spectroscopy was performed with an IR spectrometer (FT/IR-4200, JASCO Corp.) by the reflection method. Samples for IR spectroscopy were prepared by repeating the following procedure several times that toluene dispersion of CeO_2 NCs was dropped onto a reflecting metal surface and dried, resulting in transparent amber films consisting of CeO_2 NCs and organic substances. Each spectrum was obtained by accumulating signals for 500 times at a resolution of 4 cm^{-1}. Morphological observation was carried out with a transmission electron microscope (TEM; JEM2100, JEOL Ltd.) operated at 200 kV. Nanoparticle-containing toluene was dropped onto a carbon-coated Cu grid (elastic carbon supporting membrane, Okenshoji Co., Ltd.) and evaporated at room temperature. Elemental analysis of nanoparticles was performed by an energy-dispersive X-ray spectroscope (EDX) attached to TEM.

RESULTS AND DISCUSSION

Temperature Dependence of Morphology of CeO_2 Nanocrystals

Hydrothermal synthesis of CeO_2 NCs was performed at 150°C or 200°C for 6 hours using oleic acid as a stabilizing agent. Then, XRD measurement and TEM observation of the produced precipitates were carried out. Figure 2 shows XRD patterns of CeO_2 NCs synthesized at each temperature. The XRD pattern of CeO_2 nanoparticles prepared without the hydrothermal treatment is also shown in Fig. 2. Strongly-broadened diffraction patterns corresponding to the crystalline CeO_2 (JCPDS No. 34-3094) were obtained for the sample without hydrothermal treatment, indicating a low-crystallinity structure of the CeO_2 nanoparticles. On the other hand, for the samples prepared with a hydrothermal treatment at 150°C or 200°C, relatively broad but clear powder patterns of CeO_2 with the fluorite-type structure were observed. No diffraction pattern attributed to intermediate phases such as oxyhydroxide was detected for any sample. From Scherrer equation expressed as Eq. 1, crystallite size of CeO_2 NCs synthesized at each temperature was estimated using three diffraction lines at 28.355°, 33.082° and 47.479°.

$$d = \frac{0.941\lambda}{\beta \cos \theta_{\mathrm{B}}} . \qquad (1)$$

Here, d is crystallite size, λ is X-ray wavelength, β is the full width half maximum (FWHM) of the peak, θ_{B} is Bragg angle. Crystallite sizes for CeO_2 NCs for 150°C and 200°C were calculated to be 2.9 nm and 4.0 nm, respectively. Also, crystallite size for CeO_2 nanoparticles obtained without hydrothermal treatment was estimated to be much less than 2 nm which was almost the applicable lower limit of Eq. 1. It was revealed that crystalline size showed tendency to increase with the synthetic temperature increasing, corresponding to the previous observation [10,13].

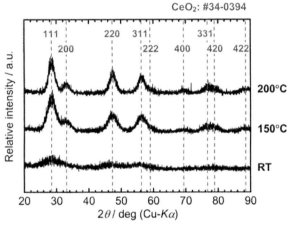

Figure 2. XRD patterns of oleate-modified CeO_2 nanoparticles prepared under the room temperature condition and CeO_2 nanocrystals synthesized with hydrothermal treatment at 150°C and 200°C for 6 hours.

All the samples were then dispersed in toluene, dropped onto TEM grids and observed by TEM. Figure 3-a shows a TEM image of the sample prepared without hydrothermal treatment. Vaguely-outlined substances, which were confirmed to contain Ce element by EXD, with 10-50 nm in size can be seen. On the other hand, no crystalline products were recognized by the TEM observation. In comparison with the XRD result, it is suggested that 10-50 nm sized substances observed in TEM should be the aggregates composed of oleate species and low-crystallinity CeO_2 nanoparticles. Figures 3-b and 3-c show TEM images of the samples after the hydrothermal treatment at 150°C and 200°C for 48 hours, respectively. Well-defined particles with the size less than 10 nm can be seen in both cases. Lattice images obtained by the magnified observation clarified that each particle was single crystal, i.e. CeO_2 NCs. For the 150°C sample,

NCs with a uniform spherical shape are mainly observed. The particle size was ca. 3 nm. On the other hand, for the 200°C sample, relatively large angular NCs with size of ca. 5 nm are visible in addition to the small spherical NCs. Projection images of the larger NCs show square shapes, indicating the cubic structure enclosed with {200} facets [13,16]. Small spherical NCs may have the truncated-octahedral shapes same as shown in the early stage of hydrothermal treatment at 200°C [13]. It can be predicted that the small spherical particles finally grow into the {200} cubic shapes by the hydrothermal treatment, because the growth rate of CeO_2 NCs in [100] direction is strongly suppressed by the organic modification compared with other directions [16]. We can conclude that {200} cubic structures observed in this study should result from the crystalline growth accelerated at high temperature.

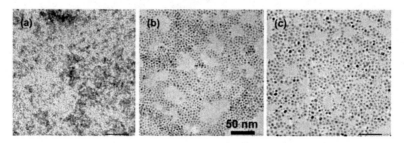

Figure 3. TEM images of oleate-modified (a) CeO_2 nanoparticles prepared under room temperature condition and CeO_2 nanocrystals synthesized with a hydrothermal treatment at (b) 150°C and (c) 200°C for 48 hours.

Effect of Molecular Structure of Carboxylic Acid

In order to clarify the effect of the structure of the organic stabilizing agents on the CeO_2-NC formation, hydrothermal synthesis of CeO_2 NCs were conducted using laureate (C_{12}) and linoleate ($C_{18:2}$) in addition to oleate ($C_{18:1}$) at 200°C for 48 hours.

XRD patterns of the obtained samples are shown in Figure 4. Only a diffraction pattern attributed to crystalline CeO_2 was detected in the pattern of each sample. Crystallite sizes calculated from Eq. 1 were almost constant at ca. 5.0 nm for all the samples, indicating that the type of carboxylate salt gives no significant effect on the crystal growth rate of CeO_2 NCs. It can also be found that the crystallite size of oleate-modified CeO_2 NCs increased with the processing time in comparison with the result of the 6-hour treatment (Fig. 2).

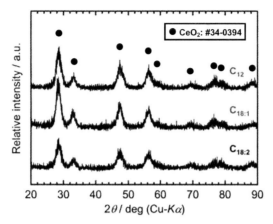

Figure 4. XRD patterns of CeO₂ nanocrystals synthesized with hydrothermal treatment at 200°C for 48 hours using laureate (C_{12}), oleate ($C_{18:1}$), and linoleate ($C_{18:2}$) as a stabilizing agent.

As depicted in Fig. 1, lauric acid (C_{12}) is a saturated fatty acid which can be in a straight form, while oleic acid ($C_{18:1}$) and linoleic acid ($C_{18:2}$) are always in crooked froms due to their *cis* double bonds. Oleic acid and linoleic acid have same carbon number, but the number of the double bond in the HC chain is different. These differences in molecular structure are expected to affect the bonding state between carboxylic acid and CeO₂ NCs. For clarifying this point, IR spectroscopy was performed.

Figure 5 shows the IR spectra for the CeO₂ NCs prepared with the carboxylate salts corresponding to the above mentioned acids. Absorption bands at around 1710 cm⁻¹, 1660 cm⁻¹, 1530 cm⁻¹ and 1440 cm⁻¹ are observed for each sample. According to the previous studies, the bands in the range 1660-1530 cm⁻¹ and that at 1440 cm⁻¹ are attributed to the asymmetric ($v_{as}(COO^-)$) and symmetric stretching vibration ($v_s(COO^-)$) of the COO⁻ bond adsorbed on metal species, respectively [12,17,18]. Therefore, it is revealed that all the carboxylic acid used in this study is chemisorbed on the CeO₂-NC surface as a carboxylate and that the COO⁻ part is coordinated to the CeO₂ surface.

The band separation, $\Delta v = v_{as}(COO^-) - v_s(COO^-)$, exhibits the coordination mode of the carboxylate to the CeO₂ surface. $\Delta v > 200$ cm⁻¹ indicates the unidentate bonding state, while $\Delta v < 110$ cm⁻¹ indicates the bidentate bond and Δv in the range of 140-200 cm⁻¹ represents the bridging bond [12,17]. All the samples in this study show a strong $v_{as}(COO^-)$ band at 1530 cm⁻¹, which corresponds to $\Delta v = 90$ cm⁻¹, revealing the bidentate coordination. It is found that the laureate (C_{12}) makes a simple bidentate coordination with CeO₂ NCs. Although the linoleate

($C_{18:2}$) sample shows a broadened spectrum, it also seems to make a bidentate coordination mainly. On the other hand, a clear absorption band at 1660 cm^{-1}, which corresponds to $\Delta v = 220$ cm^{-1}, is observed for the oleate ($C_{18:1}$) modified sample, revealing the unidentate coordination. From this spectrum, mixed bonding states between oleate species and CeO_2 NCs are expected.

The absorption band at 1710 cm^{-1} observed for the oleate and the linoleic samples is attributed to the stretching vibration mode of C=O in the free carboxylic acid form [12,17,18]. This clarifies that free carboxylic acid also remains among CeO_2 NCs without any coordination in the case of oleate- and linoleate-modified samples. Since oleate and linoleate chains have convoluted structures compared with the laureate chain, free acid may not be removed in ordinal washing process.

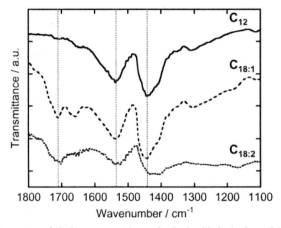

Figure 5. IR spectra of CeO_2 nanocrystals synthesized with hydrothermal treatment at 200°C for 48 hours using laureate (C_{12}), oleate ($C_{18:1}$), and linoleate ($C_{18:2}$) as a stabilizing agent.

Using the above prepared CeO_2 NCs, TEM observation was performed. Figure 6 shows the TEM images of CeO_2 NCs whose surfaces are modified with laurate, oleate and linorate. No remarkable difference in the particle size and shape is recognizable. Small round particles with size of ca. 2 nm and large square-like particles with size of ca. 5 nm can be seen in all the cases. On the other hand, the interparticle distance is clearly different from each other; almost constant distance at ca. 2 nm for laureate-modified NCs, inhomogeneous distances of 2-5 nm for oleate, and ca. 5 nm for linoleate, respectively. These distances are generally known to depend on the length of the HC chain of the stabilizing agents [19]. In the case of a carboxylate with a straight chain, interdigitation can easily takes place, while the convoluted HC chains inhibit the interdigitation. Considering the structure of the lauric acid shown in Fig. 1 and the interparticle distance in Fig. 6-a, it can be indicated that the shallow (~50%) interdigitation uniformly takes place for the laureate-modified NCs. In contrast, the interparticle distance is not constant for the oleate-modified NCs (Fig. 6-b), which indicates inhomogeneous interdigitation states, that is, deep interdigitation in the tight area and no interdigitation in the loose area. The linoleate-modified NCs (Fig. 6-c) show a wide interparticle distance compared with the size of linoleate molecules, indicating the existence of free linoleic acid or linoleate species among the NCs. This speculation is also consistent with the relatively strong absorption band at 1710 cm^{-1} observed in Fig. 5. The convoluted molecules may create complicated connections between the CeO$_2$ NCs to keep a wide interparticle distance.

These observation indicates that the selection of the stabilizing agent is important for controlling the alignment of CeO$_2$ NCs on the target surface.

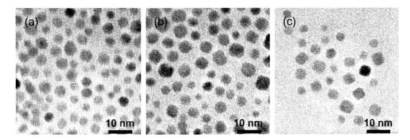

Figure 6.   TEM images of CeO$_2$ nanocrystals synthesized with hydrothermal treatment at 200°C for 48 hours using (a) laureate, (b) oleate, and (c) linoleate as a stabilizing agent.

CONCLUSIONS

This research was the first challenge to understand the effect of organic stabilizing agents on the surface states of CeO$_2$ NCs and the interaction among NCs. In the former part, NC-formation behavior in the hydrothermal process was investigated using oleic acid. Crystallization process at the high temperature was clearly shown. Then, the investigation with organic-modified CeO$_2$ NCs prepared by the hydrothermal treatment at 200°C using different stabilizing agents was performed in the latter part. XRD measurement showed the formation of nanocrystalline CeO$_2$ and indicated no remarkable difference in the crystalline size and structure. IR spectroscopy revealed the different surface conditions of CeO$_2$ NCs depending on the molecular structure of stabilizing agents. Laureate species having a short straight HC chain give a simple bidentate coordination to CeO$_2$ NCs, while oleate gives mixed coordination of bidentate and unidentate. Linoleate species having a convoluted HC chain indicated the bidentate coordination and the existence certain amount of free linoleic acid or linoleate species among NCs. TEM observation clarified the relation between the alignment of the CeO_2 NCs and the molecular structure of the modifying carboxylate species. It was also suggested that the selection of the suitable molecular structure of the stabilizing agent should be important to align CeO_2 NCs on the target substrate.

REFERENCES

1. E.P. Murray, T. Tsai and S.A. Barnett, *Nature*, **400**, 649 (1999).
2. A. Corma, P. Atienzar, H. García and J.Y. C.-Ching, *Nature Mater.*, **3**, 394 (2004).
3. R. Di Monte and J. Kaspar, *J. Mater. Chem.*, **15**, 633 (2005).
4. S. Tsunekawa, T. Fukuda and A. Kasuya, *J. Appl. Phys.*, **87**, 1318 (2000).
5. X.D. Feng, D.C. Sayle, Z.L. Wang, M.S. Paras, B. Santora, A.C. Sutorik, T.X.T. Sayle, Y. Yang, Y. Ding, X.D. Wang and Y.S. Her, *Science*, **312**, 1504 (2006).
6. X. Wang and Y. Li, *Chem. Commun.*, 2901 (2007).
7. M. Niederberger, G. Garnweitner, J. Buha, J. Polleux, J. Ba and N.J. Pinna, *Sol-Gel Sci. Technol.*, **40**, 259 (2006).
8. S.G. Kwon and T. Hyeon, *Acc. Chem. Res.*, **41**, 1696 (2008).
9. S. Yang and L. Gao, *J. Am. Chem. Soc.*, **128**, 9330 (2006).
10. T. Taniguchi, T. Watanabe, N. Sakamoto, N. Matsushita and M. Yoshimura, *Cryst. Growth Des.*, **8**, 3725 (2008).
11. T. Taniguchi, K. Katsumata, S. Omata, K. Okada and N. Matsushita, *Cryst. Growth Des.*, **11**, 3754 (2011).
12. T.-D. Nguyen and T.-O. Do, *J. Phys. Chem. C*, **113**, 11204 (2009).
13. K. Kobayashi, M. Haneda and M. Ozawa, *Adv. Mater. Res.*, **463-464**, 1501 (2012).
14. J. Zhang, S. Ohara, M. Umetsu, T. Naka, Y. Hatakeyama and T. Adschiri, *Adv. Mater.*, **19**, 203 (2007).
15. W.J. Stark, M. Maciejewski, L. Mädler, S.E. Pratsinis and A. Baiker, *J. Catal.*, **220**, 35 (2003).
16. K. Kaneko, K. Inoue, B. Freitag, A.B. Hungria, P.A. Midgley, T.W. Hansen, J. Zhang, S. Ohara and T. Adschiri, *Nano Lett.*, **7**, 421 (2007).
17. Y. Lu and J.D. Miller, *J. Colloid Interfce Sci.*, **256**, 41 (2002).
18. J.-J. Max and C. Chapados, *J. Phys. Chem. A*, **108**, 3324 (2004).
19. A. Tanaka, H. Kamikubo, M. Kataoka, Y. Hasegawa and T. Kawai, *Langmuir*, **27**, 105 (2011).

INTEGRATION OF MgO CARBONATION PROCESS TO CAPTURE CO_2 FROM A COAL-FIRED POWER PLANT

Sushant Kumar
Center for the Study of Matter at Extreme Conditions, College of Engineering and Computing
Florida International University
Miami, FL 33199

ABSTRACT

This paper analyses the practical viability of a new proposed method for post-combustion capture of CO_2 gas from a coal- fired power plant. Here, magnesium oxide (MgO) has been considered as a solid adsorbent to lock CO_2 gas. Moreover, a new retrofit design concept has also been proposed to integrate this carbonation reaction with a running coal- fired power plant. The calculation shows that the MgO based carbonation process, when integrated to a power plant, shows a better capacity to alleviate the high cost and severe energy penalty, otherwise associated with other proposed methods.

1. INTRODUCTION

CO_2 capture and storage (CCS) has received substantial attention as a potential greenhouse gas mitigation route during the last decade [1-9]. An enormous funding has been devoted to modify power plants in a way to encompass CO_2 capture technology [10]. Although at present various commercial technologies exist to capture and separate the CO_2; all these have not been demonstrated commercially at a large scale for gas or coal-fired boilers. Thus, the need of implementing these technologies at a large scale (e.g. 100 MWe) is still a challenge. The most cost effective way for CCS technology is still to be discovered.

CCS can be classified into three different sections of technology: CO_2 capture technology, CO_2 transport technology and CO_2 sequestration technology. Capture technology is believed to be the most costly affair, while sequestration is the most complex process. Coal-fired power plants use pulverized coal to run turbine and generate electricity. Coal itself contains carbon, sulfur, nitrogen, mercury and other elements. However, during combustion sulfur and nitrogen oxides, mercury and ammonia are released with CO_2. Each pollutant species requires different technology to be captured. The techniques for removal of sulfur and nitrogen oxides are well known but techniques to capture CO_2 are still experimental. Three methods are being experimented to capture CO_2: (1) Addition of capture equipment for stack gases in new or retrofitted plants; (2) burn the coal in oxygen; and (3) coal gasification.

Seifritz first introduced the concept of mineral carbonation in 1990 [11], and since then, the carbonation method has been globally accepted as one of the promising technology to capture or separate CO_2 gas from the flue gases. There have been several suggested methods based on solid adsorbents and showed ability to lower the involved expenses during the carbonation process. However, in their current state of development these technologies cannot be implemented on coal-fired power plant primarily because of these three reasons: (1) none of the technology has been demonstrated at the larger scale necessary for power plant; (2) high parasitic loads required for supporting CO_2 capture steps; and (3) if scaled-up they would not be cost effective [12]. Hence, the solution to this problem is still a challenge.

MgO-CO_2 is an exothermic reaction and forms $MgCO_3$ as a product. $MgCO_3$, when calcined around 400-500°C, liberates pure CO_2. The rest MgO is then re-injected for recarbonation reaction. Thus, the multiple cycles can be implemented to separate CO_2 from the combustion flue gas. In brief, an analysis for the post combustion capture of CO_2 from a coal-fired power plant using metal oxide, MgO has been reported here.

A large number of recent studies have estimated CCS costs based on technologies that are still immature or commercially non-existent. Thus, we hereby report our proposed method of capturing CO_2 using MgO based on our earlier experimental work [13]. A cost evaluation on the basis of our proposed process design has also been performed. Moreover, energy penalty and the overall CO_2 captured are also investigated.

2. ENERGY ANALYSIS OF THE CARBONATION PROCESS

The energy calculations have been carried out on a coal fired utility power plant generating 100 MWe. The assumed efficiency is 33%. This translates to 303 MWt energy input from coal. If coal with a calorific value of 32.727 MJ/ton is to be used, about 33.33 ton/h of coal will be required. A retrofit of the carbonation process into an existing coal-fired power plant is shown in Figure 1. This integration is so designed that it can be retrofitted with another vapor power cycle and gas turbine to get an efficient result. Molar and mass flue gas composition is shown in Table 1.

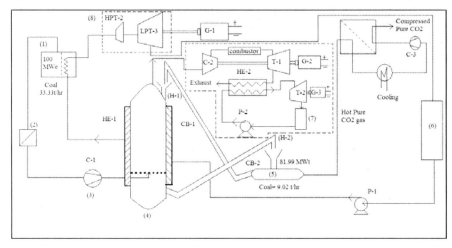

1- Boiler
2- Ash removal
3- C-1,C-2,C-3-Compressor
4- Fluidized Bed Reactor (FBR)
5- Rotary Kiln
6- Condenser Turbines
7- Condenser
8- Steam Turbine

9- P-1,P-2-Pump
10- G-1, G-2,G-3-Generator
11- HE-1,HE-2- Heat Exchangers
12- CB-1,CB-2-Conveyor Belts
13- H-1,H-2-Hoppers
14- HPT-2,LPT-3-High and Low pressure Steam
15- T-1,T-2-Gas and Steam Turbines respectively

Figure 1 Concept design for the post combustion CO_2 capture using magnesium oxide (MgO)

Table 1 mass and molar flue gas composition

Component	% Mass fraction	Molar Conc.
N_2	65	62.32
CO_2	15	9.15
H_2O	18	1.67
O_2	2	26.84

The hot flue gas (850-600^{0}C) is first cooled, cleaned and then compressed to 5 bars of pressure. A multistage reciprocating compressor needs to be employed to compress 683.33 ton/h of flue gas to 5 bars pressure. A huge power input of about 31706 hp is needed to compress such a large volume of flue gas. Hence, the idea is to use four parallel compressors each having a power of about 8000 hp. Flue gases can be fed equally into these four parallel compressors and subject to 5 bars pressure using 3-staged reciprocating compressors. A polytropic process (n=1.3) is assumed. The assumed inlet temperature and pressure of flue gas is 67°C and 1 bar respectively. Here, we have considered efficiency of compressor as 0.8. The calculation shows that to compress flue gas to 5 bars pressure, 12.71 MWe energy input will be needed [14]. This step is thus energy intensive. The final temperature of flue gas after compression will be 384.81 K (111.81°C).

2.1   Carbonation reaction and carbonation reactor

A fluidized bed reactor (FBR) has been considered to carry out MgO-CO_2 reaction. This kind of reactor permits higher mass and heat transfer rates between the interacting molecules. FBR also ensures uniform temperature gradient throughout the reactor and has the ability to be run in a continuous state. Fig.1 shows that the compressed flue gas is then routed through the FBR.

Here, the reactor contains wet MgO. Our experimental results show that this reaction can be possible even at 325°C. About 83.65 ton/ h of 325 mesh size MgO is to be fed into the FBR running at 325°C and a pressure of 5 bars. The ultimate density of MgO particles is 3580 Kg/m^3. Maximum porosity assumed is 0.2. If the assumed diameter of the reactor is 30 ft (9.144 m), then the calculated length is about 7.12 m [15].

MgO (25 C, s) + CO_2 (100 C, g) = $MgCO_3$ (325 C, s) ΔH= -91.29 KJ/mole   .... R (1)

An amount of 83.65 ton/h of MgO needs to be fed to capture 91.33 ton/h of CO_2. R (1) generates 47.37MWt energy. This liberated heat needs to be effectively utilized. If this much amount of heat will be utilized with 100 % efficiency, about 15% of coal needs not to be burnt to get desired 303 MWt. That means, instead of burning 33.33 ton/h Coal, only about 28.33 ton/h coals will be needed. To do so, the idea is to have a heat exchanger around the FBR. The incoming cold water gains heat and becomes hot at the exit. Subsequently, the amount of coal to be burnt will decrease. Water flow rate is assumed to be 2.920167x10^5 Kg/h. If the temperature of water at inlet of the heat exchanger is 41°C, it will increase by 173.94°C if 100% of the carbonation reaction heat will be utilized. However, here, only 20% heat transfer efficiency has been assumed. That translates to an electrical energy of only about 3.16 MWe.

2.2 Gas Turbine Operation

Flue gas containing reduced CO_2 is then subjected to a gas turbine [T-1] [Fig.1]. Flue gas out of FBR will be mainly N_2 and O_2 gases. The gas enters the compressor at an elevated pressure and temperature of 4 bars and 400 K (127°C), approximately. The flow rate of flue gas with the reduced amount of CO_2 is about 517.5 ton/h. About 60% of this volume of gas has been considered as an input to the gas turbine. This approximation is made to be on a safe side. Now, the task is to compress 86.25 kg/s of gas to 10 bars. The inlet temperature for the gas turbine is assumed to be 1127°C and pressure is 12 bars. Air expands through the turbines, which has an isentropic efficiency of 88%, to a pressure of 1 bar. The air then passes through the interconnecting heat exchanger and is finally discharged at 400 K (127°C) as an exhaust into the atmosphere [16].

Steam generated at HE-2 [Figure 1], then enters the turbine of the vapor power cycle at 80 bars, 400°C and expands to the condenser pressure of 8 bars (assumed conditions). The turbine and pump of the vapor cycle have isentropic efficiencies of 90 and 80% respectively. The net power that will be developed is 60 MWe, which will be in the form of electricity. About 46 MWe power will be contributed by the gas turbine while the rest from vapor power cycle.

Another possibility could be to use an air-separation unit to gain pure nitrogen gas. However, that will not be cost efficient. Thus, installation of an air-separation unit will be completely dependent on the demand of nitrogen gas.

2.3 Calcination reaction and compression of CO_2

$MgCO_3$ then needs to be fed into the calciner, which is running at a temperature in the range of 400-500°C. Here, a mixture of steam and CO_2 is produced with MgO. Steam is to be removed by the condensation process.

$$MgCO_3 \text{ (s)} = MgO(s) + CO_2 \text{ (g)} \qquad \Delta H = +158 \text{ KJ/mole} \quad \ldots\ldots \text{ R (2)}$$

81.99 MWt energy is needed for the calcination process. There are three different possibilities of achieving this much amount of energy. One of them is to burn 9.02 tons of coal. This will emit 24.35 tons of CO_2. One of the other possibility could be to use heated flue gas (between 850-600°C) to produce steam. This steam can be directly used to provide energy for the calcination process. Or, we can also use electrical energy generated by the gas turbine cycle. The electrical energy that will be needed is 27.33 MWe. The required energy for the calcination process is almost 59% of the electrical energy produced by gas turbine cycle.

The moisture free pure CO_2 is then sent to the compressor where the pressure of CO_2 is increased from 1-2 bars to 136 bar using multi-stage compressors usually equipped with inter-cooling [Figure 1]. Numerous studies revealed that the compression energy requirement of CO_2 lies in the 95-135 kWh/tons of CO_2 range [17]. Thus, for multi-stage compression of CO_2, an energy requirement of 105 kWh/ton of CO_2 is chosen. Therefore, the compression of 91.33 ton/h of CO_2 requires 9.59 MWe.

3. EFFECT OF THE INEGRATION OF THE CARBONATION PROCESS
The integration of carbonation process requires electrical energy for various unit operations and can be regarded as a "parasitic load" for any power plant. However, a justification for the

integration of any unit is needed. Our calculation is based on the account of MWe produced and spent. The efficiency by which the produced energy from the FBR is utilized will have a significant effect on the overall efficiency of the power plant. If fully utilized, this energy will have the capacity to lower the energy burden due to the compression of the flue gases. Table 2 delineates the electrical energy produced and required to operate various operations. The overall electric generation capacity of plant is 113.53 MWe. However, if the carbonation reaction will need to be carried out at a high pressure of 10 bars, the overall electric generation capacity reduces to 95.24 MWe. Thus, these figures look promising compared to that of the CCR [18] and MEA process.

4. ECONOMIC ANALYSIS OF THE CARBONATION PROCESS

The economic analysis is based on the estimated procurement cost. The total direct plant cost (TDPC) accounts for both the procurement cost as well as installation cost of equipment, piping, instrumentation and other accessories required to complete the plant construction. The total capital investment (TCI) includes charges for general facilities, contingencies, engineering, working land and miscellaneous. The calculation shows that TCI is about 57 % of TDPC calculated. The total annual variable operating costs is about $19 million; which corresponds to approximately 23% of TDPC. Based on the various literatures, fixed operating and maintenance cost is taken as 5% of the capital cost [14]. The total cost is estimated by summing up the variable operating cost, fixed charges and fixed operating cost. All the equipment costs are estimated from the web service (www.matche.com). The detailed economic analysis is shown in Table 3. The cost analysis suggests that about $57.42 should be invested per ton of CO_2 to be captured using MgO process. However, if we also consider the CO_2 that is emitted to run the calcination process then this figure becomes $83.8/ton of CO_2 captured. (Consideration: Energy to run the calcination reaction will be delivered by burning coal). While the estimates for CO_2 using MEA is about $55/ton of CO_2 captured without considering any CO_2 emitted during the regeneration process.

5.1 TRANSPORTATION

Due to the large volumes of compressed CO_2, transportation will most likely be done using pipelines rather than ground transportation. At present, we are successfully transporting about 45 million ton of CO_2 each year through a pipeline extended by about 3700 miles. This transportation medium is used for enhanced oil recovery (EOR) by the petroleum industry. These pipelines are relatively thicker and expensive as it requires pressurized CO_2 to be transported.

Here, in our study if we are assuming that the diameter of the pipeline is 0.5 meter. Then, the following relation reported by several researchers [19], is used to determine the distance within which storage site should be located to transport 91.33 ton/h of compressed CO_2:

$$(P_1-P_2) = 32Lfm^2/ (\pi^2\rho D^5) \qquad \qquad \text{................... (1)}$$

$$\text{Or, } D = (32L \ fm^2/ [(P_1-P_2) \ \pi^2\rho])^{0.2}$$

Where, P_1 and P_2 = pressures at the beginning and end of the pipe respectively

L= length of pipe

f = fanning friction factor = $(1/ (4*\log (D/E) +2.28))$,

E = roughness in the material = 0.000457

m= mass flow rate (kg/s)

ρ = density (kg/m^3)

D = diameter (m)

Using the formula (1), the calculation suggests that we can have a pipeline of a length of approximately 183 km within which we can find a safe storage site to pump liquid CO_2 underground. National Energy Technology Laboratory (NETL)[19] suggested the cost estimation procedure for CO_2 pipeline which is followed here. Table 4(a) shows the calculation. The cost incurred in $/in/mile is 65813. Here, when we are using pipeline of diameter of 19.685 inches and length of 113.71 miles, thus we need to spend about $147 million for transportation.

5.2 STORAGE

Once CO_2 is compressed, captured, and transported then the challenge left is to successfully inject it underground. A safe geologic storage site within an economical distance needs to be identified. The Department of Energy (DOE) has focused a lot of attention to do research in CO_2 storage area. The goal is to develop technologies which can safely, permanently and cost-effectively store CO_2 in the geological formations. The cost associated with storage can be categorized into five different sections: (1) Site screening and evaluation; (2) injection wells; (3) injection equipment; (4) operating and maintenance costs; (5) pore space acquisition [19]. Previous studies show that to successfully transport and store 10,000 short tons of CO_2 per day, one injection well is sufficient. Therefore, here in our case we need only one injection well. The well-depth is assumed to be 1200 meters. Table 4(b) shows the calculation for injecting 91.33 ton/h of CO_2 inside an injection well. The total cost is about $11.01 million, which is about 7.5% that of transportation cost.

6. CONCLUSION

This paper illustrates the possibility of integrating MgO carbonation unit for capturing CO_2 after the post combustion of coal. Here, we made an economic and energy analysis for MgO based carbonation unit to be integrated with the coal fired power plant. It seems to be relatively more profitable as well as eco-friendly than all other proposed technologies. Process design enables the use and integration of high quality heat available using magnesium oxide sorbent and thus makes this technology a potential market for CO_2 capture technologies.

Table 2 Electrical energy distribution for various operations

	Electrical Energy Produced	Electrical Energy Needed
Original Vapor power cycle	100	
Compression of flue gas		12.71
Carbonation reaction	3.16	
Gas power cycle	46	
Vapor power cycle-2	14	
Calcination reaction		27.33
Compression of CO_2		9.59
Overall produced electrical energy	113.53	

Table 3 Different costs for post combustion CO_2 capture using Magnesium Oxide (MgO)

Equipment	(million $)
Jacketed and Agitated reactor	0.112
Hopper for make-up sorbent	0.0262
Hopper for sorbent calcination	0.0361
Compressor	11.8636
CO_2 Compressor Cost	16.3802
Two conveyor belts	0.0744
Sorbent Inventory	0.092
Total procurement cost:	$ 28.5845

TDPC: $ 85.7535 million

General facilities	30 %	25.72605
Eng Permitting and start up	10%	8.57535
contingencies	10%	8.57535
Working capital, land and	7%	6.002745
Total capital investment		$ 134.633

Variable operating costs

Variable Non-fuel operating	1 %	0.8575
Calciner cost (Rotary Kiln)		15.956
Make up Sorbent cost		2.1983
Total annual variable		$ 19.011 million

Fixed operating and maintenance cost (5% capital cost) $6.7316 million

Capital charges (15% capital cost) $20.1949 million

Total annual cost   **$45.9384 million**

(a) If CO_2 emitted by burning coal for calcination process to be considered	
CO_2/coal (tons/tons)	2.608
Total coal consumed (tons/hr)(calcination)	9.02
CO_2 emitted (tons/hr)(calcination)	24.35
Total CO_2 captured (tons/hr)	62.57
Total CO_2 captured (tons/year)	548153.8
Cost /tons CO_2 captured	**$83.80**/tons CO_2 captured
(b) If CO_2 emitted by burning coal for calcination process not to be considered	
CO_2/coal (tons/tons)	2.608
Total CO_2 captured (tons/hr)	91.33
Total CO_2 captured (tons/year)	800050.8
Cost /tons CO_2 captured	**$57.42**/tons CO_2 captured

Table 4(a) Transportation costs for liquefied CO_2 using pipelines

Cost of Pipelines

Diameter = 19.685 inches   Length = 113.71 miles

(i)

Material	35520597.04
Labor	72670788.04
Miscellaneous	31217759.66
Right of way	5662524.423

(ii) Other Capital Costs

CO_2 surge tank	1150636
Pipeline control system	110632

(iii) Operating and Maintenance Costs

Fixed ($/year)	981544.72

Total ($/in/mile)   **65813**

Table 4(b) Geological storage costs for liquefied CO_2

Geological Storage Costs

Well depth = 1200 meter   number of injection well =1

(i)

Site screening and evaluation	4738488
Injection wells	628671.9
Injection equipment	483031.7
Liability bond	5000000

(ii) Declining Capital funds

Pore space acquisition	807.0286

(iii) Operating and Maintenance Costs

Normal daily expenses	11566
Consumables	2995
Surface Maintenance	120607.7
Subsurface Maintenance	27874.01

Total ($)   **11.01404 million**

REFERENCES

(1) Zhao, L., et al., aqueous carbonation of natural brucite: relevance to CO_2 sequestration, Environmental Science and Technology **2010**, *44,* 406–411

(2) Bhatia, S.K., Perlmutter, D. D., Effect of the product layer on the kinetics of the CO_2 – Lime Reaction. AIChE J. **1983**, 29, 79-86.

(3) Butt, D.P., et al., Kinetics of Thermal Dehydroxylation and Carbonation of Magnesium hydroxide. Journal of the American Chemical Society **1996**, 79, 1892-1898.

(4) Dunsmore, H.E. A Geological Perspective on Global Warming and the Possibility of Carbon Dioxide Removal as Calcium Carbonate Mineral. Energy Conversion and Management **1992**, 33, 565-572.

(5) Gunter, W. D., et al., Aquifer Disposal of CO_2 rich gases: Reaction Design for Added Capacity. Energy Conversion and Management **1993**, 34, 941-948.

(6) Hanchen, M., et al., Precipitation in the Mg-carbonate system- effects of temperature and CO_2 pressure. Chemical Engineering Science **2008**, 63, 1012-1028.

(7) Holloway, S. An overview of the underground disposal of carbon dioxide. Energy Conversion and Management **1997**, 38, S193-S198.

(8) Kojima, T., et al. Absorption and Fixation of Carbon Dioxide by Rock Weathering. Proceedings of the Third International Conference on Carbon Dioxide Removal, Cambridge Massachusetts, **1996**, September 9-11.

(9) Lackner, K.S., et al., Carbon Dioxide Disposal in Carbonate Minerals. Energy **1995**, 20, 1153-1170.

(10) http://sequestration.mit.edu/tools/projects/index.html

(11) Seifritz, W. CO_2 disposal by means of silicates. Nature **1990**, 345, 486.

(12) Kendra L. Zamzow, Carbon Capture & Sequestration, Center for Science in Public Participation September **2009.**

(13) Kumar S., et al., Capture of CO_2 using magnesium oxide, Under Review.

(14) Max Stone, P., Timmerhaus, K.D., Plant Design and Economics for chemical engineers, Mcgraw Hill, 4th Edition.**1991**.

(15) Mccabe, W., Smith, J., Harriott, P., Unit Operations of Chemical Engineering, McGraw Hill, 7th Edition, **2004**.

(16) Moran M.J., Shapiro H.N., Fundamentals of Engineering Thermodynamics, John Wiley and sons, 6th Edition, **2008**.

(17) Hallman, M.M. and Steinberg, M., Greenhouse Gas Carbon Dioxide Mitigation: Science and Technology, Lewis Publishers, New York, **1999**.

(18) Mahesh V. Iyer, High Temperature Reactive Separation Process for Combined Carbon dioxide and Sulfur dioxide Capture from Flue Gas and Enhanced Hydrogen Production with in-situ Carbon dioxide Capture using High Reactivity Calcium and Biomineral Sorbents, Ohio State University, **2006**.

(19) Quality Guidelines for Energy Systems Studies ,Estimating CO2 Transport, Storage & Monitoring Costs, National Energy Technological laboratory, Department of Energy, USA, March **2010**.

INVESTIGATION ON AMMONIUM PHOSPHATE MIXED SiO$_2$/POLYMER HYBRID COMPOSITE MEMBRANES

Uma Thanganathan

Research Core for Interdisciplinary Sciences (RCIS)
Okayama University
Tsushima-Naka, Kita-Ku
Okayama, 700-8530, Japan
E-mail: umthan09@cc.okayama-u.ac.jp

ABSTRACT

The key to this strategic work involves the combination of a hybrid network structure and an investigation of the corresponding properties. It represents an important part in the development of future materials. Here, a class of combination for a SiO$_2$–PVP–(NH$_4$)$_3$PO$_4$ hybrid composite electrolyte was synthesized via the sol-gel technique.

1. INTRODUCTION

The development of material technology is paying special attention to novel organic-inorganic hybrid materials as a means of obtaining electrolytes for fuel cell applications. Several researchers have reported on various kinds of proton-conducting composites as alternatives to the Nafion® membrane. This is due to such perfluorosulfonic acid polymers having many attractive properties, including a good mechanical strength, a decent chemical stability, and a high proton conductivity, and being widely used in commercial applications.[1,2] However, perfluorosulfonic acid membranes display several drawbacks such as a complex water management, CO poisoning of Pt catalyst at the anode, and a high cost.[3]

After removal of perflourosulfonic acid membranes like Nafion®, one could introduce novel non-fluorinated materials with similar properties as the later. The main idea is to develop hybrid materials based on the design of hybrid polymers with special emphasis on structural hybrid materials. The focuses for hybrid materials go further than mechanical strength, and include thermal and mechanical stability. Information for relevant SiO$_2$-mixed PVP membranes including their characterization results have been reported by several researchers.[4-7] The study of the properties of ammonium polyphosphate has been performed by Stimming et al.[8,9] It was found that NH$_4$H$_2$PO$_4$/SiO$_2$ can be used in organic transformations due to the fact that this catalyst displays several advantages such as a low toxicity,[10] low cost, ease of handling and high catalytic activity.

The development of eco-friendly synthetic methodologies have led to a ecological, facile and efficient one-pot method for the synthesis of aryl-14-*H*-dibenzo [*a,j*] xanthene derivatives catalyzed by NH$_4$H$_2$PO$_4$/SiO$_2$ under solvent-free conditions.[11] The purpose of this study has been to introduce a composite with good mechanical and thermal properties and such a material was synthesized via the sol-gel method using the water-soluble polymer poly(vinyl pyrrolidone) (PVP) and triammonium phosphate [(NH$_4$)$_3$PO$_4$.3H$_2$O]. A compound was formed by carrying out a chemical reaction between these components, and the properties of each part were eliminated to form a hybrid material with specific characteristics. The addition of (NH$_4$)$_3$PO$_4$ was expect to give rise to a composite with a high degree of homogeneity and (NH$_4$)$_3$PO$_4$ was considered to be a proton conductor serving as the supporting matrix. The other component employed in the composite, i.e., PVP, could efficiency improve the oxidative stability and mechanical properties of the membranes. Another advantage of using PVP was that it could be thermally crosslinked.[12,13] Based on a literature survey, information has yet to be recorded for the combination of SiO_2 and PVP with triammonium phosphate. It was thus a challenge to create a combination of organic-inorganic materials in order to obtain a hybrid composite electrolyte by the sol-gel method.

2.   EXPERIMENTAL SECTION

$Si(OC_2H_5)_4$ (TEOS, 99.9%, Colcote), poly(vinylpyrrolidone) (PVP) with a molecular weight of 100,000 g/mol was purchased from Nacalai Tesque, Japan. Triammonium phosphate trihydrate $[(NH_4)_3PO_4.3H_2O]$ was obtained from Cica-reagent, Kanto Chemical Co. Inc. The homogenous solution of $SiO_2–PVP–(NH_4)_3PO_4$ was synthesized by the sol-gel process with PVP, $(NH_4)_3PO_4.3H_2O$ and SiO_2 under stirring at room temperature for several hours. PVP (1g) was dissolved in a predetermined amount of water at room temperature by stirring for 1 h. This PVA solution was then mixed with tetraethoxysilane (TEOS) and triammonium phosphate under stirring at ambient conditions to produce the $SiO_2–PVP–(NH_4)_3PO_4$ (50/50 mol %–1g) hybrid composite (Figure 1). The product was analyzed and tested with varying methods such as Fourier transform infrared spectroscopy, X–ray diffraction, and thermogravimetric analysis. According to the results of the experiments, the PVP polymer generated hydrogen bonding with the –OH group in SiO_2 and with the O–H bond with H_2O in $(NH_4)_3PO_4.3H_2O$. The O–H bonds reoriented, achieving migration of the proton. PVP was homogenously dissolved in the sol. The proposed structure for the composite is shown in Figure 2.

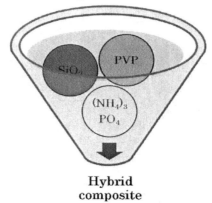

Hybrid composite

Figure 1. A new $SiO_2–PVP–(NH_4)_3PO_4$ hybrid composite synthesized by the sol–gel process.

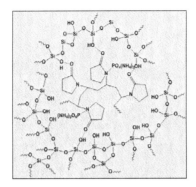

Figure 2. Systematic network structure of SiO$_2$–PVP– (NH$_4$)$_3$PO$_4$ hybrid composite.

3. RESULTS AND DISCUSSION

The structural FTIR study involved a transmittance range from 400 to 4000 cm^{-1} for the composite and the results are presented in Figure 3. It shows that PVP has a clear absorption spectrum at 1655 cm^{-1}, which corresponds to an amide. This is an excitation/absorption spectrum combined with the $>$C$|$O and C–N groups.[14] Fig. 3b illustrates broad peaks at 400–600, 845–1260, 1410–1547, and small peaks are located at 1637, 1714 and 1848 cm^{-1}. In the composite, a few of these peaks were shifted to 473, 809, 1095, 1405, 1642 cm^{-1} and other peaks disappeared completely. The clear C$|$O group absorption spectra of the PVP/SiO$_2$ hybrid were observed at different weight ratios at 1651, 1647, and 1646 cm^{-1}. As can be seen in Figure 3c, sharp peaks appeared at 1095 cm^{-1}. These we are the absorption bands for the Si–O–Si group, confirming the existence of the synthesized PVP/SiO$_2$ hybrid.[15]

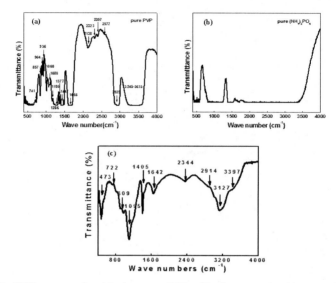

Figure 3. FTIR curves for (a) the pure PVP, (b) the pure (NH₄)₃PO₄ and (c) the SiO₂–PVP–(NH₄)₃PO₄ hybrid composite.

Because of the shifting of the absorption spectra for the PVP >C|O group between the two, on the basis of reports written by several scholars,[14-16] an identification can be made of the existence of hydrogen bonding. According to Figure 3c, the hybrid composite shows a clear C|O group absorption band at 1642 cm^{-1}, which became shifted toward the low-frequency region when adding SiO₂ and (NH₄)₃PO₄ in the composite. Generally, the information showed that the inorganic silica and organic PVP were very well dispersed at the molecular level in accord with the previous reports.[17,18] The bands at 473, 722, 809, 1095 cm^{-1} were related to the Si–O–Si bond, which confirmed the presence of Si–O–Si networks in the hybrid. The bands in the region 2914-3397 cm^{-1} and at 2344 cm^{-1} were caused by stretching of –OH. It is known that the position of the OH-absorption bands depends on the degree of strength of hydrogen bonding, and shifts to lower wave numbers when the hydrogen bonding becomes stronger.[19] The band at 2344 cm^{-1} in the composite proved the existence of intermolecular hydrogen bonds and the NH₄ species in the composite confirmed the hybrid composite. This observation pointed at the formation of a composite combining PVP, SiO₂ and (NH₄)₃PO₄.

Figure 4 shows the results from X–ray diffraction of the pure (NH₄)₃PO₄.3H₂O, the pure PVP and the SiO₂–NH₄–PVP hybrid composite. As can be seen-4, the pure PVP presented two broad peaks at 12 and 22° (Figure 4a) and the pure (NH₄)₃PO₄ showed many sharp crystalline peaks (Figure 4b). The composite exhibited a few sharp peaks at 17, 24, 29, 34 and 45° (Figure 4c), and consequently the crystallinity of the hybrid composite did not change with the addition of TEOS. The original peaks from PVP with (NH₄)₃PO₄ slightly altered their location which indicates that the original crystallinity was lower. From the final curve, a few sharp peaks can be seen for the composite. The phase analysis confirmed the formation of a hybrid composite.

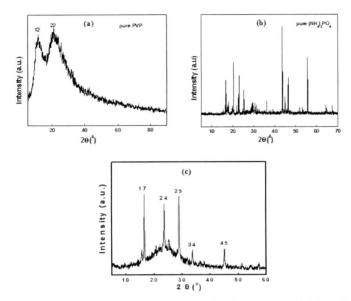

Figure 4. XRD curves for (a) the pure PVP, (b) the pure (NH₄)₃PO₄ and (c) the SiO₂–PVP–(NH₄)₃PO₄ hybrid composite.

The thermal stability was determined by thermogravimetric analysis (TGA) and the results are given in Figure 5. It seems that the hybrid composite had a high stability up to a temperature of about 410 °C. In order to analyze the thermal stability of the pure PVP and the pure salt, the TGA measurements were carried out in N_2 atmosphere. According to Figure 5a, for the pure PVP, the weight loss began after 417.3 °C, and the total weight loss was about 61.5 %. Figure 5b, displayed the weight loss for the pure salt of NH_4PO_4, amounting to a total of 37.6 % as calculated from the TGA curve. PVP and $(NH_4)_3PO_4$ were found to be stable at about 400 and 200 °C.[20] However for the composite, as can be seen in Figure 5c, the thermal stability was the result of the strengthened hydrogen-bonding interaction between PVP and $(NH_4)_3PO_4$. Three main weight losses were observed and could be calculated. The first dehydration occurred at 40 °C and corresponded to about 5.7%. It was believed to be caused by water absorption. The second dehydration was observed at 199 °C and the third at 412 °C, corresponding to 24.7%, and 33.5%, respectively. Finally after 700 °C, there occurred no further dehydration. As a conclusion, the composite was deemed thermally stable at high temperature.

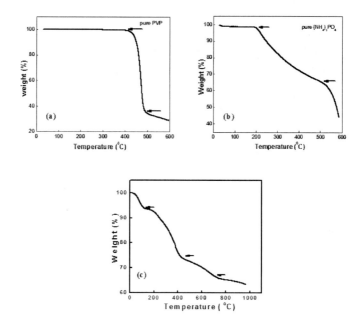

Figure 5. TGA curves for (a) the pure PVP, (b) the pure $(NH_4)_3PO_4$ and (c) the SiO_2–PVP–$(NH_4)_3PO_4$ hybrid composite.

4. CONCLUSIONS

To the best of my knowledge, there exist no other reports on the combination of SiO_2–PVP–$(NH_4)_3PO_4$ as a hybrid composite electrolyte. This is definitely an essential composite material and further work will introduce its structural and thermal properties for fuel cell applications. Such a composite will be fabricated by the same method and several properties, including the conductivity and electrochemical activities, will be studied as a function of temperature and humidity. The results of such investigations will be reported in the near future. The present study supports the idea that SiO_2–PVP–$(NH_4)_3PO_4$ hybrid composite materials are potentially suitable candidates for low/high temperature electrochemical energy conversion systems.

ACKNOWLEDGEMENT

This work was financially supported by the Ministry of Education, Sport, Culture, Science and Technology (MEXT) and the Special Coordination Funds for Promoting Sciences and Technology of Japan. Special thanks go to Prof. Kishmoto, Graduate School of Natural Science of Technology and Prof. Kimura, Graduate School of Environmental Science, Okayama University, Japan; they permitted to use their lab equipments for my research work.

REFERENCES

(1) Wahdame, B.; Candusso, D.; Francois, X.; Harel, F.; DeBernardinis, A.; Kauffmann, J. M.; Coquery, G. *Fuel Cells*, **2005**, *7*, 47.

(2) Arita, M. *Fuel Cells*, **2002**, *2*, 10.

(3) Aparicio, M.; Duran, A. *J. Sol-Gel Sci Technol.*, **2004**, 31, 103.

(4) Nakajima, H.; Kawano, K. *J. Alloys and Compounds*, **2006**, *408/412*, 343.

(5) Huang, K. S.; Lee, E. C.; Chang, Y. S. *J. Appl. Poly. Sci.*, **2006**, *100*, 2164.

(6) Hsiao, C. N.; Huang, K. S. *J. Appl. Poly. Sci.*, **2005**, *96*, 1936.

(7) Wang, H.; Tealdi, C.; Stimming, U.; Huang, K.; Chen, L. *Electrochimica Acta*, **2009**, *54*, 5257.

(8) Uma, T.; Tu, H. Y.; Warth, S.; Schneider, D.; Stimming, U. *J. Mater Sci.*, **2005**, *40*, 2059.

(9) Bigdeli, M. A.; Jafari, S.; Mahdavinia, G. H.; Hazarkhani, H. *Catal. Commun*, **2007**, *8*, 1641.

(10) Chalapathi, V. V.; Ramiah, K. V. *Curr Sci.*, **1968**, *16*, 453.

(11) Cain, B. D.; Simoni, R. D. *J. Biol. Chem.* **1989**, *264*, 3292.

(12) Zieba, J.; Zhang, Y.; Prasad, P. N.; Casstevens, M. K.; Burzynski, R. "Sol-gel-processed inorganic oxide: organic polymer composites for second order nonlinear optical applications," in *Sol-Gel Optics II*, J. D. Mackensie, (ed.), *Proc. Soc. Photo-Opt. Instrum. Eng.* 1758, 403-409 119922.

(13) Scheirs, J.; Bigger, S. W.; Then, E. T. H.; Billingham, N. C. *J. Polym. Sci. Polym. Phys. Ed.* 31, 287–297 119932.

(14) Zhang, X.; Takegoshi, K.; Hikichi, K. *Macromolecules*, **1992**, *25*, 2336.

(15) Zhang, X.; Takegoshi, K.; Hikichi, K. *Polymer*, **1992**, *33*, 712.

(16) Bovey, F. A. High-Resolution NMR of Macromolecules; Academic: New York, 1972.

(17) Wen, J.; Wilkes, G. L. *Chem. Mater.*, **1996**, *8*, 1667.

(18) Chujo, Y.; Saegusa, T. *Adv. Polym. Sci.*, **1992**, *100*, 11.

(19) Scholze, H. *Glastech. Ber.*, **1959**, *32*, 81, 142, 314.

(20) Chen, X.; Huang, Z.; Xia, C. *Solid State Ionics*, **2006**, *177*, 2413.

Nanocomposites
and Nanomaterials

NANO-HETEROGENEOUS STRUCTURING EFFECTS ON THERMAL POWER FACTOR AND THERMAL CONDUCTIVITY

Gustavo Fernandes[1], Do-Joong Lee[2], Jin Ho Kim[1], Seungwoo Jung[2], Gi-Yong Jung[2], Fazal Wahab[1], Youngok Park[1], Ki-Bum Kim[2], and Jimmy Xu[1,2]

[1]School of Engineering, Brown University, Providence, RI 02912, USA
[2]WCU Program, Seoul National University, Seoul, Korea

ABSTRACT
    We discuss our investigations on the thermoelectric properties of two novel systems, multilayer Al:ZnO films as well as holey carbon nanotube films (holey Bucky paper). Both of these systems, despite being novice materials in thermoelectric applications, share heterogeneous and nanostructuring features and offer a rich set of material properties that can be tuned in composition and feature size to enhance their thermoelectric properties. Here we explore two representative results: bi-material layer stacks as well as nanohole etching, and show that enhancements in the Seebeck coefficient as well as significant reduction in the thermal conductivity of the same material are achievable with heterogeneous nanostructuring.

INTRODUCTION
    Underlying the numerous interesting thermoelectric devices and structures is the same governing principle – the Boltzman transport equation[1]. As a result, the Seebeck coefficient and electrical conductivity are inter-locked in such a way that one goes up with the Fermi energy while the other goes down. It can be shown [2] that in a homogeneous system instead of maximizing the Seekbeck coefficient, one should ascertain an "optimal" value that is common to all materials and dimensionalities and at which the thermal power factor is maximized. The electrical and thermal conductivities are also coupled in bulk systems via the Wiedemann-Franz law [1], but not as tightly such that their ratio can be increased as demonstrated with nanostructuring in order to reduce the phonon mean-free path more than the electron's.
    Thus nano-heterogeneous structuring is suggested here as one way to introduce an extra degree of freedom into the system so as to weaken the mutual constraint, imposed by Boltzman transport of a single homogeneous system, between Seebeck coefficient, electric conductivity, and phonon scatterings. In such a way the thermoelectric figure of merit,

$$ZT = S^2\sigma / \left(\kappa_e + \kappa_L \right), \qquad (1)$$

may be enhanced. Here S is the Seebeck coefficient (V/K), σ is the electrical conductivity (S/m) and $\kappa = \kappa_e + \kappa_L$ is the thermal conductivity (W/m-K) which consists of an electronic (κ_e) and a lattice (κ_L) contributions. The numerator of equation (1) is known as the thermoelectric power factor, and ZT is related to the maximum efficiency that a perfect thermoelectric generator made of the specific material would be able to achieve.

Current available homogeneous material platforms have a maximum ZT ~ 1 at 300 K. Lead composites [3-5], in particular Na and Ta doped PbTe alloys and Na doped $PbTe_{1-x}Se_x$ have been shown to achieve ZT ~ 1.4 to 1.8 at high temperatures (>~750 K). For higher temperatures, Si-Ge alloys are found to be the best thermoelectrics, with ZT ~ 0.7 [6]. In order for thermoelectric devices to compete with modern household refrigerators and batteries, ZT larger than ~ 3 is required. Nanotechnology strategies have been used to decrease κ_L and increase S so as to improve ZT. PbTe/PbSeTe quantum dot arrays have been shown to have ZT ~ 1.5 at room temperature which is better than that of PbTe or PbTeSe alone (~ 0.5) [7]. Silicon nanowires with rough surfaces have also shown to have ZT ~ 1 and recently holey Si membranes were found to be good thermoelectric materials near room temperature [8]. Our own recent research of solution grown PbS nanoclusters has shown a large Seebeck coefficient (~ 450 μV/K), which is comparable in size to that obtained in PbTe, for example [9].

Here we will discuss two particular heterogeneous nanostructured systems of potential interest in the above context. The systems we have studied incorporate strategies aimed at maximizing thermoelectric power factor while also maximizing phonon scatterings. While our results are by no means complete at the present stage, they indicate that some of these strategies may lead to fruitful future advancements in thermoelectrics.

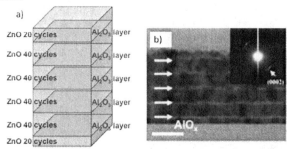

Figure 1. Stacked ZnO and Al_2O_3 layers. (a) Schematic of a representative sample. (b) TEM image showing layered composition.

THERMAL AND ELECTRICAL PROPERTIES OF MULTILAYER Al:ZnO FILMS

The first system we turn our attention to consists of a multi-layered structure based on the well known Al doped ZnO (Al:ZnO, or more concisely AZO) material platform, shown in Figure 1 [10, 11]. Homogeneous AZO has been widely studied as a transparent conductor for applications in displays and solar cells, and indeed as one potential replacement for the more ubiquitous indium-tin-oxide (ITO) [10-18]. More recently other properties of AZO have received attention, such as its IR and electromagnetic interference (EMI) shielding properties [19-24]. The thermoelectric properties of AZO remain underexplored, despite the fact that some of its material properties may be indicative of potentially good thermoelectric performance. For instance, AZO is a wide band gap semiconductor but also with a very high electron concentration, and thus highly electrically conductive. It is made of highly abundant materials on the Earth's surface, Al and Zn, and thus may offer a cheap alternative to current platforms. AZO's large band-gap indicates that it could in principle be incorporated into photovoltaic solar cells for hybrid thermal/photovoltaic energy harvesting.

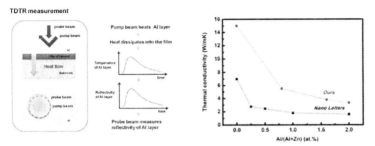

Figure 2. Thermal conductivity of AZO layered samples with increasing fractional quantity of Al_2O_3.

Nonetheless, being a semiconductor, the thermal conductivity of AZO in bulk crystalline form is expected to be large and to include a large lattice contribution. Theoretical estimates and previous experimental work have placed AZO's thermal conductivity in the neighborhood of 30 W/m-K which makes it difficult to ascertain large ZT values.

On the other hand, with recent developments in atomic layer deposition (ALD) it has been possible to obtain heterolayered AZO by depositing alternating ZnO and Al_2O_3 layers. The structure of the heterolayered AZO-based system we explore here is shown in Figure 1. Samples consisted of stacks of ZnO thin films separated by Al_2O_3 monolayers. The number of ALD cycles is indicated in the schematic drawing of Figure 1 (a). Figure 1 (b) shows a cross section transmission electron microscope (TEM) image of one such sample, with the Al_2O_3 buffer layers clearly visible.

From the thermoelectric point of view, the inclusion of multiple layers introduces two benefits. Firstly, as suggested by the TEM image in Figure 1 (b), small lattice mismatches at the interfaces between ZnO and Al_2O_3 are likely to provide surface irregularities resulting from built in strains and may, as a result, increase the scattering of both phonons and electrons at the interfaces. Because electrons and phonons experience dramatically different dynamic conditions, with different mean-free-paths and wavelengths, it can be expected that, with proper optimization of the layer distribution, the electron transport path could be spatially separated from the phonon transport so as to weaken the Boltzman transport constraint and maximize the ratio of the scattering of phonons over that of electrons. As will be discussed below, measurements of the lateral (in-plane) electrical conductivity and Seebeck coefficient of these samples, along with the vertical (cross-plane) thermal conductivity seem to suggest that we may already have achieved preferential phonon scattering in these current samples.

Secondly, the model of alternating layers also provides extra acoustic impedance as the interface between the two materials involved, ZnO and Al_2O_3, have different bonding strengths and densities. This alternating mechanical impedance variation gives rise to additional scattering of phonons.

The vertical (cross-plane) thermal conductivity was measured using the time-domain thermo reflectance (TDTR) technique. The TDTR technique has been extensively discussed in the literature [25]. Figure 2 (a) shows a brief schematic diagram of the TDTR method. Two laser beams are made to impinge onto the same portion of the sample which is covered with an aluminum film. The pump laser provides an intense laser pulse that heats up a spot on the Al film. As heat propagates away from this spot both laterally in the Al film as well as vertically into the underlying AZO film a probe laser, reflected off the same spot monitors the reflectivity of the aluminum. Because the latter changes with temperature, the reflectance measured by the probe beam is also effectively a measurement of the temperature. From this data the thermal conductivity is extracted [25].

Results for AZO films with various Al_2O_3 percentages are shown in Figure 2 (b). "Nano Letters" refers to the results of Jood et al [26] which are shown here for comparison. Their system consists of pellets of pressed powder AZO. For our films, the case of 0% Al corresponds to a pure homogeneous ZnO film. As expected from first-principles consideration above, the introduction of Al_2O_3 layers has a pronounced effect on the thermal conductivity. The film with 2% Al_2O_3 in its composition showed an almost 5-fold decrease in the thermal conductivity. These results support the hypothesis of increased phonon scatterings as more Al_2O_3 layers are added. While it is too early to say which scattering mechanism is dominant, or whether a mean-free-path (particle) is responsible, these results indicate that such decrease in the thermal conductivity may actually stem from suppression of the lattice contribution to the thermal conductivity rather than its electronic contribution.

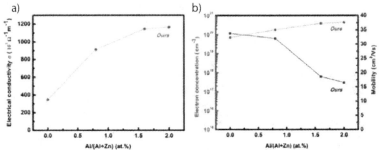

Figure 3. Lateral (in-plane) electronic properties of AZO/Al_2O_3 layered samples with increasing fractions of Al_2O_3. (a) Electrical conductivity. (b) Electron concentration and mobility.

Although we have not yet measured the cross-plane electronic properties of these samples, their measurements in the lateral direction appear to support the above conclusion. Figure 3 (a) shows that the lateral electrical conductivity increased continuously with the Al_2O_3 fraction. The same behavior was observed in the electron concentration, shown in Figure 3 (b) along with the electron mobility. The mobility is observed to decrease with increasing fraction of Al_2O_3, which suggests that in the lateral direction too, the addition of Al_2O_3 layers increases electron scattering. Such additional scattering in the lateral propagation direction is most likely introduced by surface defects and roughness at the inter-layer surfaces. From the point of view of the conductivity, however, this apparent increase in electron scattering seems to be more than compensated for by the increase in electron density. It is possible that the same pattern will be observed in the vertical direction, which would be greatly beneficial to the thermoelectric properties of this material, although further measurements will be needed in order to confirm this.

One of the strategies for achieving good thermoelectric materials is to improve the Seebeck coefficient

$$S = \frac{\pi^2}{3} \frac{k_B}{q} k_B T \left\{ \frac{d \ln[n(E)]}{dE} + \frac{d \ln[\mu(E)]}{dE} \right\}_{E=E_F}, \qquad (2)$$

where n is the carrier density and μ is the carrier mobility. According to equation (2), this can be accomplished by increasing the energy dependence of $\mu(E)$ or of $n(E)$ [3]. The former's energy dependence can be increased by scattering mechanisms that depend strongly on the energy of the charge carriers. In contrast, the energy dependence of $n(E)$ can be increased, e.g., by a local variation in the density of states.

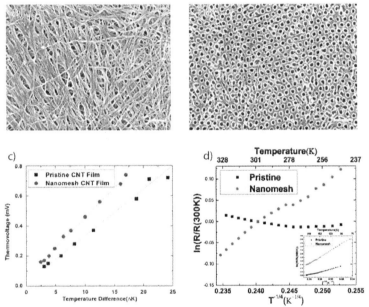

Figure 4. (a) Pristine 20 nm thick CNT film. (b) Nanopatterned 20 nm thick CNT film. (c) Thermovoltage near 300 K. (d) Temperature dependence of resistivity. The inset

graph shows the (80–250 K) temperature range and the main graph shows the (250–330 K) temperature range.

The multilayer AZO system described above could in principle satisfy both of these conditions. The multiple stacked layers may lead, for instance, to Bragg-grating effects and thus strongly energy-dependent scattering of electrons. In addition, the interlayer surfaces and the small lattice mismatch between ZnO and Al_2O_3 can be expected to present distortions in the local density of states. These could lead to increased Seebeck coefficient in the cross-plane direction, and remains to be measured.

In contrast, the in-plane Seebeck coefficient has been measured for that structure and found to be small, at around 15 μV/K. The reason may be in the high lateral conductivity observed for the structure which appears to behave like a metallic layer, with inherently low Seebeck coefficient and which also shorts whatever thermoelectric voltage that may develop in the adjacent ZnO layers. We note too that the Seebeck coefficient for pure ZnO samples as well as sol-gel AZO samples have also been measured by us and found to have relatively high values (~100-200 μV/K).

HOLEY CNT FILMS

In the same context one may view a second platform as a heterogeneous nanostructured thermoelectric: films of randomly dispersed metallic and semiconducting carbon nanotubes (CNTs) [27, 28]. More details have been recently published on this system [29]. Such films are interesting systems for basic science and engineering regarding their thermoelectric properties [29-31]. For example, metallic nanotubes are expected to exhibit no thermoelectric power (TEP), yet experiments have shown otherwise [31]. Coupling and spatial charge transfer between nanotubes can be in principle controlled via the myriad of available techniques for surface passivation as well as incorporation of polymers, nanoparticles and organic molecules, and are thus expected to lead to interesting, possibly controllable, thermoelectric properties.

To investigate some of these effects and to attempt to increase the thermopower of CNT films we adopted a novel approach consisting of etching a periodic nanoscale hole array [29], as shown in Figure 4 (a). The 20 nm thick CNT film sits atop a fused silica substrate and the holes are transferred to it via reactive ion etching using the following recipe: 30 sccm Cl_2, 5 sccm BCl_3, 50 mTorr, 100 W, 1 min. After hole transfer, the Seebeck coefficient of films were measured, and the results, shown in Figure 4 (b), indicate an enhancement of almost 34% compared to the unpatterned films. The TEPs of pristine CNT film and nanomesh CNT film are 29.54 (1.2) V/K and 39.47 (0.7) V/K, respectively, and the value for pristine CNT film agrees well with previously reported data [32].

These results can be qualitatively understood in the context of equation 2. Studies on the temperature coefficient of resistance (TCR) of such films (Figure 4 (c)) as well as micro-Raman studies [29] have indicated that the nanopatterning process introduced dangling bonds in the CNTs leading to distortions of their local density of states. In the case of the TCR behavior, it was found that the nanopatterned films exhibited a five-fold increase in the TCR (0.04 %/K for the pristine film vs 0.2 %/K for the nanopatterned one) and considerably larger activation energy when fitted to a variable-range-hopping (VRH) model [33].

At the moment, the enhancement in the Seebeck coefficient in this CNT system is accompanied by a drastic increase in the resistivity (approximately twenty-fold) and thus diminishes its practical interest for thermoelectric generation. However, given the many strategies available to alter its properties, it is conceivable that upcoming research will reveal otherwise. In addition, the Cl_2 based etching recipe employed here is rather aggressive and could possibly be substituted for an O_2-based recipe in the future which would likely introduce less damage to the CNTs. The key in this process appears to be to always strike the right balance between enhancement of the Seebeck coeffcient and excessive damages to the electronic structure that adversely affect the electrical conductivity.

SUMMARY

We have discussed two novel heterogeneous systems of potential interest in the context of thermoelectrics. In particular their ability to have their electronic and thermal properties tuned via bottom up nanofabrication approaches as well as their unique material properties makes them interesting in the context of themopower engineering. Our results presented here, although still in progress, suggest strategies to further improve the thermoelectric performance of these materials.

REFERENCES
1. C. Kittel, *Introduction to solid state physics* (Wiley, Hoboken, NJ, 2005).
2. P. Pichanusakorn, and P. R. Bandaru, "The optimal Seebeck coefficient for obtaining the maximum power factor in thermoelectrics," Appl Phys Lett **94** (2009).
3. J. P. Heremans, V. Jovovic, E. S. Toberer, A. Saramat, K. Kurosaki, A. Charoenphakdee, S. Yamanaka, and G. J. Snyder, "Enhancement of thermoelectric efficiency in PbTe by distortion of the electronic density of states," Science **321**, 554-557 (2008).
4. Y. Z. Pei, A. LaLonde, S. Iwanaga, and G. J. Snyder, "High thermoelectric figure of merit in heavy hole dominated PbTe," Energ Environ Sci **4**, 2085-2089 (2011).
5. Y. Z. Pei, X. Y. Shi, A. LaLonde, H. Wang, L. D. Chen, and G. J. Snyder, "Convergence of electronic bands for high performance bulk thermoelectrics," Nature **473**, 66-69 (2011).
6. D. M. Rowe, *CRC handbook of thermoelectrics* (CRC Press, Boca Raton, FL), 1995.
7. T. C. Harman, P. J. Taylor, M. P. Walsh, and B. E. LaForge, "Quantum dot superlattice thermoelectric materials and devices," Science **297**, 2229-2232 (2002).
8. A. I. Boukai, Y. Bunimovich, J. Tahir-Kheli, J. K. Yu, W. A. Goddard, and J. R. Heath, "Silicon nanowires as efficient thermoelectric materials," Nature **451**, 168-171 (2008).
9. T. b. published.
10. D. J. Lee, H. M. Kim, J. Y. Kwon, H. Choi, S. H. Kim, and K. B. Kim, "Structural and Electrical Properties of Atomic Layer Deposited Al-Doped ZnO Films," Adv Funct Mater **21**, 448-455 (2011).

11. D. J. Lee, J. Y. Kwon, S. H. Kim, H. M. Kim, and K. B. Kim, "Effect of Al Distribution on Carrier Generation of Atomic Layer Deposited Al-Doped ZnO Films," J Electrochem Soc **158**, D277-D281 (2011).

12. P. Banerjee, W. J. Lee, K. R. Bae, S. B. Lee, and G. W. Rubloff, "Structural, electrical, and optical properties of atomic layer deposition Al-doped ZnO films," J Appl Phys **108** (2010).

13. X. Jiang, F. L. Wong, M. K. Fung, and S. T. Lee, "Aluminum-doped zinc oxide films as transparent conductive electrode for organic light-emitting devices," Appl Phys Lett **83**, 1875-1877 (2003).

14. A. E. Jimenez-Gonzalez, J. A. S. Urueta, and R. Suarez-Parra, "Optical and electrical characteristics of aluminum-doped ZnO thin films prepared by solgel technique," J Cryst Growth **192**, 430-438 (1998).

15. H. Kim, C. M. Gilmore, J. S. Horwitz, A. Pique, H. Murata, G. P. Kushto, R. Schlaf, Z. H. Kafafi, and D. B. Chrisey, "Transparent conducting aluminum-doped zinc oxide thin films for organic light-emitting devices," Appl Phys Lett **76**, 259-261 (2000).

16. W. J. Maeng, J. W. Lee, J. H. Lee, K. B. Chung, and J. S. Park, "Studies on optical, structural and electrical properties of atomic layer deposited Al-doped ZnO thin films with various Al concentrations and deposition temperatures," J Phys D Appl Phys **44** (2011).

17. T. Minami, H. Nanto, and S. Takata, "Highly Conductive and Transparent Aluminum Doped Zinc-Oxide Thin-Films Prepared by Rf Magnetron Sputtering," Japanese Journal of Applied Physics Part 2-Letters **23**, L280-L282 (1984).

18. W. Tang, and D. C. Cameron, "Aluminum-Doped Zinc-Oxide Transparent Conductors Deposited by the Sol-Gel Process," Thin Solid Films **238**, 83-87 (1994).

19. R. Buonsanti, A. Llordes, S. Aloni, B. A. Helms, and D. J. Milliron, "Tunable Infrared Absorption and Visible Transparency of Colloidal Aluminum-Doped Zinc Oxide Nanocrystals," Nano Lett **11**, 4706-4710 (2011).

20. W. M. Kim, D. Y. Ku, I. K. Lee, Y. W. Seo, B. K. Cheong, T. S. Lee, I. H. Kim, and K. S. Lee, "The electromagnetic interference shielding effect of indium-zinc oxide/silver alloy multilayered thin films," Thin Solid Films **473**, 315-320 (2005).

21. Y. Okuhara, T. Kato, H. Matsubara, N. Isu, and M. Takata, "Near-infrared reflection from periodically aluminium-doped zinc oxide thin films," Thin Solid Films **519**, 2280-2286 (2011).

22. H. Raflayuan, and J. F. Cordaro, "Optical Reflectance of Aluminum-Doped Zinc-Oxide Powders," J Appl Phys **69**, 959-964 (1991).

23. H. Serier, M. Gaudon, and M. Menetrier, "Al-doped ZnO powdered materials: Al solubility limit and IR absorption properties," Solid State Sci **11**, 1192-1197 (2009).

24. A. Stadler, "Analyzing UV/Vis/NIR Spectra-Sputtered ZnO:Al Thin-Films III: Plasma-Parameter Dep.," Ieee T Semiconduct M **25**, 19-25 (2012).

25. D. G. Cahill, "Analysis of heat flow in layered structures for time-domain thermoreflectance," Rev Sci Instrum **75**, 5119-5122 (2004).

26. P. Jood, R. J. Mehta, Y. L. Zhang, G. Peleckis, X. L. Wang, R. W. Siegel, T. Borca-Tasciuc, S. X. Dou, and G. Ramanath, "Al-Doped Zinc Oxide Nanocomposites with Enhanced Thermoelectric Properties," Nano Lett **11**, 4337-4342 (2011).

27. E. S. Snow, J. P. Novak, P. M. Campbell, and D. Park, "Random networks of carbon nanotubes as an electronic material," Appl Phys Lett **82**, 2145-2147 (2003).

28. J. Suhr, and N. A. Koratkar, "Energy dissipation in carbon nanotube composites: a review," J Mater Sci **43**, 4370-4382 (2008).

29. S. Jung, K.-B. Kim, G. Fernandes, J. H. Kim, F. Wahab, and J. Xu, "Enhanced thermoelectric power in nanopatterned carbon nanotube film," IOP Nanotechnology **23**, 135704 (2012).

30. C. A. Hewitt, A. B. Kaiser, S. Roth, M. Craps, R. Czerw, and D. L. Carroll, "Varying the concentration of single walled carbon nanotubes in thin film polymer composites, and its effect on thermoelectric power," Appl Phys Lett **98** (2011).

31. J. Hone, I. Ellwood, M. Muno, A. Mizel, M. L. Cohen, A. Zettl, A. G. Rinzler, and R. E. Smalley, "Thermoelectric power of single-walled carbon nanotubes," Phys Rev Lett **80**, 1042-1045 (1998).

32. H. E. Romero, G. U. Sumanasekera, G. D. Mahan, and P. C. Eklund, "Thermoelectric power of single-walled carbon nanotube films," Phys Rev B **65** (2002).

33. Z. J. Han, and K. Ostrikov, "Controlled electronic transport in single-walled carbon nanotube networks: Selecting electron hopping and chemical doping mechanisms," Appl Phys Lett **96** (2010).

DEVELOPMENT OF THERMOELECTRIC DEVICES FOR STRUCTURAL COMPOSITES

Oonnittan Jacob Panachaveettil, Xinghua Si, Zahra Zamanipour, Ranji Vaidyanathan, Daryoosh Vashaee

Helmerich Advanced Technology Research Center
Oklahoma State University
Tulsa, OK 74107

ABSTRACT

In the contemporary world of energy crisis, energy harvesting is very crucial. To this effect, thermoelectrics can prove to be a very efficient form of energy generation from wasted heat. Thermoelectric devices can be incorporated onto structural composites so that waste energy can be harvested during the service of these composites components. Combining the thermoelectric principles and structural composites gives rise to a very novel and multifunctional energy harvesting method compared to the conventional methods. Simplified thick film thermoelectric devices can be manufactured for energy harvesting needs ensuring less complication compared with thin film approach. Sandwich layers of thermoelectric powders displaying high thermoelectric properties with metallic contacts were hot pressed avoiding any post processing metal deposition. This approach of thick film thermoelectric devices can be extended for other thermoelectric materials irrespective of the temperature range of operations. Moreover, tailoring n-type thermoelectric materials by doping BiSeTe with 5%vol SiGe:As led to a shift from negative Seebeck to a positive value and further annealing resulted in a shift back to negative Seebeck coefficient. Another interesting effect observed was the increase in electrical conductivity with corresponding decrease in thermal conductivity, which resulted in a higher figure of merit.

INTRODUCTION

The principle of tapping energy by utilizing the temperature difference between two surfaces is known as thermoelectric energy generation. This phenomenon of producing energy discovered in 1821 after its discoverer is known as Seebeck effect [1]. The opposite effect of producing temperature difference across two surfaces by supplying energy is known as Peltier effect [2]. Thermoelectrics can be used for both cooling and power generation purposes. Waste thermal energy being produced could be efficiently harvested and used for power generation. Peltier effect finds potential application in refrigeration and air conditioning [2]. Effectively installing thermoelectrics devices at strategic positions can help trap waste thermal heat energy produced as a byproduct of many natural or thermodynamic operations. Application of thermoelectric generators can be observed in cars, refrigerators, solar modules, satellites etc. However, in all these operations, thermoelectric generators perform only the single operation of generating electricity.

Thermoelectric generators tethered with structural composites could prove to be multifunctional in nature. Structures employing composites have increased with the advancement in material science. Composites can be replacements to conventional structural materials owing to their high specific strength, light weight, fatigue and environmental resistance. Composites are engineered from of fiber and matrix. Thermoelectric devices if incorporated onto structural composites, act as waste heat energy harvesting source in addition to providing structural integrity. These structures therefore possess multifunctional capabilities. Previous attempts into

235

fabricating these multifunctional composite have been achieved by composite reengineering altering either the matrix phase material or tailoring the interlaminar materials [3]. New or existing structures and panels employing regular composites can be replaced with multi-functional embedded thermoelectric structural composites, satiating both the structural application and energy harvesting need. Developing composites with thermoelectric materials was done by Chung et al.[3]. Processing, production scale-up difficulty, low ZT values, inadequate mechanical performance, and implementation difficulty where some of the problems associated with reengineering the composite for thermoelectric applications [3]. A unique approach to fabricate these multifunctional composites addressing the above problems is by attaching thermoelectric devices onto the surface of these composites.

Most commercially available thermoelectric modules which may be either used for energy generation or cooling purposes are bulk or to a lesser extent thin film devices. Thermoelectric devices find a wide variety of applications like automobile thermoelectric generators, heat engines, space probes and electronic equipments [4]. Bulk devices can be manufactured easily but compromise has to be made in the dimensions of the device. Thinner is the dimension of device smaller is the resistance, leading to an increase in thermoelectric voltage output [5]. Thin films and thick films devices have significantly smaller dimensions varying from a few microns to nano-scale compared to bulk devices and are used primarily in electronic equipments and sensors. Conventionally, thick and thin film devices are fabricated using deposition methods like MOCVD, PVD, CVD, electroplating etc.

The efficiency of thermoelectric devices is determined by the materials dimensionless figure-of-merit, defined by the relation (1),

$$ ZT = \left(\frac{\sigma S^2}{k}\right) T \qquad (1) $$

In which S, σ, k, and T are the Seebeck coefficient, electrical conductivity, thermal conductivity, and absolute temperature, respectively [6,7,8]. From relation (1), the figure of merit is directly proportional to electrical conductivity (σ) and inversely proportional to thermal conductivity (k).

The best figure-of-merit of 2.4 obtained for room temperature application thermoelectric material Bi_2Te_3/Sb_2Te_3 through superlattice structure deposition [9]. Thin films though efficient, their production is very complicated and expensive. Another important step is the inclusion of the metal contact layers connecting the series p-type and n-type at the top and bottom surfaces of the thermoelectric devices making it electrically in series. Although metallization in bulk devices can be achieved relatively easily, metallization in thin film is a challenge, since dealing with micron and sub-micron thickness ranges. Deposition methods of fabricating thin film device involves carrying out the production on a single substrate and carrying out subsequent etching and deposition of both p-type and n-type materials. Industrialization and commercialization of thin film thermoelectric device require enormous infrastructure.

The proposed thermoelectric module after fabrication can be mounted on to composite structures and can be used as energy harvesting structural composites. The structural composites that can be commonly used for mounting can be either low temperature polymer matrix composites or high temperature ceramic composites, since this production method can be extended to even high temperature applications. This paper discusses a unique and cost effective method of fabricating these thermoelectric devices along with metal contacts which can be attached onto composite surfaces. The device is formed by joining the p-type and n-type thermoelectric materials electrically in series and thermally in parallel. Best p-type and n-type material used for room temperature application are BiSbTe and BiSeTe alloy respectively [4,10].

Although many metals and compounds can be used for ohmic contacts, optimization of the material must be done to ensure lowest diffusion into the thermoelectric and metal contacts. Moreover, good adherence and ability to perform at operating temperatures are also criteria to select appropriate material for ohmic contact. Traditionally, nickel has been employed as the barrier material of choice for bulk thermoelectric cooling devices [7]. Nickel diffusion into the thermoelectric element has been observed, but this effect is negligible for both p-type and n-type bismuth telluride alloys, when dealing with bulk materials [11].

Different thermoelectric materials are used for energy generation or cooling purposes in different ranges of operating temperature. Bismuth telluride based alloys like bulk solid solutions including p-type $Bi_xSb_{2-x}Te_3$ and n-type $Bi_2Te_{3-y}Se_y$, still remain the best thermoelectric materials used at near room temperature [12,13]. In spite of recent progress in enhancement of the efficiency of BiSbTe thermoelectric materials, there has been little progress in developing efficient BiSeTe alloys. An interesting and potentially groundbreaking trend of shift in the sign of Seebeck coefficient along with an effect of decrease in thermal conductivity with corresponding increase in electrical conductivity was observed when arsenic doped SiGe was added to BiSeTe alloys. BiSeTe and SiGe:As are materials displaying inherently negative Seebeck coefficient characteristics. However, the annealing process following the fabrication of the thermoelectric sample revealed several interesting results. The characteristics of the sample changed from being an n-type material to a p-type material over annealing time and later returning back to higher value of negative Seebeck coefficient compared to the initial value. This paper also discusses the various trends observed in the thermoelectric properties of $Bi_2Te_{2.7}Se_{0.3}$-$Si_{80}Ge_{20}$:As composite during about 400 hours of annealing.

EXPERIMENTAL PROCEDURE

Fabrication of Nanostructured thick film thermoelectric device

In order to obtain alloyed powders, appropriate amounts of high purity elemental Bi(>99.999%), Te(>99.999%), Sb(>99.999%) and Se(>99.999%) were used. BiSbTe and BiSeTe powders were weighed according to the required stoichiometric ratio, separately mixed and sealed in a stainless steel jar with chromium stainless steel hard balls under an argon environment. In order to synthesize nanostructured alloy powders, mechanical alloying was employed using vibratory impact ball mills.

The p-type and n-type device with ohmic contacts had to be fabricated separately and later joined together with solder materials. Fabricating the device involves sandwiching the thermoelectric powder in between the contact metal powder layers.

To start with, appropriate amount of nickel powder was loaded into the graphite die. After loading each layer of powders into the graphite die in an argon atmosphere, it was cold pressed under approximately 200 MPa. The inner diameter of the graphite die used for this experiment was 2 inches (however could be extended for greater diameters). This was followed by the thermoelectric alloyed powder layer and topped off with nickel powder. Uniform and flat surface was very crucial at the interface of the thermoelectrics and metal contacts to ensure proper and uniform adhesion. Figure 1(a) and 1(b) shows a 2 inch graphite die loaded with flat surfaced nickel and thermoelectric powders. The multilayered powder combination was sintered via hot press method at ~773 K under ~100 MPa. Even though the melting temperature of the nickel powder was very high (1728.05 K) compared to the sintering temperature required to preserve the nanostructuring of the thermoelectric powder, the nickel layers were sintered because of the larger amount of pressure application along with temperature application. Figure 1(c) shows a sintered thermoelectric device after the hot pressing operation. During the fabrication of this device, copper was added as a supplement layer to above the nickel layer.

Figure 1 : (a) 2" graphite die loaded with nickel powder; (b) 2" graphite die loaded with thermoelectric powder on top of nickel powder; (c) thermoelectric device obtained after hot pressing operation(copper added as supplement layer).

The device obtained after sintering process was subjected to computer numerical controlled (CNC) milling operation. A precise CNC machine having micron level accuracy was used to create thermoelectric leg structures in the device. Figure 2, shows a miniature sample sub-device machined initially using the CNC milling machine.

Figure 2: Bulk thermoelectric thick film sub-devise
manufactured   employing innovative CNC dicing method

Employing a similar method, the complementary n-type thermoelectric can be manufactured. Using soldering materials, the p-type and n-type thermoelectrics could be joined together to form the electrically series, thermally parallel thermoelectric module.

Figure 3 shows the concept for the thermoelectric energy harvesting structural composite. These multifunctional composites have both structural composite functioning and energy harvesting functioning. These composites can potentially find applications in the areas where structural integrity combined with waste thermal energy harvesting is needed. Such applications can be found replacing composites of furnace lining, structural panels, etc. These devices can act as both generators and sensors.

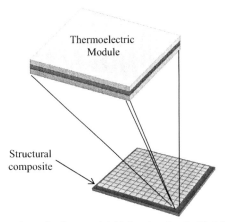

Figure 3 : Concept: Multi-functional embedded thermoelectric
structural composites; thermoelectric module attached onto
composite

BiSeTe doped with SiGe:As

In order to obtain the alloyed powders, appropriate amounts of high purity elemental Bi (>99.99%), Te (>99.99%) and Se (>99.99%) were weighted accordingly to the stoichiometric ratio of $Bi_2Te_{2.7}Se_{0.3}$. We will refer to this composition as BiSeTe in this manuscript. Similarly, high purity elemental Si(>99.9%), Ge(>99.99%) and As(>99.9%) were weighted and powders were alloyed and nanostructured following a similar milling technique to form the stoichiometric ratio of $Si_{0.78}Ge_{0.20}As_{0.2}$. We will refer to this composition as SiGe:As.

To form the final composite alloy mixture 5 vol% SiGe:As was added to BiSeTe and was further milled for about 12 hours to form the nanocrystalline alloyed powder.

The milled powder was loaded into a graphite die with an inner diameter of 1.27 cm and cold pressed at 77 MPa for 5 minutes in a Craver hydraulic press system. Following the cold pressing operation the powders were sintered to obtain the composite bulk alloy. The powders were subjected to a sintering temperature, time and pressure of 773 K, 90 seconds and 77 MPa respectively using a dc operated heating source. A cylindrical sample of diameter of 1.27 cm and 0.5 cm was obtained.

The thermoelectric properties of the sample were measured. The electrical conductivity and the Seebeck coefficient of the sample were measured by a four-point dc current switching technique and static dc method based on the slope of the voltage versus temperature-difference curves using commercial equipment (ZEM3, ULVAC), respectively. The thermal diffusivity was measured by the laser-flash method with a commercial system (LFA-457, Netzsch Instruments, Inc.). The mass density of the sample was measured by the Archimedes method.

After the initial measurement of the thermoelectric properties of the sample, the rectangular bar and the disk samples used for electrical and thermal conductivity measurements were annealed in a muffled furnace in ambient air. The annealing conditions performed during this experiment were in chronological order: 623 K for 41, 72 hours at 623 K, 68 hours at 653 K, 120 hours at 623 K and finally 96 hours at 673K.

RESULTS AND DISCUSSION

Bulk nanostrutued thick film device

A bulk nanostructured thick film thermoelectric device of 0.5mm thickness and nickel ohmic contact layer thickness of about 0.2 mm thickness was sintered and a disc shaped device was obtained. This was later machined using a precision CNC milling machine to form the thermoelectric sub-device. Figure 4(a) shows the top SEM image of such a device fabricated using this procedure. The schematic of a single thermoelectric leg with ohmic contact, along with dimensions milled during the operation is graphically represented in figure 4(b). The dimensions of the thermoelectric legs can be further reduced after optimization of the milling process.

Sub-devices after milling operation, are mounted onto ceramic plates for foundation. Ceramics, being thermal and electrical insulators, serve the purpose of maintaining the maximum temperature difference between cold and hot surface across the device. The sub-devices after fabrication can be joined together using appropriate soldering materials. Figure 4(c) shows the depicted schematic of proposed thermoelectric module.

This method of fabrication of thermoelectric device overcomes the complicated steps of thin film device fabrication and ohmic contact metal deposition employing MEMS fabrication techniques. Owing to its low cost and ease of manufacturing, these devices can be mass produced. Thermoelectric devices when attached onto composite surfaces overcome the problems previously encountered when fabricating these energy harvesting composites.

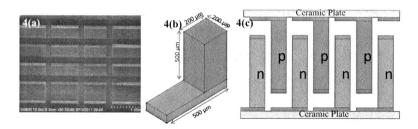

Figure 4: (a) Top SEM image of CNC milled thermoelectric sub-device; (b) Schematic representation of thermoelectric leg with dimensional features; (c) Cross section of the proposed thermoelectric device.

Fig 5 shows the scanning electron micrograph of the interface region of the thermoelectric material and the metal contact. For this device, we used higher manganese silicide (HMS) for the thermoelectric layer. The interface is dense with no micro-cracks on this scale. Fig 6(a) shows the X-ray diffraction data taken for the HMS/Ni/HMS configuration, in which Ni is the metal contact. HMS and Ni layers are marked with I and II in the diagram, respectively. The presence of elemental Ni and MnNiSi in the HMS layer indicates the diffusion of Ni into the HMS layer. It is also observed that HMS has diffused into the Ni layer.

In order to prevent the diffusion of Ni into HMS region, a thin Cr layer was applied between Ni and HMS layers. For this purpose, an HMS/Cr/Ni/Cr/HMS structure was prepared and examined with XRD. Fig 6(b) shows the results. Regions I, II, and III correspond to the HMS, Cr/Ni/Cr, and HMS layers, respectively. The XRD spectrum labeled PII was taken at the mid-point of the Ni layer and shows that a Ni-Cr alloy had been formed. It is interesting to notice no diffusion of Ni or Cr into the HMS region. Again, small diffusion of HMS into the Ni was observed similar to the Ni/HMS contact. However, this is not of concern since the electrical conductivity of Ni, even with HMS diffusion, is much higher than the thermoelectric layer and would not affect the performance of the device. It is concluded that Cr can act as an effective deterrent to diffusion of Ni into HMS in an HMS/Ni system.

Figure 5: The Scanning electron microscope (SEM) image of the TE/metal contact interface.

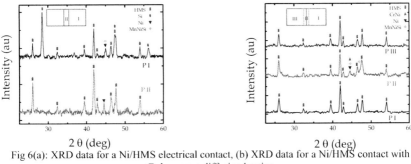

Fig 6(a): XRD data for a Ni/HMS electrical contact, (b) XRD data for a Ni/HMS contact with Cr layer as a diffusion barrier.

Anomalous thermoelectric behaviour of BiSeTe doped with SiGe:As

When 5% vol SiGe doped with As was added to BiSeTe interesting trends and effects were observed upon annealing.

The as pressed sample exhibits negative Seebeck characteristics. Figure 7 and figure 8 show the variation of the electrical conductivity and Seebeck coefficient as a function of temperature over the different annealing cycles. Electrical conductivity has increased by longer annealing time. Following 41 hours of annealing, the material changes from an n-type material to a p-type material as evidenced by the sign change of the Seebeck coefficient. The trend further proceeds to become stronger with the material displaying maximum magnitude of positive Seebeck when annealed for additional 72 hours. Upon further annealing for a total of 397 hours, the material shifts back to being an n-type material. The magnitude of the n-type material recorded after 400 hours of annealing showed a greater magnitude in negative Seebeck compared to the magnitude of the Seebeck coefficient of the as pressed sample. A large enhancement in power factor is observed at this time.

In conclusion, as it is seen in Figure.9, there is an enhancement in power factor compared to as pressed sample and also annealed samples with different annealing time. The maximum power factor at room temperature is about 0.48 at room temperature which belongs to sample annealed 113 hours.

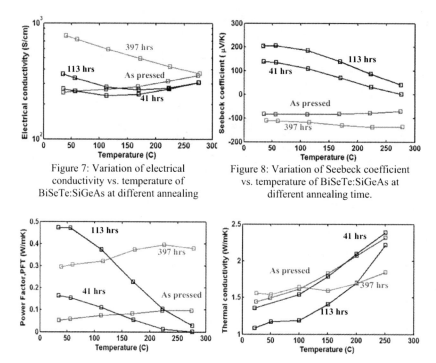

Figure 7: Variation of electrical conductivity vs. temperature of BiSeTe:SiGeAs at different annealing

Figure 8: Variation of Seebeck coefficient vs. temperature of BiSeTe:SiGeAs at different annealing time.

Figure 9: Variation of power factor vs. temperature of BiSeTe:SiGeAs at different annealing time.

Figure 10: Variation of thermal conductivity vs. temperature of BiSeTe:SiGeAs at different annealing time.

An interesting effect observed after 113 hours of annealing of the BiSeTe: SiGeAs sample was the increase in the electrical conductivity accompanied by a reduction in thermal conductivity after 373 K. Electrical conductivity at 113 hours of annealing is greater than that of 41 hours of annealing from Figure 7, but the corresponding thermal conductivity, contrary to increasing, displays the lowest thermal conductivity as shown in Figure 10. This effect attributes to the highest figure of merit displayed by sample after annealing for 113 hours as shown in Figure 11. Another interesting result is the figure of merit of the sample after 397 hours annealing time. The figure of merit of this sample has only small variation over the entire temperature range of room temperature to 523 K.

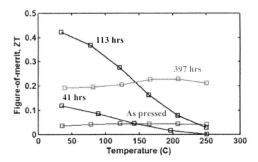

Figure 11: Variation of figure-of-merit vs.
temperature of BiSeTe:SiGeAs at different

CONCLUSION

A novel and unique method of fabrication of thermoelectric device was proposed and partially achieved. Owing to the relative ease and cost effectiveness of manufacturing thick film thermoelectric devices (in addition to the capability of harnessing the maximum waste thermal energy capitalizing on the thermoelectric effect), energy harvesting structural composites will be crucial in energy generation and find applications in different fields.

It is imperative to tailor thermoelectric materials so as to increase their figure of merit thus increasing the output energy of these materials. The trend of a negative Seebeck material displaying positive Seebeck coefficient and further shift from a positive Seebeck back to a greater magnitude of negative Seebeck compared to the original measurement was observed during the annealing of BiSeTe doped with 5% vol SiGe:As. Moreover, an interesting behavior was observed that after about 113 hours of annealing, there was a drop in thermal conductivity with a corresponding increase in electrical conductivity. This led to an enhancement in the figure-of-merit and the sample displayed the highest ZT after 113 hours of annealing during the entire experiment. This composite structure is unconventional and our result may prove to be a stepping stone to enhancing thermoelectric properties of Bi_2Te_3 based solid solution alloys.

ACKNOWLEDGMENT

This report is partially based upon work supported by Air Force Office of Scientific Research (AFOSR) High Temperature Materials program under grant no. FA9550-10-1-0010 and the National Science Foundation (NSF) under grant no. 0933763.

REFERENCES

[1] T. J. Seebeck, "Magnetische Polarisation der Metalle und Erze durch Temperatur-Differenz," *Abh. Akad. Wiss. Berlin,* vol. 1820–21, pp. 289–346, 1822.
[2] H. J. Goldsmid, *Introduction to Thermoelectricity*: Springer, 2009.
[3] D. D. L. Chung and S. Han, "multifunctional carbon fiber epoxy-matrix composites for energy harvesting."
[4] P. P. H. J. Goldsrtlid, New York,). *Thermoelectric Refrigeration*: Plenum Press, 1964.

[5] J. Laird, "SunShot takes aim at PV costs: Part two: With its SunShot program, the US Department of Energy (DoE) is trying to lead the world in research dedicated to slashing PV costs," *Renewable Energy Focus,* vol. 12, pp. 44-46, 48-49.

[6] H.J. Goldsmid, *Thermoeletric refrigeration:* Plenum Press, 1964

[7] D. M. Rowe, *CRC handbook of thermoelectrics*: CRC Press, 1995.

[8] T. Tritt, Recent Trends in Thermoelectric Materials Research: Part Three: Elsevier Science, 2000.

[9] R.Venkatasubramanian,E. Siivola, T Colpitts, B O'Quinn, "Thin-film thermoelectric devices with high room-temperature figures of merit," *Nature,* vol. 413, pp. 597-602, 2001.

[10] G. S. Nolas, J. Sharp, J. Goldsmid, *Thermoelectric: basic principles and new materials developments*: Springer, 2001.

[11] I. V. Gasenkova and T. E. Svechnikova, "Structural and Transport Properties of Sn-Doped Bi2Te3 – xSex Single Crystals," *Inorganic Materials,* vol. 40, pp. 570-575, 2004.

[12] B. Poudel, Q. Hao, Y. Ma, Y. Lan, A. Minnich, B. Yu, X. Yan, D. Wang, A. Muto, D. Vashaee, X. Chen, J. Liu, M. S. Dresselhaus, G. Chen, Z. Ren, "High-Thermoelectric Performance of Nanostructured Bismuth Antimony Telluride Bulk Alloys," *Science,* vol. 320, pp. 634-638, May 2, 2008 2008.

[13] X. Yan, B. Poudel, Y. Ma, W. S. Liu, G. Joshi, H. Wang, Y. Lan, D. Wang, G. Chen, Z. F. Ren, "Experimental Studies on Anisotropic Thermoelectric Properties and Structures of n-Type Bi2Te2.7Se0.3," *Nano Letters,* vol. 10, pp. 3373-3378, 2010.

EFFECT OF SN CONCENTRATION ON PHYSICAL PROPERTIES OF ZNO THIN FILMS
GROWN BY SPRAY PYROLYSIS ON SNO$_2$:F/GLASS

Mejda Ajili[1], Michel Castagné[2] and Najoua Kamoun Turki[1]

[1] Laboratoire de Physique de la Matière condensée, Faculté des Sciences de Tunis, Tunis El Manar (2092) Tunisie
[2] Institut Electronique du Sud, Université de Montpellier II, Place Eugène BATAILLON 34 095 Montpellier cedex 05 France.

ABSTRACT

Transparent conducting thin films of tin-doped zinc oxide (TZO) have been deposited on SnO$_2$:F/Glass by the chemical spray technique, starting from zinc acetate (CH$_3$CO$_2$)$_2$Zn$_2$H$_2$O and tin chloride SnCl$_2$. The effect of changing the tin-to-zinc ratio from 0 to 1 at.% on the structural, optical and electrical properties of ZnO:Sn thin layers, has been thoroughly investigated. It was found that X-Ray diffraction results reveal that TZO films have an hexagonale Wurtzite structure with (002) preferred orientation. The maximum transmittance is around 65% for the 0.2 at.% Sn doped thin films in the visible range. The band gap energy, calculated for all samples is around 3.25 eV. A minimum resistivity ρ of 0.10 x 10^{-2} Ω.cm is obtained for the undoped thin layers and it becomes 2.2 x 10^{-2} Ω.cm for 0.2 at.% TZO ones. The lowest surface roughness and highest grain size values are also obtained for 0.2 at.% TZO thin films.

1. INTRODUCTION

Zinc oxide (ZnO) is one of the most interesting transparent and conducting oxide (TCO) due to its electro-optical properties, high electrochemical stability, large band gap, abundance in nature and absence of toxicity. These advantages are of considerable interest for practical applications such as, gas sensors [1,2], piezoelectric devices [3], surface acoustic wave devices [4], solar cells [5-8],etc. Recently, ZnO thin films have been used as a window layer and contact layer for thin film solar cells with Cu(In,Ga)S$_2$ or Cu(In,Ga)Se$_2$ absorber material. ZnO films have been prepared by various methods of film depositions, which include radio frequency magnetron sputtering [9-11], sol–gel processing [12-16], spray pyrolysis [17,22], etc. Among these methods, the spray pyrolysis method is one of the most commonly used methods for preparation of transparent and conducting oxides owing to its simplicity, safety, non-vacuum system of deposition. Other advantage of the spray pyrolysis method is that it can be adapted easily for production of large-area films. In fact, to enhance its electrical and optical properties, ZnO is commonly doped with one of the following elements: Indium [20], Fluorine [9], Aluminum [11-13,16-19], Gallium [14], Cobalt [15] and Tin [21]. In particular, the main reason of being chosen of tin as the dopant in ZnO is to enhance the electrical conductivity. When ZnO is doped Sn, Sn^{4+} substitutes Zn^{2+} site in the ZnO crystal structure resulting in two more free electrons to contribute to the electric conduction. Furthermore, Zn can be easily substituted by Sn and does not result in a large lattice distortion due to their almost equal radii ($r_{Zn2+} = 0.074$ nm and $r_{Sn4+} = 0.069$ nm). Sn substituting Zn is therefore a good candidate as an n-type dopant in ZnO. It is well known that the physical properties of thin films depend on the substrate properties. Up to now, ZnO thin films were prepared on silicon, Al$_2$O$_3$, kapton, ZnO:Al, glass and sapphire substrates [9-22].

In this paper, we present the effect of tin dopant concentration ratio in the spray solution (y=($[Sn^{4+}]/[Zn^{2+}])_{solution}$= 0, 0.2%, 0.6% and 1%) on the structural, electrical and optical properties of ZnO thin layers deposited by chemical spray on nanostructured substrates such as fluorine doped tin oxide SnO_2:F (FTO). This SnO_2:F thin layer has a thickness of about 0,89 µm and has also been deposited by spray pyrolysis on glass substrates. Our attempt will be to use TZO thin films as optical window layer in the solar cells Au/CuInS$_2$ (p)/ZnO (n)/SnO$_2$:F (n)/glass, where Au and SnO$_2$:F will be used as an ohmic contact and CuInS$_2$ as an absorber material.

2. EXPERIMENTAL DETAILS

Spray pyrolytic system was used to obtain FTO and TZO thin films. The experimental set up has been previously described [7,8]. The thin films of FTO and TZO were sprayed onto the substrate glass kept at a temperature equal to 440°C with a spray rate of 25 ml.min^{-1} and the distance between the substrate and the nozzle is about 27 cm. For FTO and TZO, the time of spray is respectively equal to 4 and 40 min. The solution used for pulverization of FTO material consisted of tin tetrachloride (SnCl$_4$), methanol (CH$_3$O) and double distilled water (H$_2$O) [7,8]. Fluorine doping was achieved by adding fluoride NH$_4$F. The spray solution of TZO compound consisted of zinc acetate, H$_2$O and propanol-2 (C$_3$H$_8$O) [7,8]. Tin doping was achieved by adding tin chloride (SnCl$_2$).

The structure of TZO thin layers was studied by X-Ray Diffraction (XRD) which are recorded with an automated Bruker D8 advance X-Ray diffractometer with CuKα radiations for 2θ values over 20-60°. The wavelength, accelerating voltage and current were, respectively, 1.5418 Å, 40 kV and 20 mA. The morphology TZO thin films was studied using atomic force microscopy (AFM) and scanning electron microscopy (SEM). The volume composition of TZO thin films was studied by energy dispersive spectrometry (EDS) coupled with (SEM). Resistivity, surface and volume carrier concentration and mobility were determined from Hall effect measurements in the Van der Paw-configuration. The optical properties were studied according to UV-VIS-NIR spectrum with a Perkin–Elmer Lambda 950 spectrophotometer in the wavelength range of 250-2500 nm at room temperature, taking the air as reference. The average grain size D is estimated by the scherrer's formula [23]:

$$D = \frac{0.9 \times \lambda}{B \cos \theta} \qquad (1)$$

Where λ is the X-ray wavelength of CuKα radiation (λ= 1.54 Å), B is the full-width at half-maximum (FWHM) of the diffraction in radians and θ is the Bragg diffraction angle.

3. RESULTS AND DISCUSSION

3.1 Structural analysis

3.1.1 X-Ray Diffraction

In order to study the effect of tin concentration on the microstructural properties of TZO thin layers deposited on SnO$_2$:F/Glass, different values of dopant concentration were used ((y=[Sn^{4+}]/[Zn^{2+}])$_{solution}$ =0, 0.2%, 0.6% and 1%, y is the concentration ratio in the spray solution). The XRD patterns of four films for different doping concentrations are shown in figure 1. The spectra show a polycrystalline character which characterized with (100), (002), (101) and (102) principal orientations, indicating an hexagonal wurtzide structure. The reflexion peaks are very narrow showing a good crystallinity. The all TZO thin films had the highest (002) diffraction peak intensity indicating that (002) is the preferential orientation. The same result was obtained by M. R. Vaezi and al. [21]. Moreover, this preferential peak intensity increased with increasing the tin doping concentration until 0.2%. But if y exceeds this value, this intensity decreased. This reveals that an increase in tin dopant concentration more than 0.2% deteriorates the crystallinity of films. This results may be due to the formation of stresses by the difference in ion size between zinc and the tin dopant [13].

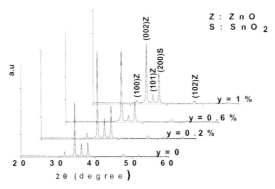

Fig. 1. X-Ray Diffraction of TZO thin layers grown on SnO_2:F/glass substrate at different tin doping concentrations ($y= [Sn^{4+}]/[Zn^{2+}]$). Peaks corresponding to FTO and TZO compound are respectively indexed by Z and S.

In addition, the peak position of the (002) plane is shifted to the high 2θ value when we increase the tin doping concentration beyond y= 0.2 % as shown in the figure.2. This result may be explained that the ionic radius of Sn^{4+} (0.069 nm) are smaller than Zn^{2+} (0.074 nm), so the subtitutional replacement of Sn leads to an increase in the (002) diffraction angle.

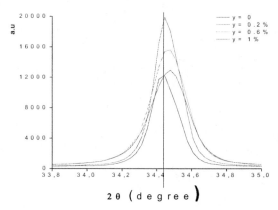

Fig. 2. Variation of the 2θ position of the (002) preferential peak of TZO/ SnO$_2$:F/Glass thin films elaborated at different concentrations of tin dopant in the spray solution (y= [Sn^{4+}]/[Zn^{2+}]).

The introduction of tin in the spray solution affected the growth of TZO thin films because the thicknesses "e" of these thin layers varied when we change the amount of tin introduced into the spray solution as shown in Figure. 3: y= 0, e= 2.53 μm; y= 0.2%, e= 3.14 μm; y= 0.6%, e= 3.03 μm and y= 1%, e= 2.74 μm. When the doping is equal to 0.2 at.%, 0.6 at.% and 1 at.%, the (002) peak intensity increases respectively by a factor of 1.62, 1.29 and 1.05 with the respect to the undoped layer. The film thickness increases also respectively by a factor of 2.05, 1.98 and 1.79. So, this disparity between the increase in the peak intensity and the increase in the film thickness, clearly indicates that the changes observed in the diffractograms are indeed the results of structural changes caused by the tin doping and not only by the variation in the film thickness. Similar results are obtained by J. L. Van Heerden and al [25].

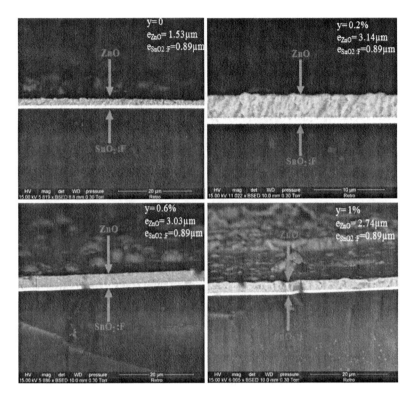

Fig. 3. Cross-sectional SEM micrograph of TZO/SnO$_2$:F thin films grown at different concentrations of tin dopant in the spray solution (y= [Sn^{4+}]/[Zn^{2+}]).

3.1.2 Surface morphology

AFM images of TZO thin films deposited on SnO$_2$:F/glass are illustrated in Figure.4. All films surfaces show a rough texture, and numerous porous surrounding the grains can be evidenced. However, 0.2 at.% tin doped ZnO thin films show dense and less rough structure.

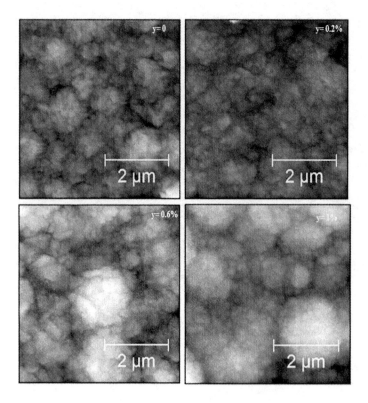

Fig. 4. Surface topography 2D of ZnO:Sn thin layers deposited on SnO_2:F/Glass for different concentrations of tin dopant in the spray solution (y= $[Sn^{4+}]/[Zn^{2+}]$).

Table 1. present the variance of height distribution (RMS roughness) calculated from AFM images and the mean grain size (D) calculated from Scherrer's formula. The maximum grain size and minimum RMS were obtained for y= 0.2%. This result is in agreement with the XRD analysis which show that the best cristallinity is obtained for 0.2 at. % TZO thin films.

Table1. Dependence of RMS and the grain size D on Sn dopant concentration in the spray solution (y= $[Sn^{4+}]/[Zn^{2+}]$) for ZnO:Sn/SnO$_2$:F/Glass.

y (%)	D (nm)	RMS (nm)
0	41.6	99.7
0.2	44.8	49.9
0.6	34.26	110.4
1	32.35	93.8

3.2 Optical properties

In order to study the effect of tin concentration on the optical properties, we plotted in Figure 5, the optical transmission and reflection of TZO thin films grown for different values of y. One can observe that the optical transmittance spectra of ZnO:Sn films represent a strong dependence with the tin doping concentration. The average transmittance in the visible region [400-800] nm is about 65% when the concentration of Sn dopant is 0.2 %. This value indicates a good quality of TZO material with high transparency of layers, predisposing them to be used as an optical window in the visible range in photovoltaic systems. Beyond y=0.2%, the transmittance decreases remarkably. The undulating shape of the reflection curves is caused by interference of the light in the film itself. This result is similar for tin doped zinc oxide elaborated by a two-stage chemical deposition (TSCD) [21]. Besides we note the presence of the small interference fringes characteristics of fairly uniform in thickness and quite homogenous layers in accordance with the morphological study. Similar results were previously reported [20,22].

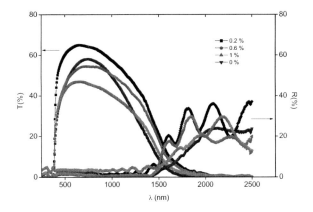

Fig. 5. Transmission and reflection spectra of TZO thin films deposited on SnO$_2$:F/Glass substrates, at different concentration ratio of tin dopant (y= $[Sn^{4+}]/[Zn^{2+}]$).

As a direct band gap semiconductor, the band of ZnO film is an important parameter. According the following formula, the band gap depend on the optical absorption coefficient and the photon energy $h\upsilon$ [25] :

$$(\alpha h\upsilon) = A \, (h\upsilon - E_g)^n \qquad (2)$$

Where A is a constant, h is planck's constant, υ is the photon frequency, E_g is the optical band gap, n is ½ for direct band gap and α is the absorption coefficient, which can be calculated from the film thickness e, transmittance T and reflectance R measurements by using the formula [26] :

$$\alpha = -\frac{1}{e} Ln[\frac{T}{(1-R)^2}] \qquad (3)$$

Figure 6 shows variations of $(\alpha h\upsilon)^2$ vs. photon energy $h\upsilon$ for ZnO :Sn (y=0.2%). The linear part of the curves for high photon energies indicates that the ZnO :Sn thin films are essentially semiconductors with direct-transtion-type. Extrapolation of the linear portions of $(\alpha h\upsilon)^2$ to zero gave the value of E_g. The estimated E_g values are in the order of 3.25 eV for all TZO thin layers. Thus, it can be concluded that Sn as a dopant of ZnO/SnO$_2$:F/Glass thin films does not contribute any significant changes to the optical band gap of TZO thin layers.

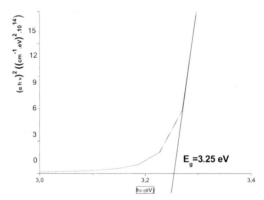

Fig. 6. $(\alpha h\upsilon)^2$ versus $h\upsilon$ plots of 0.2 at. % TZO/SnO$_2$:F/Glass thin films (y= [Sn^{4+}]/[Zn^{2+}]) (a): y= 0, (b): y= 0.2%, (c): y= 0.6% and (d): y= 1%.

3. 3 Electrical properties

It is well admitted that conduction properties of ZnO are primarily dominated by electrons generated from O^{2-} vacancies and Zn^{2+} intertitial atoms. ZnO exhibits a wide range of conductivity; its behavior varies from metallic to insulating. Its electrical characteristics can be controlled by doping with ternary elements or by adjusting process conditions. All our Hall measurement results are listed in table 2. It is shown that an undoped film exhibits a resistivity of 0.10×10^{-2} Ω.cm and the resistivities of 0.2, 0.6 and

1 at.% Sn-doped ZnO thin films were 2.16×10^{-2}, 3.5×10^{-2} and 14.22×10^{-2} $\Omega.cm$, i.e. when the Sn dopant introduced in the spray solution, the resistivity of film greatly changed. The resistivity of doped films increased with increasing dopant concentration, which may be due to a decrease in carrier concentration caused by carrier traps at the grain boundaries [27]. So, it is apparent from table 2, that the surface and volume charge carrier concentration (N_s and N_v) of the films show a gradual decrease with the indroduce of tin dopant in the spray soution. In contrast, The highest value of carrier mobility ($\mu = 9.50$ cm^{-2} .V^{-1} .s^{-1}) is obtained for y= 0.2 %. This result can be correlated by the XRD analysis which showed that the best cristallinity is obtained for 0.2 at. % TZO thin layers.

Table 2. Hall data of TZO thin layers deposited on SnO_2:F grown at different tin doping concentrations (y= $[Sn^{4+}]/[Zn^{2+}]$= 0, 0.2%, 0.6% and 1%).

y (%)	ρ (10^{-2} .Ω.cm^{-1})	N_s (10^{15} .cm^{-2})	N_v (10^{19} .cm^{-3})	μ (cm^{-2} .V^{-1} .s^{-1})
y= 0	0.10	100	6.54	8.79
y= 0.2	2.16	9.22	5.04	**9.50**
y= 0.6	3.50	5.21	3.03	4.22
y= 1	14.22	3.16	1.88	1.03

4. CONCLUSION

In summary, TZO/SnO_2:F/Glass thin layers have been deposited by chemical spray pyrolysis. The obtained films were polycrystalline with the hexagonal structure and had a (002) preferred orientation. The best cristallinity for TZO thin films is obtained for y concentration ratio equal to 0.2% (y= $[Sn^{4+}]/[Zn^{2+}]$). The introduction of tin in the spray solution affected the growth of TZO/SnO_2:F/Glass thin films because the thickness of these thin layers varied when we change the amount of tin introduced into the spray solution. On the other hand, The highest value of carrier mobility ($\mu = 9.5$ cm^{-2}.V^{-1}.s^{-1}) is obtained for y= 0.2 % and band gap (E_g= 3.25 eV) were also achieved for all TZO/SnO_2:F/Glass and the maximum transmittance is obtained for y= 0.2%. This will allow us to use TZO thin layers grown for y equal to 0.2 at. % TZO thin films as optical window in photovoltaic cell such as SnO_2:F/TZO/$CuInS_2$/Au where the growth by spray of $CuInS_2$ thin films is well controlled in our laboratory [7,8,28,29].

REFERENCES

1. M.-W. Ahn, K.-S. Park, J. Heo, D.-W. Kim, K.J.Choi, J,-G. Park, Sensors and Actuators B **138** (2009) 168-173.
2. Kuwei Liu, Makoto Sakurai, Masakazu Aono, Sensors and Actuators B **157** (2011) 98-102.
3. S.H. Jeong, B.N. Park, S.-B. Lee, J,-H. Boo, Thin Solid Films, **516** (2008) 5586-5589.
4. Tsung-Tsong Wu and Wei-Shan Wang, Journal of Applied Physics, 96 number **9** (2004) 5249-5253.
5. B. Asengo, A.M. Chaparro, M.T. Gutierrez, J. Herrero, J. Klaer, Solar Energy Materials and solar cells **87** (2005) 647-656.
6. M. Krunks, A.Katerski, T. Dedova, I. Oja Acik, A; Mere, Solar Energy Materials and Solar Cells **92** (2008) 1016-1019.
7. N. Jebbari, B. Ouertani, M. Ramonda, C. Guasch, N. Kamoun Turki and R. Bennaceur, Energy Procedia **2** (2010) 79-89.

8. N. Kamoun Allouche, T. Ben Nasr, N. Kamoun Turki and M. Castagne, Energy Procedia **2** (2010) 91-101.
9. Yu-Zen Tsai, Na-Fu Wang, Chun-Lung Tsai, Thin Solid films **518** (2010) 4955-4959.
10. Petronela Prepelita, R. Medianu, Beatrice Sbarcea, F. Garoi, Mihaela Filipescu, Applied surface science **256** (2010) 1807-1811.
11. J.J Ding, H.X. Chen, S. Y. Ma, Applied surface science, **256** (2010) 4304-4309.
12. S. W. Xue, X. T. Zu, W.G. Zheng, M. Y. Chen, X. Xiang, Physica B **382** (2006) 201-204.
13. Xu Zi-qiang, Deng Hong, Li Yang, Cheng Hang,Materials Science in Semiconductor Processing **9** (2006) 132-135.
14. K. Y. Cheong, Norani Muti, S. Ramanan, Thin Solid Films **410** (2002) 142-146.
15. Nupur Bahadur, A.K. Srivastava, Sushil Kumar, M. Deepa, Bhavya Nag, Thin Solid Films **518** (2010) 5257-5264.
16. Young-Sung Kim, Weon-Pil Tai, Applied Surface Science **253** (2007) 4911-4916.
17. H. Mondragon -Suarez, A. Maldonado, M. De la L. Olvera, A. Reyes, R. Castanedo-Pérez, G. Torres-Delgado, R. Asomozo, Applied Surface Science **193** (2002) 52-59.
18. M. de la L. Olveraa, A. Maldonadoa, J. Vega-Péreza, O. Solorza-Feriab, Materials Science and Engineering B **174** (2010) 42–45.
19. L. Castaneda, R. Silva-Gonzalez, J. M. Gracia-Jiménez, M. E. Hernandez-Torres, M. Avendano-Alejo, César Marquez-Beltran, M. De la L. Vega-Pérez, A. Maldonado, Materials Science in Semiconductor Processing **13** (2010) 80-85.
20. M. Amlouk, S. Belgacem, N. Kamoun, H. Elhouichet, R. Bennaceur, Semi-Conductors and thin films, Ann. Chim. Fr., 1994, **19**, pp. 469-472.
21. M. R. Vaezi, S. K. Sadnezhaad, Materaials Sciences and Engineering B **141** (2007).
22. M. Mekhnache, A. Drici, L. Saad Hamideche, H. Benzarouk, A. Amara, L. Cattin, J. C. Bernède, M. Guerioune, Superlattices and Microstructures **49** (2011) 510-518.
23. Buerger MJ (1960) X- ray crystallography. Wiley, New York, p 23.
24. J.L. van Heerden, R. Swanepoel, Thin Solid Films **299** (1997) 72-77.
25. M. Girtan, G. Folcher, Surf. Coat. Technol. **172** (2003) 242.
26. S. Belgacem et R. Bennaceur, Revue Phys. Appl. **25** (1990) 1245-1258.
27. Chien-Yie Tsay, Hua-Chi Cheng, Yen-Ting Tung, Wei-Hsing Tuan, Chung-Kwei Lin, Thin Solid Films **517** (2008) 1032–1036.
28. N. Kamoun, R. Bennaceur et J. M. Frigerio, J. Phys. III France **4** (1994) 983-996.
29. N. Kamoun, N. Jebbari, S. Belgacem, and R. Bennaceur, Journal of Applied Physics **91** (2002) 1952-1955.

EFFECT OF HEAT TREATMENT ON THE PHYSICAL PROPERTIES OF In$_2$S$_3$ PREPARED BY CHEMICAL BATH DEPOSITION

Mouna Kilani[a], Cathy Gguasch[b], Michel Castagné [b] and Najoua Kamoun-Turki[a]

(a) Laboratoire de Physique de la Matière Condensée, Faculté des Sciences de Tunis El Manar, Tunisie (2092), Tunisia.
(b) Institut d'Electronique du Sud, Unité Mixte de Recherche 5214 UM2-CNRS - Université MontpellierII, Place Eugène Bataillon cc083 F-34095 Montpellier Cedex 05 France.
* Corresponding author: mouna_kilani@yahoo.fr

ABSTRACT
    Indium sulfide thin films are successfully synthesized on Pyrex by chemical bath deposition and annealed in nitrogen for one hour at temperature (T_a) in the range 200-400°C. The properties of the films are investigated by X-ray diffraction (XRD), scanning electron microscopy and optical spectrophotometer. XRD analysis reveals that for the as-prepared film only the cubic In$_2$S$_3$ phase is observed, while from T_a= 200 ° C, we observe the appearance of the tetragonal phase. The surface morphology of the annealed films at 400°C showed uniform, densely- packed and continuous. The optical band gap value of the In$_2$S$_3$ practically does not change (Eg ~2.3Ev). These results are discussed to introduce the effect of the annealing temperatures in the physical properties of In$_2$S$_3$ material.

1. INTRODUCTION
  Over the last decade metal chalcogenides have received much attention in the field of materials science, thanks to their vast potential for use in thin films solar photovoltaic and other optoelectronic devices like holographic, recording system and optical imaging [1-3]. Among them, In$_2$S$_3$ thin films appear to be promising candidates owing to its stability, band gap energy (2-3eV), transparency and photoconductor behavior. Indium sulfide (In2S3) buffer layer have been used as substitute for cadmium sulfide (CdS) in Cu(InGa)(SSe)$_2$ (CIGS)thin film solar cells to avoid toxic cadmium and obtain more environment friendly photovoltaic technology[4]. Various methods such as spray pyrolysis [5], ultrasonic dispersion [6], chemical bath deposition (CBD) [7-8], physical vapor deposition [9], etc. have been used to prepare In$_2$S$_3$. Among these methods, chemical bath deposition (CBD) is selected in the present study because of its simplicity, inexpensive equipments and low deposition temperatures, which together results in low cost processes and large area of scientific and industrial applications.
  In$_2$S$_3$ is an n type semiconductor exhibiting low conductivity [10] and exists in three crystallographic phases α, β and γ. Among these structures, ß-In$_2$S$_3$ which is the most stable phase at room temperature crystallizes in a normal spinel structure with a high degree of tetrahedral and octahedral vacancy sites [11-12].
  Therefore studies concerning effect of heat treatment on physical properties play significant role in enhancing the device efficiency in the solar cell. In the present work, we have deposited polycrystalline In$_2$S$_3$ films by chemical bath deposition. The films are annealed in nitrogen at 200, 300 and 400°C. Detailed analysis is done with an aim to see whether the nitrogen-annealed samples become more suitable for photovoltaic application. The effect of nitrogen annealing on In$_2$S$_3$ prepared by chemical bath deposition has not yet been studied, to the best of our knowledge.

2. EXPERIMENTAL DETAILS

In_2S_3 thin films are deposited on Pyrex via the CBD process by means of an aqueous solution containing indium trichloride, $InCl_3$ (0.025 M) and thioacetamide, TA (0.10 M) as indium and sulphur precursors respectively [13-15]. Then the films are annealed at different temperatures 200, 300 and 400°C for 1h in nitrogen. First, the film cristallography is studied by X-ray diffraction (XRD) using an automated Bruker D8 advanced X-ray diffractometer with CuKα radiation for 2θ values ranging from 20 to 70°. Experimental results are treated using the program X-Pert High Score. Second, all thin layers were investigated in cross section by scanning electronic microscopy (SEM, XL 30 (S.E)) in order to study the morphology of films and by associated EDS (energy dispersive spectra) for film composition. The optical transmittance was obtained using Perkin Elmer lambda 950 spectrophotometer in the wavelength range 250-2500 nm.

3. RESULTS AND DISCUSSION

3.1. Structural properties

Figure1 depicts the XRD patterns of the indium sulfide thin film CBD prepared and annealed in nitrogen at different heating temperatures. For as deposited film, only two peaks are observed, at approximately 33.6° and 48° assigned respectively to (400) and (440) reticular planes which characterize the cubic phase of the β-In_2S_3 compound (JCPDS#32-0456). The as-grown layer showed cubic β-In_2S_3 structure with the (400) peak as the preferred orientation. The annealing of the films at 200, 300 and 400°C produces the appearance of a new peak at 27.4 assigned to (109) tetragonal phases (JCPDS#73-1366). With increase of T_a to 400°C, the intensity of the (109) peak is found to increase. The best crystallinity has been selected for the film annealed at 400°C, the preferred orientation (109) becomes more intense and narrow indicating an improvement of the crystallinity of the layers.

The average crystallite size 'D' is calculated using the Scherrer formula [16]

$$D = \frac{0.9 \times \lambda}{\beta \cos(\theta)},$$

where D is the diameter of the crystallites corresponding to the preferred orientation at $2\theta = 48°$, λ is the wavelength of the CuKα line, β is the full width at half maximum in radians and θ is the Bragg angle. It is clear to note that the grain size of the as- deposited films and the films annealed at 200°C is practically constant of about 13 nm. However, after annealing at 300°C and 400°c, the crystallite size D is found to be respectively 28nm and 43 nm. This increase of crystallite size with the increase of annealing temperature is mainly due to the coalescence of the neighboring particles. At higher temperatures, the atoms have sufficient thermal energy to move into stable positions, leading to a significant increase in the intensity of (109) plane. A similar crystallite grain change is observed by Sandoval-Paz et al. [17] for annealed In_2S_3 films prepared by chemical bath deposition.

3.2. Optical properties

Optical transmission measurements in the wavelength range, 250–2500 nm is also performed in order to investigate the effect of heat treatment on the optical performances of the window layer In_2S_3. Figure 2 shows the optical transmittance of the In_2S_3 thin films annealed at different temperature. As-deposited films exhibit a good transparency in the visible and infrared regions

(65-75%). In fact, we note a decrease in the transmission values for the In_2S_3 annealed at 300 and 400°C films compared to the as deposited layer. This decrease of the transmission values can be explained by the presence of secondary phase. Based on the optical transmission and reflection measurements, $(\alpha h\nu)^2$ is plotted as a function of photon energy ($h\nu$) in Fig. 3 for β-In_2S_3 annealed at different temperature in nitrogen. The following equation can be applied for a direct band transition [18]: $(\alpha h\nu)^2 = A(h\nu - E_g)$ where A is a constant. The band gap energy is obtained by extrapolating the linear portion of the plot to the crossing with $h\nu$ axis. There is no remarkable change in the value of band gap after annealing it increases from 2.2 to 2.3 eV.

3.3. SEM and EDX studies

Indium sulphide composition has been investigated by means of EDAX in order to determine the In-to-S ratios at different annealing temperatures. Fig. 4 shows the typical EDAX spectra of In_2S_3 layers annealed at 400 °C in nitrogen during one hour. The EDAX analyses reveal that the film composition has the same value before and after annealing (S/In ratio of 1.43). Fig.5 shows SEM micrographs of the In_2S_3 as prepared and annealed at different temperatures. It is clear that the surface morphology of the In_2S_3 strongly depend of the temperature of annealing. The surface of as prepared film shows a sphere like In_2S_3 are formed on Pyrex substrate in an irregular spatial distribution. However, after annealing at 300°C, a substantial change in surface architecture is observed, it can be seen an obvious appearance of big island. At 400°C the surface relief shows more compact, uniform and continuous structure without any gaps or cracks.

CONCLUSION

In summary, indium sulphide thin films have been annealed in nitrogen at temperatures that vary in the range 200-400°C. The crystallinity and average grain size of the films increased with in annealing temperature. The EDAX analysis showed that the ratio [S]/[In] is in the range of 1.43 which is near to the theoretical value 1.5 of the stoechiometric value. The optical transmittance was found to decrease with annealing temperatures. It is important to conclude that the CBD technique allows the synthesis of good β-$In_2 S_3$ thin films. Moreover, we count to ameliorate the electrical properties of In_2S_3 dposited on pyrex and on SnO_2 without affecting the optical properties by doping of the layers with an appropriate dopant element followed by a heat treatment in order to reach the best performance of window based on doped In_2S_3 binary materials for future thin film-solar cell. Further studies in this direction are under progress.

REFERENCES
[1] M.Lajnef, H. Ezzouia, The Open Applied Pysics Journal. 2, (2009) 23-26.
[2] N.Barreau , S. Marsillac, J.C. Bernède, Vaccum. 56, (2000) 101 -106.
[3] T.T. John, S Bini, Y. Kashiawaba, T Abe, Y. Yasuhiro, C. S. Kartha and K. P. Vijayakumar, Semiconductor science and technology. 18, (2003) 491-500.
[4] D. Hariskos, M. Ruckh, U. Rqhle, T. Walter, H.W. Schock, J. Hedstrfm, L. Stolt, Solar Energy Materials and Solar Cells. 345, (1996) 41-42.
[5] T. T. John, M. Mathew, C. S. Kartha, K.P. Vijayakumara, T. Abeb, Y. Kashiwaba, Solar Energy Materials & Solar Cells. 89 (1), (2005) 27–36.
[6] Z. Li , X. Tao, Z. Wu, P. Zhang, Z. Zhang, UltrasonicsSonochemistry. 16 (2009) 221– 224.
[7] B. Asenjo, C. Sanz, C. Guillén, A. M. Chaparro, M. T. Gutiérrez, J. Herrero, Thin Solid Films. 515, (2007) 6041.
[8] C.D. Lokhande, A. Ennaoui, P.S. Patil, M. Giersig, K. Diesner, M. Muller, H. Tributsch, Thin Solid   Films. 340, (1999) 18–23.

[9] N. Barreau, J.C.Bernede , C. Deudon, L.Brohen, S.Marsillac, Thin Solid Films. 241, (2002) 4–14.

[10] S. Becker, T. Zheng, J. Elton and M. Saeki, Solar Energy Materials 13 (1986) 97-107.

[11] M. Mathew, C. S. Kartha, K. P. Vijayakumar, J. Mater Science. 20 (2009) 294-298.

[12] I. Puspitasari, T. P . Gujar, K. Jung, O. Joo, Journal of Materials Pocessing Technolology. 201, (2008) 775- 779.

[13] B. Yahmadi, N. Kamoun, C. Guasch and R. Bennaceur, Materials Chemistry and Physics. 127, (2011) 239–247.

[14] M.Kilani. B.Yahmadi, N. Kamoun and M. Castagné, J Mater Sci. J Mater Sci. 46, (2011) 6293–6300.

[15] M.Kilani. C.Guasch, M. Castagné and N. Kamoun, J Mater Sci. J Mater Sci. 47, (2012) 3198-3203.

[16] J.P.Eberhart, analyse structural et chimique des matériaux. p.231.
     Films. 472, (2005) 5 – 10.

[18] M.G. Sandoval-Paz, M. Sotelo-Lerma, J.J. Valenzuela-Jàuregui, M. Flores-Acosta, R. Ramrez-Bon, Thin Solid   Films. 472 (2005) 5 – 10.

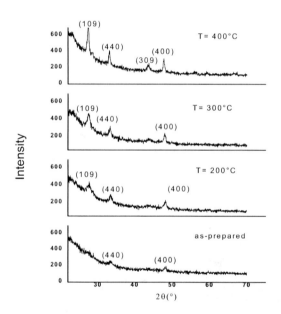

Fig. 1 X-ray spectra of In₂S₃ annealed in nitrogen at different temperature.

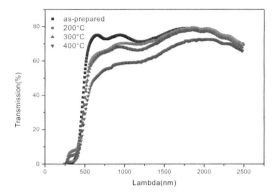

Fig.2. Transmittance (T) spectra of of In_2S_3 annealed in nitrogen at different temperature.

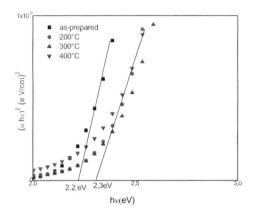

Fig.3. $(\alpha h\nu)^2$ versus (hν) spectra of In_2S_3 annealed in nitrogen at different temperature.

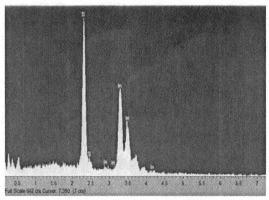

Fig. 4 EDAX spectrum of In$_2$S$_3$ layer annealed at 400°C.

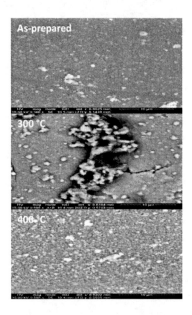

Fig. 5 Scanning electron micrograph of of In$_2$S$_3$ as prepared and annealed in nitrogen at different temperature.

EFFECT OF INDIUM CONCENTRATION ON THE PHYSICAL PROPERTIES OF In$_2$O$_3$ NANOMATERIALS GROWN BY SPRAY PYROLYSIS

Nasreddine Beji , Mejda Ajili ,Zeineb Sboui and Najoua Kamoun-Turki

Laboratoire de Physique de la Matière condensée, Faculté des Sciences de Tunis, Université Tunis El Manar (2092) Tunisie

Abstract

In$_2$O$_3$ thin films were deposited on glass substrate, by chemical spray pyrolysis, starting from indium chloride (InCl$_3$) and bidistilled water (H$_2$O). The structural and optical properties of In$_2$O$_3$ thin films as a function of different concentration of   In (y= [In^{3+}]= 0.02moll^{-1}, 0.025moll^{-1}, 0.03moll^{-1} , 0.035moll^{-1} and 0.04moll^{-1}, y is the indium concentration in the spray solution) were investigated . X-Ray diffraction pattern of indium oxide thin layers reveals the presence of cubic structure with (222) preferential orientation. The maximum transmittance is equal to 88% for y= 0.04 moll^{-1} . Optical analysis by means of transmission T(λ) and reflection R(λ) measurements allow to determine the direct band gap energy value which is in the order of 3.18 eV. The In$_2$O$_3$ thin films grown by spray pyrolysis have a big potential use as a window material for photovoltaic devices[1] such as Au/CuInS$_2$/In$_2$O$_3$/SnO$_2$:F, in which CuInS$_2$ is used as an absorber material [2] and SnO$_2$:F as an ohmic contact [3] and In$_2$O$_3$ is found to be one of the promising materials that can be used as a transparent electrode especially in solar cells [4] , along with many potential applications. Besides, Transparent conducting oxide (TCO) materials have received a great deal of interest due to their favourable physical properties such as low resistivity, good adherence to substrates and high transmittance in the visible and near infrared regions, for variety of technological applications [5–6]. Conducting indium oxide films on glass substrates can be used as either conducting or trans parent electrodes in optoelectronic devices [7].

Keywords : Indium oxide thin films, Optical and structural properties, chemical spray pyrolysis

1 Introduction:

Indium oxide (In$_2$O$_3$) thin film is technologically an important transparent conducting oxide (TCO) material. It is used in different fields like: photovoltaic devices, transparent windows in liquid crystal displays, sensors, antireflection coat-ings [8], and electrochromic devices [9]. Thin films of In$_2$O$_3$ can be prepared by a variety of techniques such as chemical vapour deposition [10], spray pyrolysis [11], evaporation of indium followed by oxidation [12], vacuum evaporation [13], and magnetron sputtering [14]. Among these techniques, spray pyrolysis provides an easy route to fabricate thin films at low cost. It can be easily modified for mass production and device-quality oxide films can be obtained over a large area.

In order to obtain optimal characteristics of In$_2$O$_3$ thin films, the parameters such as substrate temperature , thickness, concentrations and other deposition conditions have to be optimized.In$_2$O$_3$ films have been prepared by various deposition techniques such as vacuum evaporation ,dipcoating, sputtering, spray pyrolysis, sol-gel , etc. In this paper, results are described by preparing coatings by spray pyrolysis while varying indium concentration. The spray technique is one of the most commonly used techniques for preparation of transparent and conducting oxides owing to its simplicity, non-vacuum system of deposition and hence inexpensive method for large area coatings.

2 Experimental details :

Spray pyrolysis system was used to synthesize In$_2$O$_3$ thin films. The thin films of In$_2$O$_3$ were sprayed on to the glass substrate kept at a temperature 500°C with a spray rate of 2.5 ml.min^{-1} by varying indium concentrations, The experimental set up has been previously described[15-16]. The distance samples-nozzle is about 28 cm and the time of spray is 100 min. The solution used for pulverization consisted of indium chloride (InCl$_3$), and double distilled water (H$_2$O).

The structure of the layers was studied by X-Ray Diffraction (XRD) which are recorded with an automated Bruker D8 advance X-Ray diffractometer with CuKα radiations for 2θ values over 20-80°. The wavelength, accelerating voltage and current were, respectively, 1.5418 Å, 40 kV and 20 mA. The optical properties were studied according to UV-VIS-NIR spectrum with a Perkin–Elmer Lambda 950 spectrophotometer in the wavelength range of 250-2500nm at room temperature, taking the air as reference.

3 Results and discussion:

3.1 Structural properties:

3.1.1 X-Ray Diffraction

In order to study the effect of indium concentration on the structural properties of In$_2$O$_3$ thin layers deposited on Glass substrate. The XRD patterns of In$_2$O$_3$ thin layers grown for different indium concentrations are shown in Figure 1. The spectra show a polycristalline character with the typical cubic structure of In$_2$O$_3$. The same thin layers deposited on glass substrates had the highest (222) diffraction peak intensity indicating that (222) is the preferential orientation, The same behaviour was observed by A. Moses Ezhil Raj et col [7] , J. Joseph Prince et col[1] and Sung-Jei Hong et col [18]. Secondary peaks are (400) and (440).The highest intensity of the priviligied peak is obtained for y=0.04mol/l.

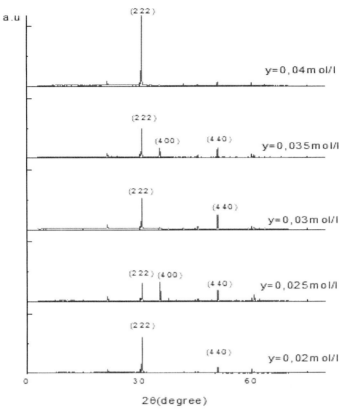

Figure 1 XRD spectra of the In₂O₃ thin films grown by spray pyrolysis on glass substrate for different concentrations of indium in the spray solution (y=[In^{3+}])

3.2 Optical properties:

In order to study the effect of indium concentrations on the optical properties, we have plotted in Figure 2, the optical transmission and reflection. we can observe that the optical transmittance spectra of In₂O₃ thin films represent a high tranmittance. The average transmittance in the visible range is about 85%, practically the same result was obtained by Shu-Guang Chen et col [19] . The highest transmittance is noted for y=0.04mol.l^{-1}, in fact it is about 90%. This value indicates a good quality of transparency of layers, predisposing them to be used as an optical window in visible range in photovoltaic systems . Besides we note the presence of the interference fringes characteristics of fairly uniform in thickness and quite homogenous layers.

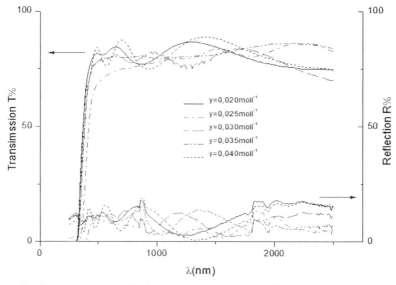

Figure 2. Transmission and reflection spectra of In$_2$O$_3$ thin films deposited on glass substrates, for different concentration of indium (y= [In^{3+}]) in the spray solution

As a direct band gap semiconductor, the band of In$_2$O$_3$ film is an important parameter. According the following formula, the band gap depend on the optical absorption coefficient and the photon energy hυ [17]:

$$(\alpha h\upsilon)= A (h\upsilon - E_g)^n \quad (1)$$

Where A is a constant, h is planck's constant, υ is the photon frequency, E_g is the optical band gap, n is ½ for direct band gap and α is the absorption coefficient, which can be calculated from the film thickness e, transmittance T and reflectance R measurements by using the formula [3]:

$$\alpha = -\frac{1}{e} Ln[\frac{T}{(1-R)^2}] \quad (2)$$

Figure 3 shows the variations of $(\alpha h\upsilon)^2$ vs Photon energy (hυ) for In$_2$O$_3$ thin layers grown for concentration of indium in the spray solution equal to 0.04 mol.l^{-1}. The linear part of the curves for high photon energies indicates that the In$_2$O$_3$ thin films are essentially semiconductors with direct-transition-type. Extrapolation of the linear portions of $(\alpha h\upsilon)^2$ to zero gave the value of E_g. The In$_2$O$_3$ thin films exhibit a direct band gap ranging in the order of 3.17 eV .

The variation of Eg as a function of indium concentration values can be attributed to changes in intensities of the peaks (222) , (400) and (440) when y varies , as shown by XRD analysis (figure 1).
Likewise we plotted the curves $(\alpha h\upsilon)^2$ vs Photon energy (hυ) for all other indium concentration values ,the results of gap are summarized in the following table :

y=[In](mol.l^{-1})	0.02	0.025	0.03	0.035	0.04
Eg(eV)	3.21	3.04	3.19	3.18	3.17

Table 1. Optical band gap energy (E$_g$) of indium oxide (In$_2$O$_3$) thin layers grown for different values of indium concentrations y in the spray solution

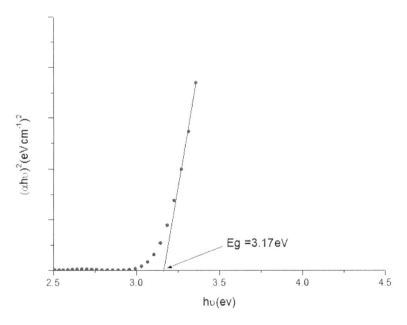

Figure 3 . Variation $(\alpha h\upsilon)^2$= f(hυ) of In$_2$O$_3$ thin films grown for indium concentration in the spray solutionequal to y= [In^{3+}]=0.04 mol.l^{-1}

The refractive index and n and the extinction coefficient k were calculated from transmittance T and reflectance R measurements by using these formulas [3] :

$$n = \frac{1 + [1 - (\frac{1-R}{1+R})^2 (1+k^2)]^{\frac{1}{2}}}{(\frac{1-R}{1+R})} \qquad (3)$$

$$k = \frac{\lambda}{4\pi e} Ln[\frac{(1-R)^2}{T}] \qquad (4)$$

Where λ is the wavelength and e is the thickness of the thin layer

(a)

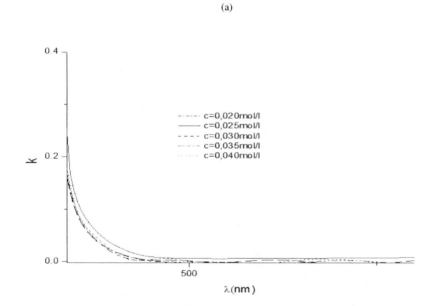

λ(nm)

(b)

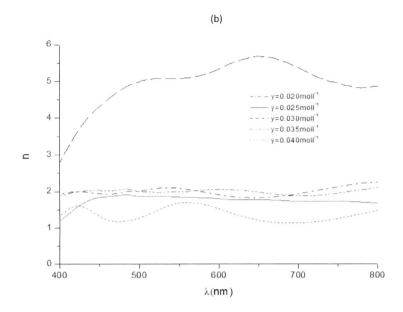

Figure 4. . Extinction coefficient k and refractive index n as a function of the wavelength , of In_2O_3 grown for different concentration of indium in the spray solution

Spectra 4-a shows that   in the transparency region [400; 800] nm the extinction coefficient k is pratically equal to zero and the variation of indium concentration don't contribute to ny significally changes to the extinction coefficient . In the absorption domain [250-400] nm, the variation of k can be attribute to the variations of band gap energy with y values , as shown in the table 1

The variations of the refractive index n with indium concentration y in the spray solution can be attributed to the improvement of the cristallinity of In_2O_3 , as shown in figure 1

The using of MAUD software allow us to determine the thickness t and lattice parameter a of In_2O_3 thin films elaborated for different indium concentrations in the spray solution ($0.020 \leq y \leq 0.040$ mol.l^{-1}). The results of this analysis are given in this table :

[In](mol/l)	0.020	0.025	0.030	0.035	0.040
t(Å) (maud)	3580	10722	5202	5180	3751
t(Å)(double weighing method)	3498	10613	5321	5294	3802
a(Å)(maud)	10.02	10.12	11.2	10.05	10.10

Table 2. lattice parameter and thickness of In_2O_3 grown for different indium concentrations calculated with MAUD software and double weighing method

The thickness of In_2O_3 thin films varies with indium concentration, in fact when y increases from 0.020 to 00.025 mol.l^{-1} t reach a maximum value 1.0722μm.

We note that the when y varies from 0.025 to 0.040 mol.l^{-1} t decrease to 0.3751 μm (table 2) and the lattice parameter a varies , in fact , when y increases from 0.020 to 0.030 the lattice parameter increases to a maximum value a=11.2 Å, it can be due for the expansion of the cubic lattice of In_2O_3 thin layers, then when y exceeds 0.030mol.l^{-1} the lattice parameter "a" decreases and reach 10.02 Å, in theory the lattice parameter is in the order of 10.11 Å

Conclusion: In summary, In_2O_3 thin layers have been deposited by chemical spray pyrolysis on glass substrates. The obtained films were polycrystalline with the cubic structure and had a (222) preferred orientation . The best cristallinity of In_2O_3 thin films is obtained for concentration (y= [In^{3+}]) of indium in the spray solution equal to 0.04moll^{-1} (T_S= 500°C).for which the band gap is in the order of 3.17 eV .

This will allow us to use In_2O_3 thin films as optical window in photovoltaic cell such as Au/CuInS$_2$/In$_2$O$_3$/SnO$_2$:F where the growth by spray of SnO$_2$:F and CuInS$_2$ thin films is well controlled in our laboratory [2-3].

References:

[1] J. Joseph Prince, S. Ramamurthy, B. Subramanian, C. Sanjeeviraja and M. Jayachandran

Journal of Crystal Growth 240 (2002) 142–151

[2] N. kamoun, N. Jebbari, S. Belgacem, R. Bennaceur, J. Bonnet, F. Touhari, and

L. Lassabatere, Journal of applied physics 91 (2002) 1952-1956

[3] S. Belgacem and R. Bennaceur, Revue Phys. Appl. 25 (1990) 1245-1258.

[4] P. Prathap, Y.P.V. Subbaiah, M. Devika and K.T. Ramakrishna Reddy, Materials Chemistry and Physics 100 (2006) 375–379

[5] Z.M. Jarzebski, Phys. Stat. Sol. (A) 71 (1982) 13.

[6] A.K. Hana, J. Photochem. Photobiol. A Chem. 132 (2000) 1.

[7] A. Moses Ezhil Raj, K.C. Lalithambika, V.S. Vidhya, G. Rajagopal,A. Thayumanavan,

M. Jayachandran and C. Sanjeeviraja, Physica B 403 (2008) 544 – 554.

[8] K.L. Chopra, S. Major, D.K. Pandya, Thin Solid Films 102 (1983) 1.

[9] C.G. Granquist, Sol. Energy Mater. Sol. Cells 60 (2000) 2301.

[10] J. , H.P. Schweizer, Thin Solid Films 29 (1975) 155.

[11] J.C. Manifacier, L. Szepessy, J.F. Bresse, M. Peroten, R. Stuck, Mater. Res. Bull. 14 (1979) 109.

[12] S. Naguchi, H. Sakai, J. Phys. D. 13 (1980) 1129.

[13] K.R. Murali, V. Sambasivam, M. Jayachandran, M.J. Chockalingam, N. Rangarajan, V.K. Venkatesan, Surf. Coat. Technol. 35 (1988) 297.

[14] W.G. Haines, R.H. Bube, J. Appl. Phys. 49 (1978) 304.

[15] N. Jebbari, B. Ouertani, M. Ramonda, C. Guasch, N. Kamoun Turki and R. Bennaceur, Energy Procedia 2 (2010) 79-89.

[16] N. Kamoun Allouche, T. Ben Nasr, N. Kamoun Turki and M. Castagne, Energy Procedia 2 (2010) 91-101.

[17] M. Girtan, G. Folcher, Surf. Coat. Technol. 172 (2003) 242.

[18] Sung-Jei Hong, Jeong-In Han , Current Applied Physics 6S1 (2006) e206–e210.

[19] Shu-Guang Chen, Chen-Hui Li, Wei-Hao Xiong, Lang-Ming Liu, Hui Wang, Materials Letters 59 (2005) 1342 – 1346

ONE-POT SYNTHESIS OF FUNCTIONALIZED FEW-WALLED CARBON NANOTUBE/MnO$_2$
COMPOSITE FOR HIGH PERFORMANCE ELECTROCHEMICAL SUPERCAPACITORS

Yingwen Cheng,[1,2] Hongbo Zhang,[1,2] Songtao Lu,[1,3] Shutong Zhan,[1] Chakrapani Varanasi [1,4] and Jie Liu [1,2,*]

[1] Department of Chemistry, Duke University, Durham, NC 27708
[2] Center for the Environmental Implication of NanoTechnology, Duke University, Durham, NC 27708
[3] Department of Chemistry, Harbin Institute of Technology, Harbin P.R.China 150001
[4] U.S. Army Research Office, P.O. Box 12211, Research Triangle Park, NC 27709-2211

Corresponding Author: Prof. Jie Liu, E-mail: j.liu@duke.edu

ABSTRACT
    In this paper, we described a one-pot approach to synthesize functionalized few-walled carbon nanotubes (fFWNTs)/MnO$_2$ composite for high performance electrochemical supercapacitor by mixing raw FWNTs with KMnO$_4$ under acidic condition. This approach combines the purification and functionalization of FWNTs and the synthesis of fFWNTs/MnO$_2$ composite in one step. The electrochemical performance of the composite was evaluated by cyclic voltammetry and galvanostatic charge-discharge techniques. The results indicate that the composite with 51wt% of MnO$_2$ has the high specific capacitance of 253 F/g (496 F/g based on MnO$_2$) at 10 mV/s and 176 F/g at the high scan rate of 500 mV/s (66% retention), indicating improved specific capacitance and excellent rate capability. Furthermore, the composite show superior electrochemical stability over 2000 charge/discharge cycles at the current density of 5 A/g. Therefore, this simplified one-pot approach of making the fFWNT/MnO$_2$ composite show high promises towards practical application in supercapacitors.

    The growing energy crisis requires both advanced renewable energy production and efficient storage systems to meet the diverse, yet rigorous, requirements.[1] Supercapacitors, also known as electrochemical capacitors, are energy storage devices that possess high power density, excellent reversibility and long cycle life and thus have received tremendous research interests.[2-4] They store energy by either electrolyte ion adsorption (electrochemical double layer supercapacitor) or fast reversible redox reaction (pseudo-capacitor).[5] Materials that have been widely used for supercapacitors include various forms of carbon, metal oxides and conducting polymers.[3] Within these materials, MnO$_2$ is particularly interesting because of its high theoretical specific capacitance up to 1300 F/g, low cost, natural abundance and environmental benign properties. [6,7] However, the poor conductivity of MnO$_2$ significantly limits its performance experimentally both in specific capacitance and rate capability.[8] In fact, most of MnO$_2$-supercapacitor systems designed by previous works only deliver specific capacitance in the range of 125~250 F/g at low current densities.[8-12] Hence, improving its performance is critical prior any practical applications.
    Inspired by the observations of the abnormally high capacitance (>700 F/g) acquired from the ultrathin MnO$_2$ film deposited on flat current collectors,[13-15] significant progresses in electrode design toward translating the current collector from flat to 3-dimensional have been made.[16] In this approach, the 3-D current collectors define the internal pore structure of the electrode and the ultrathin MnO$_2$ coated on them provide the pseudo-capacitance. As a result, the electrode could achieve high mass

loading without sacrificing its performance. Typically, the 3-D current collectors being studied are various forms of carbon, including mesoporous carbon[17,18], carbon nanotubes (CNTs)[19-21] and graphene.[22] Among them, CNTs have received intense interests because of their high electrical conductivity, superior mechanical and chemical stability.[23] In particular, the entangled structure formed by CNTs provides a conducting and readily accessible network for the electrolyte ions that could decrease the internal resistance of the electrode.[24] With the incorporation of CNTs, the specific capacitance and especially the rate capability of MnO₂ improved dramatically. In fact, most of the previous work focused on multi-walled carbon nanotube have documented that the specific capacitance of MnO₂ is generally higher than 250 F/g with the incorporation of CNTs.[25-29]

Few-walled carbon nanotubes (FWNTs) are unique nanotubes with 2~6 walls.[30] Compared with MWNTs, they have much better graphitization structure and remarkable electronic and mechanical properties.[30-33] More importantly, FWNTs have good tolerance to surface functionalization procedures by keeping the structural integrity of their inner tubes and hence, could disperse well in their composites while preserve the superior properties that originate from their tubular carbon structure.[32] Recently, FWNTs have been successfully applied in MnO₂ based supercapacitors and showed specific capacitance up to 427 F/g.[19] Unfortunately, the FWNTs produced with the chemical vapor deposition (CVD) method contain carbaonaceous and non-carbonaceous impurities that need to be removed prior the fabrication of the fFWNT/MnO₂ composite.[30] Hence, purification processes are necessary and usually involve selective oxidation of carbonaceous impurities and acid treatment to remove the metallic species.[34] Additionally, a separate functionalization step in strong acids (such as concentrated HNO₃) is also required to improve their dispersity to ensure good interaction with MnO₂.[24] These two processes (purification and functionalization) not only are labor intensive and time-consuming, but also involve dangerous and corrosive chemicals. Therefore, developing a simple and cost effective process is still necessary.

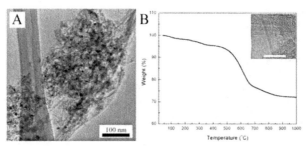

Figure 1: TEM image (A) and TGA-DTA plot (B) of the raw FWNTs obtained by chemical vapor deposition. Both of them show that the raw material contains a large amount of carbonaceous impurities and catalyst particles and support. The insert picture in (B) is a typical TEM image of the FWNTs (scale bar: 10nm)

In this paper, we reported a simple but efficient one-pot process that purify, functionalize FWNTs and produce the MnO₂/fFWNTs composite in one step. Given the fact that the impurities in the raw FWNTs are mainly metallic species and carbonaceous impurities with less graphitization perfection (Figure 1), we used a combined solution of KMnO₄ and H₂SO₄ to treat the raw material. During the reaction, H₂SO₄ ionizes and dissolves metallic impurities (Co/Mo catalyst and MgO catalyst support) while KMnO$_4$ preferably react with the carbonaceous impurities at room temperature given their less chemical stability compared with nanotubes. In the meantime, KMnO$_4$ could also react with the outer wall of nanotubes and lead to outer-wall selectively functionalized FWNTs (fFWNT). The reaction between KMnO$_4$ and carbon produces MnO$_2$ grown on nanotubes as nanoparticles through the electroless coating process.[17] Hence, this process combines the purification and functionalization of FWNT and the synthesis of fFWMT/MnO$_2$ composite in one step.

The amount of MnO$_2$ loaded onto fFWNT depends on the amount of KMnO$_4$ with respect to raw FWNTs. With the same amount of 100 mg raw FWNTs, the weight percentages of MnO$_2$ in the composites made from 50, 100, 150 and 200 mg of KMnO$_4$ were 8%, 51%, 70% and 73%, respectively (Table 1). Additionally, 70% seems to be the limiting percentage, as the composites obtained from more than 200 mg of KMnO$_4$ all have similar MnO$_2$ percentage (data not shown). Nevertheless, The BET specific surface areas of these composites measured were all higher than 200 m^2/g.

Table 1: Summary of the experimental details and characterization results: (1) and (2): amount of raw FWMT material and KMnO$_4$ used to synthesize the composites; (3): weight percentage of MnO$_2$ in the composite measured by TGA; (4) BET specific surface area; (5) and (6): specific capacitance calculated based on the mass of the composite and the MnO$_2$, respectively; (7): capacitance retention, defined as the ratio of the specific capacitance measured at 500 mV/s and 10 mV/s.

Sample	m$_{Raw FWNT}$ (mg)[1]	m$_{KMnO4}$ (mg)[2]	MnO$_2$ wt%[3]	BET-SSA (m^2/g)[4]	C$_{sp}$@10 mV/s (F/g) Composite[5]	MnO$_2$[6]	CR (%)[7]
fFWNT/ 8%MnO$_2$	100	50	8%	276	65	812	68%
fFWNT/ 51%MnO$_2$	100	100	51%	236	253	496	66%
fFWNT/ 70%MnO$_2$	100	150	70%	212	232	331	65%
fFWNT/ 73%MnO$_2$	100	200	73%	207	212	290	63%

The effectiveness of the combined solution of KMnO$_4$ and H$_2$SO$_4$ in purifying and functionalizing FWNT was monitored by Raman spectroscopy. After the reaction, the fFWNT/MnO$_2$ composites were boiled in 6M HCl to dissolve the MnO$_2$, followed by membrane filtration to collect the nanotubes. The Raman spectra acquired from these fFWNTs are presented in Figure 2. As this figure shows, the raw FWNT exhibits a high D-band (disorder-related mode) at ~1300 cm^{-1} that originates from the carbonaceous impurities,[35] whose intensity gradually decreases first and then increases with the increase of the concentration of KMnO$_4$. It is established that the carbonaceous impurities are less chemically stable than carbon nanotubes, hence they are more vulnerable for KMnO$_4$ etching and result in the gradual decrease of D-band.[36] However, FWNTs start to involve in the reaction at higher concentration of KMnO$_4$ (>150 mg). In this reaction, the outer wall of the FWNTs is oxidized and functionalized with hydroxyl and carboxyl groups as defective sites that lead to the increase of the D-band. Meanwhile, intense functionalization of the FWNTs will lead to comprised conductivity and stability. Hence, a proper balance between removing the carbonaceous impurities and functionalizing the FWNTs must be achieved through a well-controlled process. At the same time, the process should also be able to yield acceptable MnO_2 coatings on fFWNTs for supercapacitor application. Thus, based on Figure 2 and Table 1, we concluded that composites made from 100mg (fFWNT/51%MnO₂) and 150mg (fFWNT/70%MnO₂) of KMnO₄ could be good candidates for our purpose. In fact, the fFWNTs obtained from these two composites are almost free from carbonaceous impurities and have excellent water solubility.

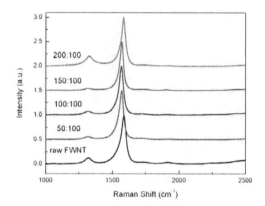

Figure 2: Raman spectrum of the fFWNTs obtained after acid dissolving the MnO_2 in the composite. The caption denotes the mass of the KMnO₄ and the mass of the raw FWNT used in the synthesis. (For example, 50:100 means 50mg of KMnO₄ and 100mg of raw FWNT).

The composition of the composite CNT/51%MnO₂ was studied with X-ray photoelectron spectroscopy (XPS, Figure 3) and energy dispersive X-ray spectroscopy (EDS, Figure 4C). Both of them reveal that the composite is free from metallic impurities, indicating the effectiveness of the purification process. The survey scan XPS spectrum acquired from this composite shows that it is mainly composed by manganese, carbon and oxygen. The average oxidization state of the Mn was analyzed with the Mn 3s core level XPS spectrum that has doublet peaks with separation depends linearly with the Mn oxidization state.[19,37] Since the peak separation for our synthesized composite is 4.78 eV, the average oxidization state of Mn is estimated to be 3.75. This indicates MnO_2 as the dominant oxide.

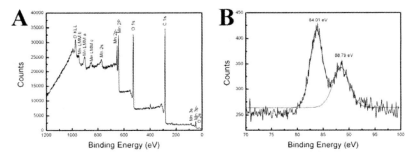

Figure 3: XPS spectrum of the composite (A) survey scan that indicate the material is free from the metallic impurities. (B) Mn 3s core-level spectrum that used to analysis the average oxidation state.

The crystal structure of the composite was studied with X-ray diffraction (XRD) and a typical pattern is presented in Figure 4A. As shown in this figure, no obvious peak was observed and hence the composite is considered as amorphous. The nitrogen adsorption/desorption isotherm acquired from the composite is shown in Figure 4B, with the corresponding BJH pore size distribution presented as the inserted image. This composite displays a typical type-IV isotherm, with the pore size peaked at ~4 nm and BET-SSA being 236 m^2/g (Table 1). Figure 4C and 4D show the morphology and microstructure of the composite revealed by SEM and TEM. Interesetingly, the MnO$_2$ particles grew uniformly on the FWNTs and no obvious large-sized aggregates were present. Furthermore, it is also clear that the composite preserves the porous structure of a entangled FWNT film, thus easily accessible for the electrolyte ions. The TEM image reveal that the manganese oxide particles were grown on the carbon nanotube surface as flate-like structures (Figure 4D).

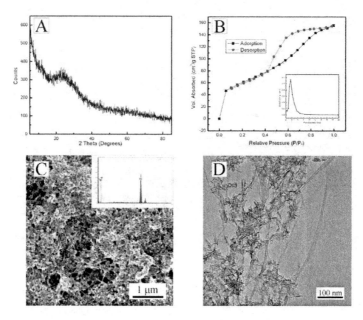

Figure 4: Set of characterization results for the fFWNTs/MnO₂ composite: (A) X-ray diffraction pattern shows that the composite is amorphous; (B) nitrogen adsorption/desorption isotherm shows that the composite mesoporous material; (C) SEM and (D) TEM images that show the MnO2 grow relatively unformly on fFWNT as flake-like structure.

As mentioned above, the poor conductivity of MnO₂ creates significant limitation to its electrochemical performance. With the highly conductive fFWNT dispersed uniformly in the MnO₂ matrix by the one-pot process, the conductivity of the composite improved dramatically and reached as high as 4.24 S/cm. The electrochemical properties of the composite were evaulated in 1M Na₂SO₄ based on the three-electrode system,with a plantinum wire and Ag/AgCl (4M KCl) electrode serving as the counter and reference electrode. The working electrode was fabricated by casting the compsite on a gold electrode without the addition of any conducting additives and binder.[25,38] A set of typical cyclic voltametry (CV) curves of the fFWNT/51%MnO₂ composite acquired at increasing scan rates from 10 mV/s to 500 mV/s were ploted in Figure 5A. All these voltammetric curves exhibit a nealy symmetrical rectangular shape, indicating highly capacitive nature with good response. The C_{sp} is caluclated from the CV curves using equation $Csp = (\int IdV)/vmV$, where C_{sp} is the specific capacitance (F/g), I is the response current, V is the potential window, v is the CV scan rate (mV/s), and m is the mass of the composite material rather than the electroactive MnO₂ unless otherwise noted. The composite delivers a C_{sp} of 253 F/g at the scan rate of 10 mV/s. Surprisingly, this composite still retains 167 F/g (66%) at the high scan rate of 500 mV/s (Table 1), implying the superior performance of the composite for high power operations.

The variation in the specific capacitances of the composites (both based on the composite and MnO$_2$) as a function of the MnO$_2$ percentage is listed in Table 1. All of these composite show CV curves with ideal rectangluar shape at increasing scan rates. However, the highest overall C_{sp} of 253 F/g (497 F/g based on MnO$_2$) was observed from the composite fFWNT/51%MnO$_2$. It is believed that the better performance was from the optimization of the composition of the material. Normally, higher content of fFWNT would lead to better electrochemical utilization of MnO$_2$ by increasing the effective contact area between MnO$_2$ and fFWNTs and providing higher electric conductivty.[25] However, more fFWNTs also tend to lower the overall C_{sp} because their contribution to the capacitance by double-layer charge storage is only trivial due to their low surface area. This is particualrly evident for the composite fFWNT/9%MnO$_2$, where the MnO$_2$ has the C_{sp} of 812 F/g, but the overall composite only delivers 65 F/g.

Furthermore, it is worth noting that more than 65% of the C_{sp} measured at 10 mV/s was retained at 500 mV/s for all these composites studied, which is ~5% higher than the composite prepared through a regular FWNT purification, functionalization and a separate composite synthesis process.[39] The better performance for the composite synthesized with this one-pot apporach might due to the better dispersion of fFWNT in the MnO$_2$. Usually, the regular FWNTs purification process through selective air-burning of the carbonaceous impurities creates a lot of FWNTs bundles that are hard to disperse into individual tubes. These bundles comprised the efficiency of FWNTs because MnO$_2$ are unable to grew on each individual tube. Our one-pot process avoids the high temperature burning step and the FWNTs are functionalized in-situ in solution, hence lead to the dramatic reduction of the bundle formation. In fact, the TEM images of the fFWNTs after MnO$_2$ removed reveal that most of fFWNTs are individual tubes. This further confirms the advantage of the simple one-pot reaction process.

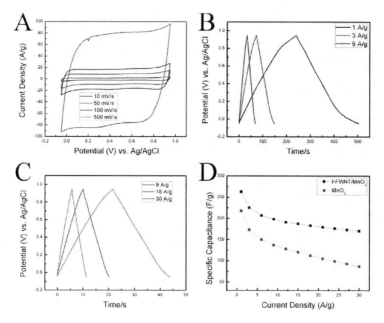

Figure 5: Electrochemical testing results for the fFWNT/51%MnO$_2$ composite: (A) CV curves acquired from 10 mV/s to 500 mV/s; (B) and (C) galvanostatic charge/discharge curves of the composite at different current densities; (D) comparions of the C_{sp} of the composite with pure MnO$_2$.

Rate capacity is an important factor for evaluating the performance of supercapacitors. A good electrochemical energy storage system is required to provide high energy density at high charge-discharge rates. Hence, in addition to the CV test, the behavior of the fFWNT/51%MnO$_2$ composite was also evauuated by the galvanostatic charging/discharging technique at different current densities from 1 A/g to 30 A/g. The results are ploted in Figure 5B and 5C and all these curves show highly linear and symmetrical shape that is typical for ideal supercapacitor. This implies that the composite has excellent electrochemical reversibilty and charge-discharge properties.[22] Additionally, the composite show much smaller iR drop compared to the pure MnO$_2$ at the same current densities. The C_{sp} of the composite at different current densities were calculated and compared with pure MnO$_2$ in Figure 5D. The compsite exhibited a C_{sp} of 263 F/g (based on the composite) at 1 A/g and 170 F/g at 30 A/g. In contrast, the C_{sp} of the MnO$_2$ electrode fabricated at similar mass loadings decreased sharply from 218 F/g to 87 F/g at the same current density range. The superior rate capability for the composite electrode can be attributed to its high-surface area and porous network structure that allows a high rate of solution infiltration and reduced diffusion path of ions, with the improved electrical conductivity that provides electron transport across the film.

Long cycling life is an important requirement for supercapacitor electrodes. In this paper, the electrochemical stability of the composite material was examined by a 2000 cycles charge-discharge test at a high current density of 5 A/g. The capacitance retention ratio as a function of cycle numbers is presented in Figure 6. It is observed that the composite showed ~5% fluctuation in the C_{sp}, however, no significant capacitance fade was observed and the composite remained relatively stable. The inserted figure presented 9 typical curves showing very stable charge-discharge behavior. This stability further indicates the superior mechanical stability of fFWNT/MnO$_2$ composite for electrochemical supercapacitors.

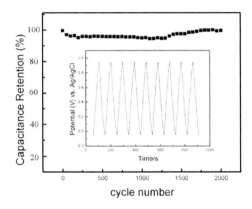

Figure 6: Variation of the specific capacitance as a function of charge-discharge cycle numbers measured at 5 A/g in 1 M Na$_2$SO$_4$. The inserted picture shows typical charge-discharge curves at this current density.

In summary, a simple but cost-effective approach is developed to fabricate fFWNT/MnO$_2$ composite for high performance electrochemical supercapacitors. This approach is based on raw FWNT materials and combines the purification and functionalization of FWNTs and the synthesis of fFWNT/MnO$_2$ composite in one step. The obtained composites are amorphous and have high surface area. Electrochemical testing results show that the composites exhibit improved specific capacitance, excellent rate capability and long cycling stability that originate from the highly conductive functionalized FWNTs. Additionally, the one-pot process also enables uniform disperse of fFWNT in the MnO$_2$ matrix that also ensures superior supercapacitor performance. We believe this greatly simplified process could readily applicable to other type of CNTs and enable cost-effective production of the fFWNT/MnO$_2$ composite toward their application in practical high performance supercapacitors.

ACKNOWLEDGEMENTS

This work is in part supported by a research grant from Army Research Office (ARO) under contract W911NF-04-D-0001 and the National Science Foundation (NSF) and the Environmental Protection Agency (EPA) under NSF Cooperative Agreement EF-0830093, Center for the Environmental Implications of NanoTechnology (CEINT). Any opinions, findings, conclusions or recommendations expressed in this material are those of the author(s) and do not necessarily reflect the views of the ARO, NSF or the EPA. This work has not been subjected to EPA review and no official endorsement should be inferred. The authors also acknowledge the support from Duke SMIF (Shared Materials Instrumentation Facilities). S. Z. and J. L. also acknowledge support from Lord Foundation for supporting undergraduate student in research activities.

REFERENCES

(1)   Yang, Z. G.; Zhang, J. L.; Kintner-Meyer, M. C. W.; Lu, X. C.; Choi, D. W.; Lemmon, J. P.; Liu, J. *Chem Rev* **2011**, *111*, 3577.

(2)   Winter, M.; Brodd, R. J. *Chem Rev* **2004**, *104*, 4245.

(3)   Simon, P.; Gogotsi, Y. *Nat Mater* **2008**, *7*, 845.

(4)   Miller, J. R.; Simon, P. *Science* **2008**, *321*, 651.

(5)   Conway, B. E. *Electrochemical Supercapacitors: scientific fundamentals and Technological Applications*; Plenum Publishers: New York, 1999.

(6)   Xu, C. J.; Kang, F. Y.; Li, B. H.; Du, H. D. *J Mater Res* **2010**, *25*, 1421.

(7)   Lang, X.; Hirata, A.; Fujita, T.; Chen, M. *Nature Nanotechnology* **2011**.

(8)   Wei, W. F.; Cui, X. W.; Chen, W. X.; Ivey, D. G. *Chem Soc Rev* **2011**, *40*, 1697.

(9)   Lee, H. Y.; Goodenough, J. B. *J Solid State Chem* **1999**, *144*, 220.

(10)   Toupin, M.; Brousse, T.; Belanger, D. *Chem Mater* **2002**, *14*, 3946.

(11)   Subramanian, V.; Zhu, H. W.; Wei, B. Q. *J Power Sources* **2006**, *159*, 361.

(12)   Reddy, R. N.; Reddy, R. G. *J Power Sources* **2004**, *132*, 315.

(13)   Toupin, M.; Brousse, T.; Belanger, D. *Chem Mater* **2004**, *16*, 3184.

(14)   Pang, S. C.; Anderson, M. A.; Chapman, T. W. *J Electrochem Soc* **2000**, *147*, 444.

(15)   Broughton, J. N.; Brett, M. J. *Electrochim Acta* **2004**, *49*, 4439.

(16)   Belanger, D.; Brousse, T.; Long, J. W. *Electrochem. Soc. Interf.* **2008**, 49.

(17)   Fischer, A. E.; Pettigrew, K. A.; Rolison, D. R.; Stroud, R. M.; Long, J. W. *Nano Lett* **2007**, *7*, 281.

(18)   Zhang, L. L.; Wei, T. X.; Wang, W. J.; Zhao, X. S. *Micropor Mesopor Mat* **2009**, *123*, 260.

(19)   Hou, Y.; Cheng, Y. W.; Hobson, T.; Liu, J. *Nano Lett* **2010**, *10*, 2727.

(20)   Zhang, H.; Cao, G. P.; Wang, Z. Y.; Yang, Y. S.; Shi, Z. J.; Gu, Z. N. *Nano Lett* **2008**, *8*, 2664.

(21)   Kim, J.-H.; Lee, K. H.; Overzet, L. J.; Lee, G. S. *Nano Lett* **2011**, *11*, 2611.

(22)   Yan, J.; Fan, Z. J.; Wei, T.; Qian, W. Z.; Zhang, M. L.; Wei, F. *Carbon* **2010**, *48*, 3825.

(23)   Dillon, A. C. *Chem Rev* **2010**, *110*, 6856.

(24)   Lota, G.; Fic, K.; Frackowiak, E. *Energ Environ Sci* **2011**, *4*, 1592.

(25)   Yan, J.; Fan, Z. J.; Wei, T.; Cheng, J.; Shao, B.; Wang, K.; Song, L. P.; Zhang, M. L. *J Power Sources* **2009**, *194*, 1202.

(26)   Ma, S. B.; Nam, K. W.; Yoon, W. S.; Yang, X. Q.; Ahn, K. Y.; Oh, K. H.; Kim, K. B. *J Power Sources* **2008**, *178*, 483.

(27)   Gao, L.; Xie, X. F. *Carbon* **2007**, *45*, 2365.

(28)   Zhai, L.; Li, Q. A.; Liu, J. H.; Zou, J. H.; Chunder, A.; Chen, Y. Q. *J Power Sources* **2011**, *196*, 565.

(29)   Jiang, R. R.; Huang, T.; Tang, Y.; Liu, J.; Xue, L. G.; Zhuang, J. H.; Yu, A. H. *Electrochim Acta* **2009**, *54*, 7173.

(30)   Qi, H.; Qian, C.; Liu, J. *Chem Mater* **2006**, *18*, 5691.

(31)   Hou, Y.; Tang, J.; Zhang, H. B.; Qian, C.; Feng, Y. Y.; Liu, J. *Acs Nano* **2009**, *3*, 1057.

(32)   Kumar, N. A.; Jeon, I. Y.; Sohn, G. J.; Jain, R.; Kumar, S.; Baek, J. B. *Acs Nano* **2011**, *5*, 2324.

(33) Qian, C.; Qi, H.; Liu, J. *J Phys Chem C* **2007**, *111*, 131.
(34) Feng, Y. Y.; Zhang, H. B.; Hou, Y.; McNicholas, T. P.; Yuan, D. N.; Yang, S. W.; Ding, L.; Feng, W.; Liu, J. *Acs Nano* **2008**, *2*, 1634.
(35) Dresselhaus, M. S.; Dresselhaus, G.; Saito, R.; Jorio, A. *Phys Rep* **2005**, *409*, 47.
(36) Haddon, R. C.; Hu, H.; Zhao, B.; Itkis, M. E. *J Phys Chem B* **2003**, *107*, 13838.
(37) Chigane, M.; Ishikawa, M. *J Electrochem Soc* **2000**, *147*, 2246.
(38) Chen, S.; Zhu, J. W.; Wu, X. D.; Han, Q. F.; Wang, X. *Acs Nano* **2010**, *4*, 2822.
(39) Cheng, Y. W.; Zhang, H. B.; Cordova, I.; Liu, J. *manuscript in preparation* **2011**.

Solar

THE USE OF INEXPENSIVE, ALL-NATURAL ORGANIC MATERIALS IN
DYE-SENSITIZED SOLAR CELLS

J. Whitehead[1], J. Tannaci[2] & M. C. Shaw[3,*]
[1]Bioengineering Department
[2]Chemistry Department
[3]Physics Department
60 West Olsen Road, #3750
California Lutheran University, USA
* Corresponding author

Abstract
The Dye-Sensitized Solar Cell (DSSC) is a photovoltaic cell which utilizes organic material to produce a small voltage, between 0.4V to 0.5V, in a process similar to photosynthesis. Our research explores the effects of using all-natural organic materials, specifically blackberries, blueberries, raspberries, and acai berries as four sources of dye for DSSCs. The acai berries were specifically selected based on their high levels of anthocyanidins, pigment molecules which adhere well to the titanium dioxide nanostructure of the DSSC; coupled with their increased absorbency as compared to other berries. In the present investigation, we have measured and compared the optical absorbance of dyes from each of the four sources to investigate how absorbance relates to electrical output. As part of this work, we have synthesized 12 cells utilizing various berry dyes, and demonstrated an electrical output of 0.3V to 0.5V from a white light source with an average intensity of 38,900 lux. Significantly, we have demonstrated the equivalent, if not superior, electrical output of the DSSCs containing acai berry dye as compared to DSSCs using blackberry dye. These findings are especially significant as blackberries are known to exhibit among the most optimal electrical performance of their class.

Introduction & Background
    The dye-sensitized solar cell is a photovoltaic device whose principal of operation is similar to the natural electron transfer process which occurs during photosynthesis. Both processes rely upon dye pigmentation to act as electron donors when excited by photons of light. The dye-sensitized solar cell (DSSC), however, utilizes a porous titanium dioxide (TiO_2) nanostructure, to which the dyes adhere, as an electron carrier for the cell (Fig. 1). These electrons are then carried by the nanostructure to an tin coated, and thus conductive, glass plate which acts as the electron accepter of the device (negative terminal).[1,2,3] From here the electrons travel through the external electrical load, thus comprising an electric current and corresponding electrical potential, or voltage. Opposite the dyed, TiO_2 nanostructure, a second tin-coated glass plate acts as a counter electrode to the system (positive terminal; Fig. 1). Evenly dispersed between the two plates is an electrolyte solution containing iodide/ tri-iodide which functions to replace the electron lost to the titanium dioxide nanostructure through oxidation (Fig. 1). Consequently, the electron lost can be regenerated by obtaining an electron from the layer of graphite on the counter electrode which acts as a catalyst. *The voltage produced by the system is the difference between the chemical potential of the titanium dioxide and the redox potential of the iodide/ tri-iodide electrolyte.*[1]

285

The critically necessary dye molecules are anthocyanidins. These plant pigments are the sugar-free counterpart to anthocyanins which are prevalent in berries, and are responsible for their color, taste, texture, and antioxidant properties.[4] An anthocyanidin molecule is pictured in Fig. 2, along with a table depicting of the various functional groups which make up the different types of anthocyanidins.

In order to obtain the best attachment of an organic molecule such as an anthocyanidin to an inorganic nanocrystalline structure of titanium dioxide a dehydration reaction must occur.[5] Thus cyanidin is the most favorable of the anthocyanidins owing to its two –OH functional groups in the R^1, R^2, or R^3 position (Fig. 2). Delphinidin and Pelargonidin would also attach to the titanium dioxide nanostructure of a DSSC, however, the attachment would not be as well established due to an additional and a missing –OH group respectively. In order to form the ideal bond between an anthocyanidin, preferably cyanidin, and the TiO_2 nanostructure, the titania will bond with the single bonded oxygen and the double bonded oxygen of the anthocyanidin[1,2] will donate the electron, as illustrated schematically in Figure 3.

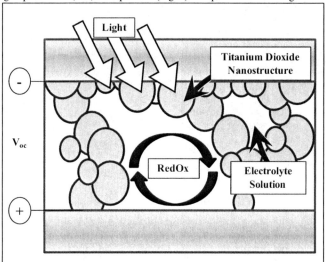

Fig. 1. Schematic of a dye-sensitized solar cell (DSSC). The DSSC utilizes a porous titanium dioxide (TiO₂) nanostructure, to which the berry dyes adhere, as an electron carrier for the cell. These electrons are then carried by the nanostructure to a tin coated, and thus conductive, glass plate (top) which acts as the electron accepter of the device (negative terminal). From here the electrons travel through the external electrical load. Opposite (bottom) is a tin-coated glass plate that acts as a counter electrode to the system (positive terminal). Evenly dispersed between the two plates is an electrolyte solution containing iodide/ tri-iodide which functions to replace the electrons lost to the titanium dioxide nanostructure through oxidation.

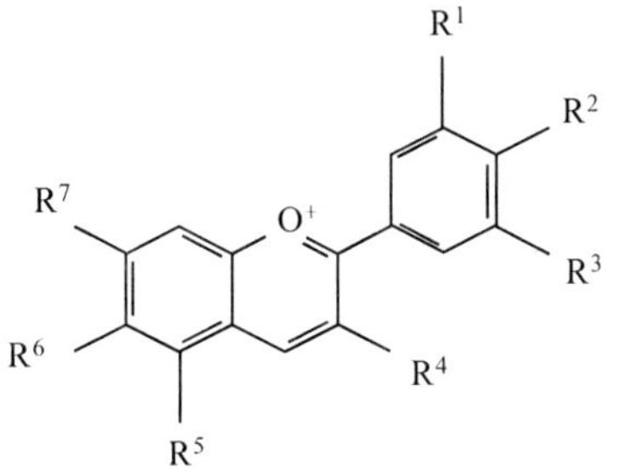

Anthocyanidin	R¹	R²	R³	R⁴	R⁵	R⁶	R⁷
Cyanidin	OH	OH	H	OH	OH	H	OH
Pelargonidin	H	OH	H	OH	OH	H	OH
Delphinidin	OH	OH	OH	OH	OH	H	OH
Malvinidin	OCH₃	OH	OCH₃	OH	OH	H	OH
Peonidin	OCH₃	OH	H	OH	OH	H	OH
Petunidin	OH	OH	OCH₃	OH	OH	H	OH

Fig. 2. Schematic (top) of the organic model of an anthocyanidin dye molecule. Table (bottom) of the different functional groups that distinguish the different anthocyanidins.

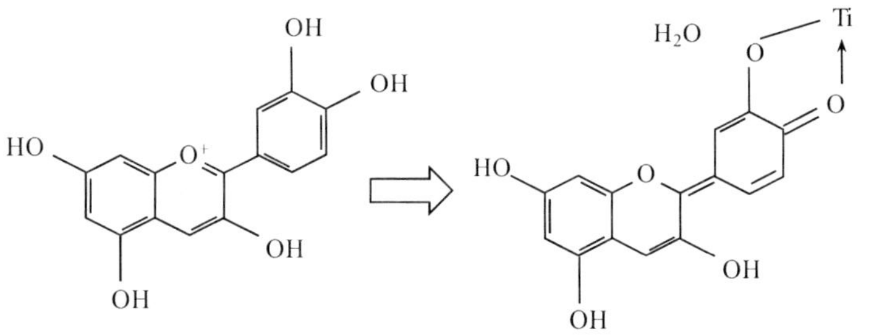

Fig. 3. Schematic illustrating the attachment of the titania molecule to the single bonded oxygen, where the electron is donated from the double bonded oxygen of the anthocyanidin.

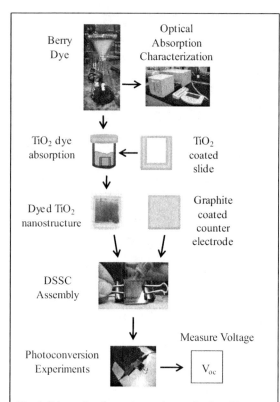

Fig. 4. Schematic of experimental organization. First, dyes were extracted from each of four types of natural berry sources. The optical absorption of each of the dye solutions was then obtained. Next a nanostructured TiO_2 coating was synthesized on a glass plate, and soaked in the dye solution to activate the TiO_2 coating. The dyed TiO_2-coated glass plate was then clamped to a graphite coated counter electrode glass plate, and an iodide/tri-iodide electrolyte solution was introduced, thus completing the DSSC assembly. Following assembly, the electrical characteristics (voltage and current) were characterized against a controlled output light source.

Blackberries were chosen as our benchmark dye source due to their abundance of cyanidins[4,6,7] and previous work demonstrating their superior characteristics as natural organic material for DSSCs, specifically at the DSSC's inception in the late 1990's.[1,2] For the present investigation, however, *acai berries were chosen as a potential alternative dye source*, due to their antioxidant rich properties[8,9] which we hypothesized would lead to superior attachment to the nanostructure of a DSSC; as well as their extremely dark pigmentation which we also hypothesized would lead to greater light absorption and therefore greater electron excitation in a DSSC. Blueberries and raspberries were investigated for comparison purposes due to the presence of favorable anthocyanidins in these two dyes, just not in as great abundance as in blackberries and acai berries.[6,7] The various anthocyanidins known to exist in each of the various berry sources are shown below in Table 1.[4,6, 7,8,9]

Table 1. Summary of Anthocyanidins Present in Berry Sources

Berry Source	Acai Berry	Blackberry	Blueberry	Raspberry
Cyanidin	✓	✓	✓	✓
Pelargonidin		✓		✓
Delphinidin			✓	
Malvinidin			✓	
Peonidin	✓	✓	✓	
Petunidin			✓	

Finally, all-natural organic materials (naturally occurring berries) were selected due to their environmentally and economically benign characteristics especially as compared to ruthenium dyes[3] used in more synthesized and more complex dye-sensitized solar cells. To summarize, the following paper describes the design, development and assembly of a dye-sensitized solar cell, the acquisition of the necessary berry dye pigments, and the experimental characterization of optical absorbance and electrical output for each of the berry dyed DSSCs.

Experimental

The organization of the experimental portion of this investigation is illustrated schematically in Fig. 4. As a summary, the experiments began by extracting dyes from each of four types of natural berry sources. The optical absorption of each of the dye solutions was then obtained. Next a nanostructured TiO_2 coating was synthesized on a glass plate, and soaked in the dye solution to activate the TiO_2 coating. The dyed TiO_2 -coated glass plate was then clamped to a graphite coated counter electrode glass plate, and an iodide/tri-iodide electrolyte solution was introduced, thus completing the DSSC assembly. Following assembly, the electrical characteristics (voltage and current) were characterized against a controlled output light source. Each of these experimental steps is described in detail below.

Experimental - Preparation of Titanium Dioxide Nanostructure Coated Glass Plates

The titanium dioxide nanostructure that acts as the electron carrier was created from commercial colloidal TiO_2 powder[1, 2, 10] using a procedure found in the literature.[1, 2] Namely, 6 g of TiO_2, 8 mL of acetic acid and 1 mL of de-ionized water were ground with a mortar and pestle until a thick white paste-like substance was created. Separately, adhesive tape was applied to three of the sides of a conductive glass plate to allow for a clear contact point for subsequent electrical and structural connections. A droplet (5 μL) of the TiO_2 solution was applied to the surface of the glass plate and then distributed as a thin layer of uniform thickness by spreading with a glass rod. The TiO_2 coating was first air dried for 30 minutes at room temperature. Next, the tape was removed and the TiO_2 coated glass was placed on a hotplate at 200°C – 250°C for an hour. This procedure allowed the TiO_2 to anneal and adhere to the glass. The TiO_2 coated glass plate was then allowed to cool to room temperature prior to being placed in the dye solution.

Experimental - Dye Solution Preparation

A mixture of anthocyanidins (plant pigmentation) was obtained from each berry type using an extraction solution of methanol/acetic acid/water (25:4:21).[1] After first crushing and thoroughly blending the berry mixture, gravity filtration was utilized to obtain the desired anthocyanidins: cyanidin, pelargonidin, and delphinidin. After preparation, dye solutions were permanently stored in brown bottles at 5°C to minimize photo-degradation due to exposure to light.

Experimental - Ultraviolet-Visible Light Spectroscopy

Each solution was analyzed using ultraviolent-visible light spectrometer Agilent Model 8453 Value. Square test tubes containing a controlled concentration of each berry dye solution (1:10, berry dye: methanol), as well as a blank methanol negative control were used to achieve the highest precision absorption spectroscopy possible.

Experimental - TiO_2 Dye Absorption

To prepare the TiO_2 coated glass plates for DSSC assembly, they were placed in the appropriate dye solution for a minimum of twelve hours at 5°C in order for the titanium dioxide nanostructure to absorb the dye pigments; specifically, to allow the anthocyanidins to saturate the nanostructure of the DSSC. For this step, the TiO_2 coated glass plates were again placed in clear glass bottles, which were wrapped in aluminum foil to minimize photo-degradation due to light exposure.

Experimental - Dye-Sensitized Solar Cell Assembly

Once the glass plate with the TiO_2 had been dyed overnight, it was ready for assembly. First, a separate counter electrode was prepared from the same glass plate material used for the TiO_2 coating. To prepare the counter electrode, the uncoated glass plate was treated with a thin carbon coating which acted as the redox catalyst. This layer was applied using the graphite from a pencil, as specified by the literature.[2] The counter electrode was then placed on top of the plate containing the titanium dioxide nanostructure and secured with binder clips, one on each side of the two plates (Fig 4). The two plates were offset by a distance of approximately 2 mm in order to allow for an attachment point for the alligator clips. Two drops of the electrolyte solution iodide/tri-iodide, which acts as the electron mediator for the system, was applied on the ridge created by the offset of the two glass plates (Fig. 4). In order to evenly distribute the electrolyte across the nanostructure of the cell the binder clips used to secure the device were alternately opened and closed, and thus, through a combination of gentle pressure cycling and capillary action, the iodide/ tri-iodide electrolyte solution permeated the TiO_2 nanostructure of the DSSC. This step completed the preparation and assembly of the DSSC.

Experimental - Electrical Characterization of the Dye-Sensitized Solar Cells

The DSSC, assembled as described above, was attached to a stage set a consistent distance from a controlled white light source of a 360 W overhead projector. Using a light sensor and automated data acquisition (DAQ) software, we determined that the light intensity at the point of testing was measured to be, on average, 38,940 lux; well within the illumination range of natural sunlight (25,000 to 130,000 lux).

The DSSC was mounted so that the plate containing the TiO_2 nanostructure faced the light source. This plate was the positive terminal; the counter electrode was the negative terminal. The DSSC was then attached to a simple series circuit (Fig. 5) with both manual and automated ammeters connected in series with a 500 Ω load resistor; as well as both manual and automated voltmeters connected in parallel across the resistor. Each DSSC was tested for a minimum of 120 minutes. Data was collected manually every 10 seconds for the first five minutes and then every five minutes after that for a total of two hours.

After testing, a dye sensitized solar cell could be reused and reactivated by lightly drying the nanostructure and placing the cell back in its dye solution storage container. If the nanostructure had not been discolored due to photo-degradation, then the cell can be reactivated by simply adding additional drops of the electrolyte solution of iodide/tri-iodide.[2]

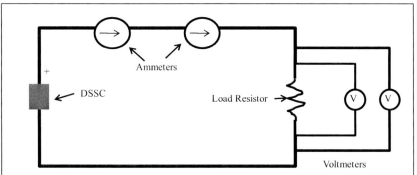

Fig. 5. Schematic of electrical characterization: The DSSC was attached to a simple circuit with both manual and automated ammeters connected in series with a 500 Ω load resistor; as well as both manual and automated voltmeters connected in parallel across the resistor. Each DSSC was tested for a minimum of 120 minutes. Data was collected manually every 10 seconds for the first five minutes and then every five minutes after that for a total of two hours.

Results - Optical Absorption

The ultraviolet- visible light spectroscopy data are presented in Fig. 6 across a wavelength range of 400 – 700 nm. The absorption data shown were first normalized by the optical absorption of the acai berry at 400 nm. The acai berry dye consistently had the highest absorption factor across all wavelengths. Blackberry dye also had high absorption, although lower than acai berry dye. Blueberry dye followed, with raspberry dye found to have the lowest absorption of the four berry dyes of interest (Figure 6).

Results - Electrical Output

The open circuit voltage was measured over a two hour period for each DSSC; as well as for a blank DSSC (negative control) containing no berry pigmentation on the titanium dioxide nanostructure. Example data for the open circuit voltage of an acai berry DSSC and a blackberry DSSC are represented in Figure 7. The blank control DSSC exhibited an initial open circuit voltage of only ~0.25 V which decreased to 0V after approximately 90 minutes, and is not shown in Fig. 7.

As can be seen from Fig. 7, the acai berry DSSC exhibited a higher output voltage as compared to the blackberry DSSC over the entire time range of interest. No statistically significant difference was detected between the output current of any of the DSSCs when connected to a load resistor, so only open circuit voltage data are described below.

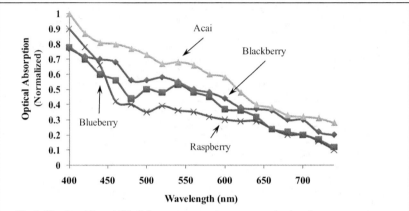

Fig. 6. The ultraviolet- visible light spectroscopy data across a wavelength range of 400 – 700 nm. The absorption data shown were first normalized to eliminate concentration effects. The acai berry dye consistently had the highest absorption factor across all wavelengths. Blackberry dye also had high absorption, although lower than acai berry dye. Blueberry dye followed with raspberry dye found to have the lowest absorption of the four berry dyes of interest.

Discussion

The goal of the present investigation was to determine whether an increase in open circuit voltage of DSSCs prepared with acai berry dye would correlate with a higher berry dye optical absorbance owing to the increased concentration of dye anthocyanidins. This is the case for acai berry[8,9] and blackberry dyes,[4] despite the greater number of anthocyanidins found in blueberry and raspberry dyes[6,7] (cf. Table 1), which have a lower concentration of the most favorable anthocyanidins, namely, cyanidin.

After measuring the optical absorption of each of the berry dyes, we observed an absorption peak around 550 nm. This is relatively close to the published peak of cyanidin (520 nm).[1] Therefore, we used the 550 nm wavelength as our point of comparison for absorption among the four berries of interest, since each of the four berries contain cyanidin.[4,8,9] Furthermore, cyanidin is known to have superior attachment properties to the titanium dioxide nanostructure of the DSSC among the known anthocyanidins and therefore acts as the best electron donator to the system.[1,2] The best electron donor therefore should correspond to the highest open circuit voltage, our means of comparison.

Indeed, at the 550 nm wavelength, the acai berry dye was found to have a higher absorption as compared to the blackberries. Blackberries are used as our benchmark owing to their established status as the best type of natural organic material to use in a dye-sensitized solar cell.[1,2] Furthermore; the open circuit voltage data exhibited two broad regimes; (i) initial electron excitation and (ii) steady-state plateau region. In the first, the majority of all available electrons in the dye are excited to higher energy levels; whereas in the second, the electrolyte solution becomes saturated and can only fill some of the holes left behind by the excited electrons. Thus, the voltage plateaus owing to the redox / electron replenishment rates equilibrate. Finally, the decrease in open-circuit voltage for the blank, non-dye containing DSSC resulted from the only the electrolyte solution acting as the donor, rather than the mediator, and thus being limited in the number of available electrons. This finite supply of electrons and no source of electron replenishment thus led to exhaustion of the DSSC.

To explore the measured increase in open circuit voltage for the acai berry DSSCs as compared to the blackberry DSSCs in more detail, the output voltage after 20 minutes for each of three acai berry dye and three blackberry dye DSSCs was averaged and plotted as a function of the normalized optical absorption at 550 nm in Fig. 8. As can be seen, the increase in the open circuit voltage correlates with the increased optical absorption. *This key finding confirms our hypothesis that the DSSC containing the largest number of electron donors, in addition to having the highest absorbance, would make for the DSSC with the highest open circuit voltage.* Thus, we conclude from this investigation that acai berry dye DSSCs offer superior open circuit voltage characteristics in dye-sensitized solar cells.

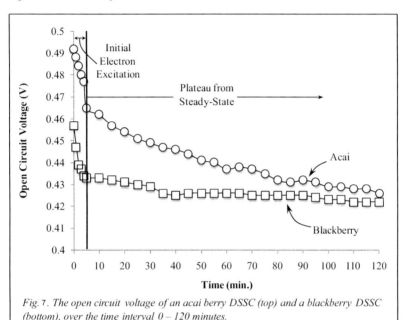

Fig. 7. *The open circuit voltage of an acai berry DSSC (top) and a blackberry DSSC (bottom), over the time interval 0 – 120 minutes.*

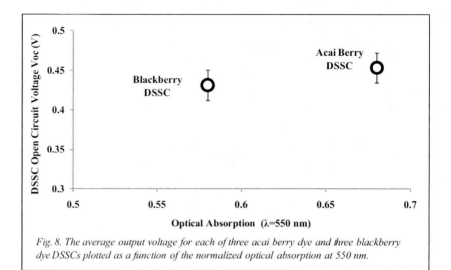

Fig. 8. The average output voltage for each of three acai berry dye and three blackberry dye DSSCs plotted as a function of the normalized optical absorption at 550 nm.

Conclusions

We have demonstrated a novel dye-sensitized solar cell chemistry based on the dye obtained from acai berries. The open circuit voltage of the acai dye DSSC was measured to be higher than that obtained from a blackberry dye DSSC. Furthermore, this higher open circuit voltage of the acai berry DSSC was found to correlate with an increase in optical absorbance by the berry dye at a wavelength of 550 nm. Finally, the increased optical absorbance of the acai berry dye was found to correlate with the prevalence of dye anthocyanidins in acai berries.

References

1.   Cherepy, N. J.; G.P. Smestad, M. Graztel, J.Z. Zhang; *Am. Chem. Soc.* 1997, 101, 9342-9351.

2.   Gratzel, M and G.P. Smestad; *Journal of Chemical Education.* 1998, 75, 752-756.

3   Gratzel, M.; *Accounts of Chemical Research.,* 2009, 42, 1788-1798.

4   Cuevas-Rodriguez, E.O. et al.; *J .Agric Food Chem.,* 2010, 58, 7458-7464.

5   Coakley K. M. and M. D. McGehee; *Am. Chem. Soc.,*2004, 16, 4533-4542.

6   Schauss, A.G., et al.; *J .Agric Food Chem.,* 2006, 54, 8598-8603.

7   Wu, X., Prior R.L.; *J .Agric Food Chem.,* 2005, 53, 2589-2599.

8   Schauss, A.G ., *G.S. Jensen, X. Wu; ACS Symposium Series.,* 2010, 1035, 213-223.

9   Jensen, G.S., et al.; *J. Agric. Food Chem.,* 2008, 56, 8326-8333.

10   Yang H.G. et al.; *Nature,* 2008, 453, 638-642.

DEEP LEVEL DEFECTS IN N$^+$-CdS/P-CdTe SOLAR CELLS

P. Kharangarh, Z. Cheng, G. Liu, G. E. Georgiou and K. K. Chin
New Jersey Institute of Technology, Dept. of Physics and Apollo CdTe Solar Energy Center,
University Heights, Newark, NJ, 07102, USA,

ABSTRACT:
    We characterize the thin film n$^+$-CdS/p-CdTe solar cells made with evaporated Cu as a primary back contact, using the temperature dependence of the reverse bias diode current (J-V-T) to determine the energy levels of deep defects. Since the solar cell quickly degrades (probably because of the well-established Cu diffusion from the back ohmic contact into CdTe) with measurement at temperatures greater than ~100°C, measurements are done below this temperature (~100°C). The results of our J-V-T measurements on solar cells made at NJIT show that while modest amounts of Cu enhance cell performance, an excessive high temperature process step degrades device quality and reduces efficiency. Results identify the physical trap though the energy (activation) level. The location of and the amount of trap are derived from the voltage dependence of diode leakage using a Shockley-Read-Hall (SRH) recombination model.

INTRODUCTION:
    Thin film solar cell technology has relatively low production cost because of the reduced material purity and the relatively simple in-line processing steps and equipment. Thin film hetero-structure n$^+$–CdS/p–CdTe solar cells have demonstrated 17.3% under one Sun Illumination [1]. However, much improvement is theoretically possible as CdTe (cadmium telluride) has a near-optimal band gap ~ 1.5eV corresponding to an ideal efficiency of 29% [2].
    There are certain limitations which limits the efficiency of CdTe solar cells. The shallow "doping", material defect in CdTe solar cell plays the role of both dopant as well as trap so we must be very careful with applying simplifying assumption which is completely valid for Si. There is no real distinction between Defects and Impurities –Both act as beneficial "dopant" or bad "trap" unlike Si with intentional shallow dopants and no intentional traps. Also, for contact quality with low doped p-CdTe –Fundamental Questions arose–
What is "dopant" species and concentration? What is "trap" species and concentration?
    However, this Cu$_{Cd}^-$ energy level is controversial with a wide range of activation energies being obtained from different techniques. Electrical data from Hall Effect and Deep Level Transient Spectroscopy measurements which ignore the temperature dependence of R-G lifetime, report 0.3-0.4 eV [3, 4] while photoluminescence measurements report ~0.146eV. This compares to the theoretical calculations ~0.22eV [5]. Interstitial copper also has a majority carrier trapping defect as a deep level with activation energy ~ 0.55eV [6]. Therefore we must understand the role of defect states on diode current and the effect of processing on defect states.
    In this paper, we describe J-V data from a baseline process in our laboratory (Apollo CdTe Solar Cell Research Center, NJIT) but with variations in the back contact with ~5-10%.

EXPERIMENTAL DETAILS:
    n$^+$-CdS/p-CdTe solar cell diodes were fabricated with a process similar to that described by [7]. The fabrication process and the solar cell physical structure was explained in brief [8]. Three different types of CdTe solar cells, namely cells 1, 2 with an area of 0.18cm^2 by Apollo CdTe Solar Cell Center at NJIT and cell 3 with an area of 0.25 cm^2 were studied. The difference in cells[1, 3] is a variation of back contact process and thickness of CdTe layer. In cell1, The thickness of CdTe layer was 12μm and the back contact process involves Cu evaporation

annealed at 280°C and J-V measurements were done from 25°C to 80°C; also cell 2 has same back contact process, but J-V measurement were done in reverse order (from 110°C to 25°C), while cell 3 processed with annealing of Cu powder at 160°C, thickness of CdTe was chosen 9μm and J-V measurements were done form 120°C to 25°C. All cells were processed at Apollo CdTe Solar Research Center, NJIT (4" x 4" glass).

All solar cells have a Cu containing back contacts. Our back contact process involves (1) evaporating and reacting a thin layer of copper on CdTe, (2) applying a carbon conducting paste containing mixed micron-size Cu particles and ZnTe powder and (3) annealing at 160°C(Cell 3)/280°C(Cells 1, 2) with 100sccm helium flow for 30 minutes. The J-V characteristics were analyzed in dark and under 100mWcm^{-2} ~ 1 sun illumination (filtered xenon lamp) from the side through the "transparent" glass substrate. J-V measurements are done by using a Keithley 236 Source Meter.

J-V MEASUREMENTS AND THEORY:

Figure 2 shows J-V under dark and 1sun illuminated conditions for each cell, the parameters for which are summarized in Table 1. Different J-V curves were observed for the three cells. This is expected because of the processing differences for each cell.

The diode current-voltage characteristics in dark for model – n+-CdS /p-CdTe solar cells

$$J(\lambda, P_\lambda, V) = \{J_{0,SRH} [\exp (^{qV/AkT}) - 1] + J_L(\lambda, P_\lambda)\} \; f(V) \qquad [1]$$

where J_o is the reverse bias "saturation"" current (saturates for the ideal diode but is a function of applied voltage or SRH R-G), V is the applied bias, A is the ideality factor (A=1 for the ideal diode and A=2 for SRH R-G). J_L is the light current and $f(V)$ is collection transfer function (holes from CdTe to CdS contact and electrons CdS to CdTe contact).

In this paper, Reverse bias diode current is dominated by Shockley-Read-Hall recombination $J_{0,SRH}$ (N_t, E_t, σ_n, σ_p, $V^{1/2}$, T) when measurements were done at large reverse bias $V \geq -1V$ without illumination between the temperature range from 25°C - 100°C.

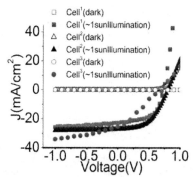

Figure 2. J-V Characteristics of each cell in dark and illumination condition

TABLE I. Extraction of different parameters from Figure2:

Cells (NJIT)	Cell 1	Cell 2	Cell 3
$J_{sc}(mA/cm^2)$	22.8	25	23.7
$V_{oc}(V)$	0.73	0.81	0.48
FF(%)	53.2%	53.2%	49.5%
$\eta(Eff)(\%)$	9.4%	10.8%	4.79%
A	3.09	4.44	3.35
$J_0(dark)(mA/cm^2)$	7.6E-7	2.65E-5	1.94E-5

RESULTS AND DISCUSSION:
Temperature-Dependence of Dark Reverse Bias Diode Current for Each Cell:

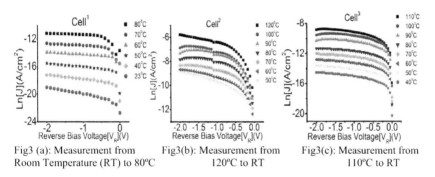

Fig3 (a): Measurement from Room Temperature (RT) to 80°C   Fig3(b): Measurement from 120°C to RT   Fig3(c): Measurement from 110°C to RT

The semi-log plot of reverse J– V characteristics of the heterojunction is shown in Fig. 3(a, b, c) for each cell. The leakage current density (J_o) was obtained by extrapolating the J – V characteristics at high reverse bias voltage (-1V).

Theory and Measurements:
Since we measure ideality factor A>2(measured from 0 to V_{oc} in dark J-V curve for each cell) and see the reverse bias leakage current increasing with reverse bias, we explore SRH R-G in the core diode. Here, the "saturation" generation current is proportional to the number of traps in the core diode depletion region.

The generation current in the depletion layer in CdTe is given by [9]

$$J_{gen} = \int_0^W qGdx \qquad [2]$$

The net generation rate G (in steady state, equal to the negative of the recombination rate –U) for two trap levels is

$$G = \sum_{j=1}^{2} \frac{\sigma_{nj}\sigma_{pj}v_{th}N_{tj}(pn - n_i^2)}{\sigma_{nj}[n + n_i \exp(\frac{E_{tj} - E_i}{kT})] + \sigma_{pj}[p + n_i \exp(\frac{E_i - E_{tj}}{kT})]} \qquad [3]$$

where q is the electronic charge, W is the space charge depletion width with a trap density N_t and energy level E_t relative to the intrinsic energy level E_i, n_i is the intrinsic carrier concentration, σ_n and σ_p are the electron and hole capture cross-sections and v_{th} is the thermal velocity calculated using the effective mass.

For large reverse bias (V >>kT/q), equation [4][16] reduces to

$$G = \frac{n_i}{2\sqrt{\tau_{n0}\tau_{p0}}\ \cosh[\dfrac{(E_t - E_i)}{kT} + \dfrac{1}{2}\ln\dfrac{\tau_{p0}}{\tau_{n0}}]} \tag{4}$$

Where $\tau_{n0} \sim 1/\sigma_n\ v_{thn}\ N_t$ and $\tau_{po} \sim 1/\sigma_p\ v_{thp}\ N_t$ are the electron and hole lifetimes is the depletion region. This result detailed in reference [16] indicates that the traps are most effective when

$$\frac{(E_t - E_i)}{kT} + \frac{1}{2}\ln\frac{\tau_{p0}}{\tau_{n0}} = 0 \tag{5}$$

Thus, the lifetime effect on generation current can be ignored for a certain range of lifetime ratio range. With the assumption that $|(E_t-E_i)/kT| \gg \ln(\tau_{p0}/\tau_{n0})$, the cosh reduces to an exponential and, in this case, the trap energy level can be simply obtained from Arrhenius plot of the reverse bias generation current. Note that the assumption that τ_{p0} and τ_{n0} are not too different may not be true in CdTe/CdS solar cell.

With the simplifying assumption that $|(E_t-E_i)/kT| \gg \ln(\tau_{p0}/\tau_{n0})$, the generation current can be written as [9]

$$J_{gen} = \frac{qn_iW}{\tau_g} \tag{6}$$

where q is the charge, n_i is the intrinsic carrier concentration, W is the depletion width and τ_g is the generation life time. The diode current is then the form of equation [1] with $J_0 = J_{gen}$ and ideality factor A=2. If we assume that τ_g (i.e. σv_{th}) does not strongly depend on temperature, then the temperature dependence of J_{gen} that of n_i, $\sim T^{3/2}$ exp (-E_g/2kT). However, neglecting the temperature dependence of σ can introduce errors for CdTe.

Assuming SRH R-G is the dominant reverse bias leakage mechanism, a slope of Ln (J_0T^{-2}) vs. 1000/T should yield the trap energy level. Such a plot at -1V is shown in Fig. 4(a, b, c) for our solar cells. If recombination through localized mid-gap states within the CdTe depletion region is dominant, then a plot of Ln (J_0T^{-2}) vs. 1000/T should yield activation energy approximately equal to half of the *CdTe* band gap [14, 15]. This is true in cells 2, 3.

For one trap, a single slope was measured. For more than one trap, multiple slopes were obtained since different traps dominate in different temperature ranges. Cells 2, 3 do not have shallow traps and fit more into the discussion of SRH R-G. However, two distinct slopes are observed. The shallower level difference of cells 2, 3 may be a result of the variation in the Cu-containing back contact process.

Identification of defects using J-V-T Measurements:

Devices were analyzed at a reverse bias of -1V. Several trap levels were found in all cells in the temperature range of 40^0C-120^0C. The measurement of J-V-T suggests the difference between cell 2 and cell 3 is the diffusion of Cu at higher temperature.

To identify the observed activation energies from J-V-T with a specific deep level is difficult. Published data for deep levels for CdTe give different activation energies from different techniques. Our simple technique cannot distinguish if more than one deep level is operative as a generation center, the measured activation energy may be average of several levels, further complicating the analysis. Further, an analysis of the J-V at various temperatures provides the information on the distribution of generation centers.

Cu used in CdTe solar cell is well known to be a fast diffuser. Assuming that it is a positive ion (Cu_i^+), however, the junction field of the cell will resist the forward diffusion. Hence, the rate of diffusion would be expected to be dependent on the bias across the cell. Recently, capacitance transients were used by Enzenroth, et al. [19] and Lyubomirsky et al. [20] determine the diffusion parameters of mobile Cu_i^+ ions in CdTe and materials based upon a transient ion drift (TID) method was first developed by Heiser and Mesli [21]. In particular, Enzenroth used the TID approach to quantify an increase in mobile Cu_i^+ as a function of increased Cu added during cell fabrication. The presence of mobile charge, in particular, Cu_i^+ could be the cause to get deep traps in cells 2, 3.

The activation energy (E_T) for Arrhenius plot for cell 1 (which was measured from Room Temperature to 80°C) of 1.47eV which is equal to the band gap of CdTe solar cell. At temperature below 100^0C, the diffusion current from outside of the depletion region dominates.

By comparing with the published data [22], the activation energy obtained for cell 2 (Measured with aging at 110°C to RT); E_{t1} 0.33eV is possibly due to substitutional Cu_{Cd}^- (effect of σ (T) proper); E_{t2} = 0.60eV possibly V_{Cd}^{2-} or Cu interstitial? At higher temperature, the junction reverse current is dominated by generation current produced with in the depletion region.

The activation energy E_{t1} =0.67eV and E_{t2} = 0.74eV obtained for Cell 3 (Measure with aging at 120°C to RT is possibly due to V_{Cd}^{2-}, V_{Cd}^- Cu interstitial.

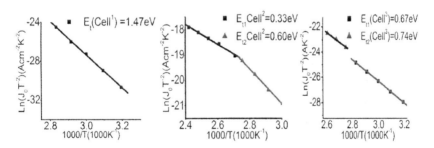

Figure 4(a, b, c): Experimental results for Ln (J_0T^{-2}) versus 1000/T for different CdTe/CdS solar cells.

CONCLUSIONS

We have presented J-V-T measurement at reverse bias (-1V) generation current (SRH R-G theory) to correlate the various traps with experimentally measured activation energies from plot of Ln (J_0T^{-2}) versus 1000/T of variously processed n⁺-CdS/p-CdTe solar cells. Since J-V-T theory is difficult to interpret trap energy level E_T but can be a good tool for process reproducibility. By J-V-T theory it was observed that Copper containing contacts initially perform well (copper forms a shallow acceptor in CdTe), but their high temperature performance can be a problem, as the copper under circumstances diffuses through the CdTe layer and forms Cu related substitutional or interstitial defects.

ACKNOWLEDGMENTS
The authors gratefully acknowledge the support of the Apollo Solar Energy, based in Chengdu, People's Republic of China.

REFERENCES:
1   A. Burger, *First Solar Sets Thin Film CdTe Solar Cell Efficiency World Record*, July 27, 2011.
2   W. Shockley, H.J. Queisser, *J. Appl. Phys.*, 32, 510 (1961).
3   A. E. Rakhshani, *J. Phys., Condens. Matter* 11 9115(1999).
4   F. Seymour, V. Kaydanov and T. R. Ohno, *Applied Physics Letter* 87, p-153507(1-3) (2005).
5   S. H. Wei and S. B. Zhang, *Phys .Rev .B.*, 66, 155211(2002).
6   J. V. Li, R. S. Crandall, I. L. Repins, A. M. Nardes and D. H. Levi ,*37th IEEE Photovoltaic Conference Proceedings*, p.1-4 (2011).
7   D. H. Rose, F. S. Hasoon, R. G. Dhere , D. S. Albin, R. M. Ribelin, X. S. Li,Y. Mahathongdy, T. A. Gessert And P. Sheldon, *Progress in Photovoltaics , Research and Applications, Res. Appl.* 7, 331-340(1999).
8   P. Kharangarh, D. Misra, G. E. Georgiou and K. K. Chin, *ECS Transactions*, 41 (4) p. 233-240 (2011).
9   S. M. Sze. Physics of Semiconductor Devices (2nd Edition).
10   S. A. Ringel, A.W. Smith, M. H. MacDougal, A. Rohatgi, *J. Appl. Phys.*, 70, p. 881-890(1991).
11   J. B Yoo, A. L. Fabrenbruch, Bube, *J. Appl. Phys.*, 68, p. 4694-4700(1990).
12   Su-Huai and S.B. Zhang, *Physical Review B* 66, 155211(2002).
13   Robert W. Birkmire and Erten Eser, *Annu.Rev.Mater.Sci.*, p. 625-53(1997).
14   M. Wimbor, A. Romeo and M. Igalson , Opto Electronics Review 8(4), p. 375-377 (2000).
15   V. P. Singh, H. Braf-man, J. Makwana, and J. C. McClure, Solar cells, 31, p.23, (1991).
16   C. T. Sah, Robert N. Noyce and William Shockley, *Proc. of the IRE*, p. 1228-1243(1957).
17   V. P. Singh, D. L. Linam, D. W. Dils, J. C. McClure, G. B. Lush, *Solar Energy Materials & Solar Cells* 63 p. 445-466 (2000).
18   V. Komin, V. Viswanathan, B. Tetali, D. L. Morel and C. S. Ferekides, *31th IEEE Photovoltaic Conference Proceedings*, p.736-739 (2002).
19   R. A. Enzenroth, K. L. Barth, and W. S. Sampath, *4th IEEE Photovoltaic Conference Proceedings*, p.449-452 (2006).
20   I. Lyubomirsky, M. K. Rabinal, and D. Cahen, *J. Appl. Phys.*, 81, p. 6648-6691(1997).
21   T. Heiser and A. Mesli, *Appl. Phys. A*, 57, p. 325-328(1993).
22   X. Mathew, *Solar Energy Materials & Solar Cells*, 76, p. 225–242(2003).

GROWTH DYNAMICS IN THIN FILMS OF COPPER INDIUM GALLIUM DISELENIDE SPUTTERED FROM A QUATERNARY TARGET

Authors: J.A. Frantz,[a] R.Y. Bekele,[b] J. D. Myers,[a] V.Q. Nguyen,[a] A. Bruce,[c] S.V. Frolov,[c] M. Cyrus,[c] J.S. Sanghera[a]

[a]U.S. Naval Research Laboratory, 4555 Overlook Avenue, SW, Washington, DC 20375, USA
[b]University Research Foundation, 6411 Ivy Lane #110, Greenbelt, MD 20770, USA
[c]Sunlight Photonics, 600 Corporate Court, South Plainfield, NJ 07080, USA

ABSTRACT

Sputtering of copper indium gallium diselenide (CIGS) offers a potential route to highly-manufacturable thin film solar cells. Thin films of CIGS are deposited from a quaternary sputtering target, and their composition, structure, and device performance are compared to those of state-of-the-art evaporated CIGS. SEM and TEM analysis reveal that the sputtered films have a unique morphology – differing significantly from that of evaporated films – with small grains near the bottom contact that coalesce into larger grains as film growth progresses. We examine the temperature dependence of grain growth by depositing films while varying deposition temperature from 85-600 °C. Devices that employ sputtered CIGS films as the absorber are fabricated. Light IV curves are obtained, and conversion efficiency as high as 10.5% is observed. Quantum efficiency measurements indicate that the morphology may have a strong impact on device performance.

INTRODUCTION

Thin film copper indium gallium diselenide (CIGS) has rapidly become an important photovoltaic material, competing for market share with silicon and other more established technologies. The most well-established method of depositing CIGS is by co-evaporation from four independent elemental sources. This technique has resulted in laboratory devices with record efficiencies of approximately 20%.[1,2] While this method has produced excellent devices, there are several disadvantages associated with it. As a result of the point source nature of typical evaporation sources, uniform deposition over large areas is difficult. Copper, with the highest evaporation temperature, presents the greatest challenge. Simultaneous control of all four sources can be challenging, with variations in relative deposition rate potentially leading to incorrect stoichiometry and diminished device performance.

Alternate methods for vacuum deposition of CIGS films include sputtering the metals with, or followed by, treatment in a selenium environment. The selenium environment is typically generated by treating the sample in H_2Se gas or by heating a solid Se source in proximity to the film. These deposition techniques exploit the advantages of sputtering – good uniformity over large areas, high deposition rates, and efficient material usage. Recent reports of 15.7% for a module[3] and 17.8% for a 30x30 cm submodule[4] have used these methods. It is interesting to note that, while the record small-area laboratory devices are based on evaporated films, these record larger-area modules and submodules employ a sputtered CIGS layer. Despite their advantages, however, these methods also have drawbacks. H_2Se gas is highly toxic, and supplying Se vapor from a solid source complicates the deposition process, requiring control of another deposition source and potentially additional processing steps.

More recently, there has been interest in deposition by sputtering from a quaternary target – a single sputtering target that contains Se as well as the metals.[5-8] Some reports of films deposited using this method require extra processing steps – such as selenization or etching in potassium cyanide to remove an unwanted Cu_2Se phase – in order to make working devices. The method described here, however, uses RF magnetron sputtering from a single quaternary target without additional selenization or the need for extra processing steps.[5,6] We have fabricated working devices using CIGS films that are sputtered in a single step and have demonstrated conversion efficiency as high as 10.5%. We compare the composition, structure, and electronic properties of these films to state-of the-art evaporated films. While the films are similar in composition, we observe important differences in morphology that may limit device performance. We speculate that improving the films' morphology is critical to further improvements in device performance and may result in sputtered films with device performance comparable to that of evaporated films.

EXPERIMENT

Bulk CIGS was formed by heating high purity (>99.999%) precursors in a vacuum-sealed quartz ampoule. The bulk material was ground into a powder inside a nitrogen-purged glovebox, and sputtering targets were formed by hot pressing the powder into 3" diameter disks. The disks were machined to the proper dimensions and indium bonded to a copper backing plate. Sputtering targets with a variety of compositions were fabricated. A near-stoichiometric composition of $CuIn_{0.7}Ga_{0.34}Se_{2.06}$ was used. This composition was chosen so that even with a small amount of Se lost to the vapor phase during target fabrication or sputter deposition, there is still sufficient Se present. The excess Ga is added in stoichiometric proportions for Ga_2Se_3, in order to incorporate the Se into a secondary phase that is more stable during heating.

Pieces of soda lime glass (SLG) with a 500 nm to 1 μm thick layer of sputtered Mo were used as substrates. CIGS films, approximately 2 μm in thickness, were prepared by RF magnetron sputtering in a sputter-up geometry. This process was carried out in an Ar atmosphere, typically at a pressure of 3 mT with an RF sputter power of 200 W. The target composition, substrate temperature, RF sputtering power, and the excess selenium content in the deposition chamber were varied. After removal from the deposition chamber, some samples were used for CIGS film characterization, and others remained in the process line for device fabrication. For device samples, approximately 50 nm of CdS was deposited via chemical bath deposition, and 100 nm of ZnO was deposited by sputtering. A 200 nm thick layer of Al_2O_3:ZnO was deposited by sputtering to serve as the top contact. Ni/Al grids, composed of 50 nm of Ni followed by 400 nm of aluminum, were deposited via evaporation. Samples were scribed by hand to obtain individual cells, each with an area of approximately 0.5 cm^2.

Energy dispersive spectroscopy (EDS) with standardless calibration was employed to determine the films' composition. Composition as a function of depth was obtained by use of secondary ion mass spectroscopy (SIMS) with ion beam milling. Samples for cross sectional scanning electron microscope (SEM) analysis were obtained by mechanically breaking samples. Samples for transmission electron microscope (TEM) analysis were capped with Pt and prepared for imaging by the focused ion beam (FIB) lift-out technique. Light I-V curves were obtained in a solar simulator under one sun, AM 1.5 illumination. Quantum efficiency (QE) was measured in a QE test bed.

The composition and morphology of sputtered CIGS films were compared to those of evaporated CIGS films. The record laboratory evaporated devices use a three-stage evaporation process in which substrate temperature and the relative evaporation rates of the constituents differ for each stage.[1,2] The evaporated films chosen for comparison here, however, were deposited in a single-stage deposition process, with a substrate temperature of 550 °C, because their deposition process is more analogous to that of the quaternary sputtered CIGS.

RESULTS AND DISCUSSION

A series of films was deposited, starting with a new target, and the films' composition was measured via EDS with standardless calibration. The composition as a function of cumulative deposited thickness is shown in Figure 1. The dotted lines indicate the composition of the reference evaporated film. Note that the values for both sputtered and evaporated films do not correspond exactly to those of stoichiometric CIGS. For example, the measured Se composition is approximately 45% rather than 50%. This apparent non-stoichiometry is a reflection of error in the standardless calibration rather than an indication of incorrect composition. This plot, therefore, should be seen as a qualitative comparison. The most prominent feature of the plot is a decrease in In concentration and an increase in Cu concentration during the first ~10 μm of deposition. After this point, the composition remained relatively constant, with slight variations within the uncertainty of the measurement. This "burn-in" behavior is typical and well-known for multicomponent sputtering targets. Preferential sputtering of one or more components occurs initially. After the formation of a subsurface region that is depleted of the preferentially sputtered species a steady-state condition emerges, and relative sputter rates stabilize.[9]

In order to obtain information about the variation of composition with depth, SIMS measurements with ion beam depth profiling were carried out. The results, shown in Figure 2, indicate that composition is nearly constant with respect to depth. The Na content is higher near the Mo/CIGS interface, consistent with diffusion of Na from the SLG substrate. Cu content is lower near the top surface, with a concentration of approximately 19 at.% compared to a concentration of 29 at.% through most of the film. The fact that the film is Cu-poor near the surface is consistent with the absence of a significant amount of $CuSe_2$, a phase that is present after many other CIGS deposition processes and necessitates etching in KCN. The low Cu region near the top surface is similar to that observed in three-stage evaporated CIGS. For evaporated CIGS, this Cu-poor region is created intentionally by turning off the copper source during the final deposition stage. It is not yet entirely clear why this same effect is observed in the quaternary sputtered films, but it may be advantageous.

Figure 3 shows SEM images of a top view (a) and cross section (b) of evaporated CIGS as well as a top view (c) and cross section (d) of a sputtered film. Both deposition methods produce dense films, but there are some important structural differences. The evaporated film exhibits ~1 μm scale grains throughout its thickness. The sputtered film, however, consists of small ~100 nm scale grains near the Mo/CIGS interface that coalesce into larger grains that have dimensions of ~1 μm near the top surface of the film.

TEM images reveal further details about the films' structure. Figure 4 shows a bright field high resolution TEM (HRTEM) image of the cross section of a film near the Mo/CIGS interface. The Mo delaminated during sample preparation, leaving only the CIGS. Small, ~100 nm grains are clearly visible. Figure 5 shows a larger scale TEM image of a FIB-thinned sample. Stacking faults and screw dislocations are visible in the image. The inset shows a high resolution TEM image of the Mo-CIGS interface. The presence of a $MoSe_2$ layer is evident at atomic resolution. The existence of this layer was confirmed with cross-sectional EDS (not shown). This

MoSe$_2$ layer is similar to that found in evaporated films, and its presence here is significant because this layer is important for adhesion and the formation of an ohmic contact.

Deposition temperature was varied to examine its effects on film properties. Figure 6 shows grain size, estimated from SEM images, as a function of substrate temperature during deposition. The insets show representative top view SEM images for several data points. The maximum temperature used was approximately 600 °C, higher than the 550 °C maximum working temperature of the glass. While these substrates were warped after deposition, it was still possible to obtain images of the films. Grain size exhibits an exponential dependence on substrate temperature, similar to the behavior observed in evaporated CIGS [8]. When plotted on an Arrhenius plot, shown in Figure 7, the activation energy can be extracted using the equation

$$E_A = -R\frac{\partial(\ln d)}{\partial(1/T)}$$

where d is the grain size and R is the gas constant with the value of 8.31 J/(mol·K). The resulting activation energy for grain growth in this case is 22 kJ/mol, or 230 meV.

Device samples were fabricated using sputtered CIGS samples that were deposited with a substrate temperature of 550 °C and a deposition power of 200 W. After CIGS deposition, no further processing steps were carried out before deposition of the rest of the stack. Figure 8 shows light I-V curves for such a device with a conversion efficiency of 10.5%, and the inset shows measured QE. The QE is degraded at the base as evidenced by the drop in efficiency for low energy photons. This behavior may be due to the presence of the small grains near the Mo-CIGS interface evident in the SEM and TEM images. Taken together, these results suggest that the greatest gains in performance for these devices may be achieved by understanding the nucleation and grain growth during deposition and finding ways to mitigate the formation of small grains near the base.

CONCLUSIONS

We describe the structure and composition of CIGS films deposited from a single sputtering target by RF magnetron sputtering. The films have several features in common with state-of-the-art evaporated CIGS – their composition, the presence of Cu-poor region near the surface, a dense structure with ~1 μm scale grains, and the presence of a MoSe$_2$ layer at the Mo-CIGS interface. The main structural feature that distinguishes the quaternary sputtered films from evaporated films is the small, ~100 nm, grains that form near the base of the films.

When used to make photovoltaic devices the sputtered films yield efficiencies as high as 10.5% without any additional processing. The small grains near the base appear to be harmful to device performance, as indicated by the drop in efficiency for low energy photons observed in the QE measurement. Further improvements in efficiency may be obtained by developing deposition techniques that reduce the prevalence of these small grains. With improved efficiency, quaternary sputtered CIGS might offer a more manufacturable alternative to evaporated CIGS.

REFERENCES
1. I. Repins et al., "19.9%-Efficient ZnO/CdS/CuInGaSe$_2$ Solar Cell with 81.2% Fill Factor," *Prog. Photovolt.: Res. Appl.* **16**, pp. 235–239 (2008).

2. P. Jackson et al., "New World Record Efficiency for Cu(In,Ga)Se$_2$ Thin-Film Solar Cells Beyond 20%," *Prog. Photovolt.: Res. Appl.* **19**, pp. 894-897 (2011).

3. http://www.miasole.com/news-archive, press release.

4. http://www.solar-frontier.com/news/179, press release.

5. J.A. Frantz et al., "Characterization of Cu(In,Ga)Se$_2$ Thin Films and Devices Sputtered From a Single Target Without Additional Selenization," *37th IEEE PVSC* (2011).

6. J.A. Frantz et al., "Cu(In,Ga)Se$_2$ Thin Films and Devices Sputtered From a Single Target Without Additional Selenization," *Thin Solid Films* **519**, pp. 7763-7765 (2011).

7. J. H. Shi et al., "Fabrication of Cu(In, Ga)Se$_2$ Thin Films by Sputtering from a Single Quaternary Chalcogenide Target," *Prog. Photovolt.: Res. Appl.* **19**, pp. 160-164 (2011).

8. C.-H. Chen et al., "A Straightforward Method to Prepare Chalcopyrite CIGS Films by One-Step Sputtering Process Without Extra Se Supply," *37th IEEE PVSC* (2011).

9. A. Oliva et al., "Sputtering of Multicomponent Materials: The Diffusion Limit," *Surf. Sci.* **166**, pp. 403–418 (1986).

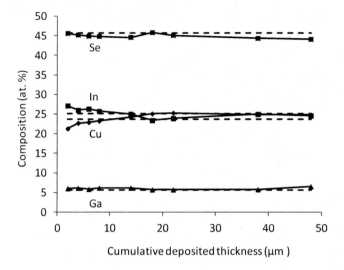

Figure 1: The effect of target burn-in on composition as measured by EDS. The dotted lines indicate the composition of the reference evaporated film.

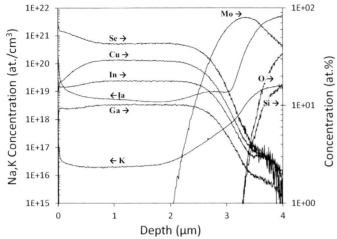

Figure 2: SIMS profile of a CIGS film. The concentration of major constituents is shown on the right axis and trace constituents on the left.

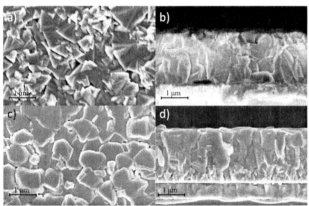

Figure 3: SEM images showing a top view (a) and cross section (b) of evaporated CIGS; and a top view (c) and cross section (d) of a sputtered film.

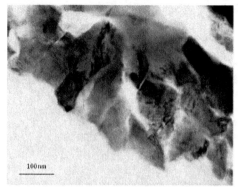

Figure 4: A bright field HRTEM of grains in film at Mo/CIGS interface. Mo delaminated during sample preparation leaving only the CIGS film.

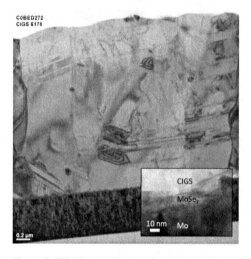

Figure 5: TEM image showing the cross section of a CIGS device. The inset shows a HRTEM profile of the MoSe₂ layer at the Mo-CIGS interface.

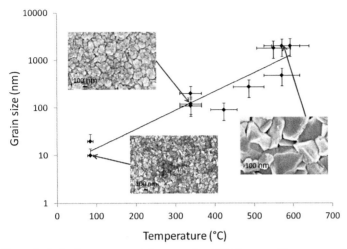

Figure 6: Grain size as measured from top-view SEM images with an exponential fit. The insets are representative SEM images for data points indicated by the arrows.

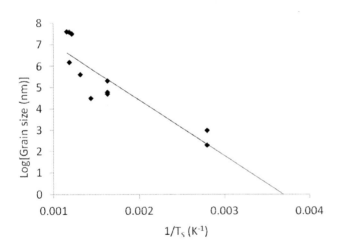

Figure 7: Arrhenius plot using the data shown in Figure 6 that can be used to extract the activation energy for grain growth.

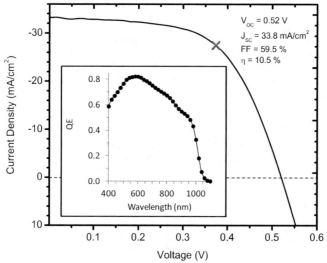

Figure 8: Light-IV of a sputtered CIGS device measured under AM 1.5 illumination. The "x" indicates the point of maximum power, and the inset shows measured QE.

Wind

NOVEL FLEXIBLE MEMBRANE FOR STRUCTURAL HEALTH MONITORING OF WIND TURBINE BLADES - FEASIBILITY STUDY

S. Laflamme[1], H. Saleem[1], B. Vasan[2], R. Geiger[2], S. Chaudhary[2], M. Kollosche[3]
K. Rajan[4] and G. Kofod[3]

[1] Dept. of Civil, Constr., and Env. Engineering, Iowa State University, Ames, IA, USA;
[2] Dept. of Electrical and Computer Engineering, Iowa State University, Ames, IA, USA;
[3] Institut für Physik und Astronomie, Potsdam University, Potsdam, Germany;
[4] Dept. of Materials Science and Engineering, Iowa State University, Ames, IA, USA

ABSTRACT

Several sensing systems have been developed over the last decade for structural health monitoring (SHM) of wind turbine blades, and some have shown promise at damage detection. Yet, there exist several challenges in establishing links between sensor signal, damage state (diagnosis), and residual life (prognosis) of a wind turbine blade. In order to create a complete SHM system capable of damage diagnosis, localization, and prognosis for wind turbine blades, the authors propose a novel sensing method capable of distributed sensing. The sensor is a flexible membrane composed of several soft capacitors arranged in a matrix, with the advantages of being robust, easy to install, inexpensive, capable of covering large surfaces, and involving relatively simple signal processing. By measuring changes in capacitance, it is possible to determine local strains on a structural member at pre-defined levels of precisions. To demonstrate the promise of the smart material at SHM of blades, the sensing system is tested for quasi-static and dynamic loads. Test results show the great potential of the proposed sensing technology at damage diagnosis, localization, and prognosis of wind turbine blades. The latest state of research on the sensing solution is also discussed, and a new measurement method is demonstrated on a fiberglass plate.

Keywords: structural health monitoring, wind turbine blades, flexible membrane sensor, area strain gauges, soft capacitors.

1 INTRODUCTION

Blade components of wind turbines are associated with the highest failure rates and repair costs.[1,2] Currently, the vast majority of structural health monitoring (SHM) of wind turbine blades is conducted offline, using a combination of visual inspections and nondestructive evaluation techniques (NDT). NDT, such as real-time X-rays, optical coherence tomography, Eddy current, and laser strain, have been thoroughly researched and evaluated.[3-5] Those methods may be expensive, and depend on maintenance scheduling and on the judgment of inspectors.[6] Also, the major disadvantage of those techniques is that most of those methods cannot be used for continuous or real time monitoring of large-scale systems. There is a need to automate the process of damage diagnosis, localization, and prognosis of wind turbine blades. Some sensing methods have been showed successful at such automation, including acoustic emission using piezoelectric sensors, strain and displacement measurements using fiber optic, and inertial sensing using accelerometers.[7] Benefits of automating the health monitoring process of wind turbine blades are substantial:[2,8]

- Avoidance of premature breakdown by allowing timely preventive maintenance.

- Improved maintenance scheduling directly resulting in reduced costs.

- Remote monitoring and diagnosis, especially beneficial for offshore wind farms.

- Improvement of the energy production capacity factor by reducing conservatism in blade design.

- Support for further development of turbines by gathering data on blade behaviors.

This paper proposes to develop a large-scale flexible membrane that can be deployed on blade beams (spars / shear beams), and behaves like a matrix of capacitance-based strain gauges. The proposed method is analogous to biological skin, in the sense that it is capable of global monitoring by deploying sensors over large regions. Each strain gauge is a soft capacitor, which is fabricated using a highly sensitive elastomer sandwiched between electrodes. Changes in the material geometry correspond to changes in the material capacitance, which are directly related to local deformations. The proposed sensory membrane differs from other smart materials proposed for SHM of large-scale systems[9–12] by combining all of the following advantages: 1) inexpensive material; 2) low voltage consumption (1-2.5 volts); 3) easy to install; 4) customizable surface monitoring; and 5) robust with respect to physical damages.

The paper is organized as follow. Section 2 presents the proposed sensing method. Section 3 reviews results obtained from previous laboratory verifications to demonstrate the sensing method. Section 4 discusses the latest advances on the technology development, including a field application to study the behavior of the material in a harsh environment, and a new measurement method demonstrated for frequency identification of a fiberglass plate. Section 5 concludes the paper.

2 SENSING PRINCIPLE

The sensing method is a flexible membrane constructed with a matrix of individual capacitive-based sensors, termed sensing patches, where each patch can be customized in shape and size. A sensing patch is constituted from a thermoplastic elastomer sandwiched between two layers of highly compliant electrodes. The value of the capacitance C of a sensing patch is obtained using the surface area A, thickness d, vacuum permittivity ϵ_0, and permittivity of the polymer ϵ_r:

$$C = \epsilon_0 \epsilon_r \frac{A}{d} \qquad (1)$$

Assuming that for a small strain the thickness remains constant, a strain produces a change in the area ΔA, which directly results in a change in capacitance ΔC. Fig. 1a schematizes the sensing principle. The sensor is adhered to the monitored surface by a bonding agent. A strain (bottom red arrows), for example bending or a crack, produces a change in the sensor geometry (top red arrows), which in turn results in a change in capacitance.

Thus, by using a differential measurement setup, where the change in capacitance ΔC between two patches is measured,[14] the sensitivity of the sensor can be increased substantially by raising its permittivity ϵ_r. This concept has been demonstrated in Ref.[15] Here, the permittivity of the sensor has been adjusted by adding 15%vol high permittivity nanoparticles consisting of titanium dioxide (TiO_2), to a TPE material SEBS (Dryflex 500120, Elastoteknik), constituting the sensing material. The sensing material is sandwiched between stretchable electrodes fabricated from the same SEBS material, mixed with 10%vol carbon black (Printex XE 2-B). Fig. 1b is a picture of a sensing patch.

The sensitivity of a sensing patch is shown in Fig. 2. In this experiment, a free standing sensor has been subjected to quasi-static strain in a tensile tester (Zwick Roell Z005). The sample sensor was pre-strained at 10%, and measured from this level. Results show that the strain history can be detected above 0.2 ppm, which corresponds to approximately 1.5 μm. Note that there is a drift in the capacitance data, which may be caused by material relaxation. To minimize this effect

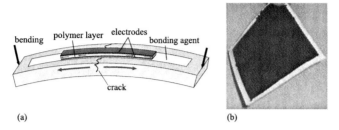

(a)   (b)

Figure 1: a) schematic of the sensing principle; and b) a sensing patch, 75 x 75 mm (3 x 3 in).[13]

along with the impact of environmental changes (e.g. temperature and humidity), all laboratory verifications showed in the next section used the differential measurement setup discussed above, where ΔC is measured instead of C.

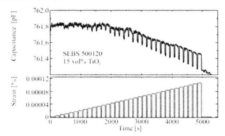

Figure 2: sensitivity of a sensing patch with respect to strain. Capacitance time history (above) and strain time history (below).[13]

3 LABORATORY VERIFICATION

The performance of the sensing solution has been demonstrated for detecting cracks,[14, 16] quasi-static strain,[17] and dynamic strain.[13] This section gives examples of these laboratory verifications.

3.1 Quasi-Static Load

The crack detection and strain measurement capability of the sensor was evaluated using a three-point load setup (Fig. 3b) using two sensing patches. One of the patches was located at mid-span (Sensor 2) under the specimen and the second (reference) located next to the specimen. A quasi-static load was applied using an Instron servo-hydraulic testing machine at a constant displacement rate of 0.500 mm/min until failure. Fig. 3a shows a filtered time history of the differential capacitance versus the applied load on the specimen. There is a strong correlation (99.8%) between both measurements during the first 300 seconds, before the first crack. Note that in the case of the three-point load setup, the relationship between load and curvature is approximatively linear. Therefore, the sensor strain corresponds directly to a change in the load. In Fig. 3a, the formation of the crack (Fig. 3c) is observed from the the first jump in differential capacitance accompanied by a drop in the load. The later expansion of the crack caused a failure, denoted in the second jump in the measurements. Results show that the sensor is capable of measuring strains and detecting cracks.

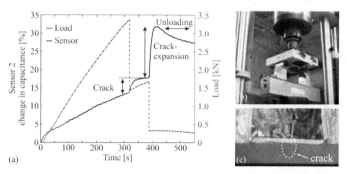

(a)

Figure 3: a) relative capacitance and applied load against time; b) typical three-point load setup for the wood specimens; and c) cracked specimen (bottom flange view).[13]

3.2 Dynamic Load

The dynamic performance of the sensor was evaluated on a 5.5 m (18 ft)-long HP10x42 steel beam. A 4000 rpm capacity shaker was installed on the top flange of the beam to produce a dynamic excitation (chirp signal) along the strong axis of the beam, at 2.85 m (9.3 ft) from the right extremity of the beam. Two sensing patches were installed 1.8 m (6 ft) away from the right support, 280 mm (11 in) apart. The experimental setup is shown in Fig. 4a. Data were sampled at 200 Hz for approximately 35 seconds. A plot of the power spectral density is shown in Fig. 4b. The plot was obtained after reconstructing the signal from the first 68% singular values from a principal component analysis. Table I compares the results obtained from the experimental verification against a finite element model. The model was built in SAP2000, using soft boundary conditions in the weak direction to match the first modal result. The first 4 modes were identified in the frequency response, but with a larger error for the 4th mode, which can be explained by noise and/or modelling inaccuracies.

Fig. 4c shows a wavelet transform using morlet wavelets on a section of the original signal extracted using a Tukey windowing function to reduce frequency leakage. The chirp signal is denoted by the increasing frequency response over time. The plateau at 23.8 Hz is the fundamental frequency of the torsional mode, which strongly resonates in its neighborhood. A second increasing frequency response can be observed at the lower middle of the plot. This might be caused by the shaker producing different excitation frequencies along the weak axis of the beam. Results show that the sensor is capable of detecting frequency responses.

4 STATE OF RESEARCH

The proposed technology is at an early development stage. Current research is focused on material testing using a field application and amelioration of the electronics for data measurements. These efforts are described respectively in the next subsections.

Table I: comparison of modal properties.

mode number	1	2	3	4
mode type (axis)	weak	polar	strong	weak
FEM (Hz)	12.3	26.5	42.9	55.5
experimental (Hz)	12.0	23.8	47.2	71.1
difference (%)	-2.44	-10.2	10.0	28.1

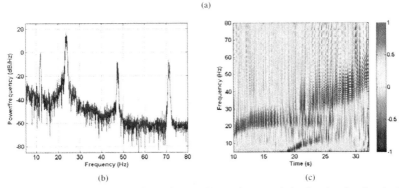

(a)

(b)　　　　　(c)

Figure 4: a) picture of the experimental setup; b) pseudo spectral density plot showing the first four frequencies; and c) normalized wavelet transform of the response.[13]

4.1 Field Application

A field application is currently being studied to evaluate the behavior of the sensor in a harsh winter environment. Two patches were installed on the South Skunk River Bridge (eastbound direction), located in Ames, IA. The bridge is a three-span two-lane highway overpass, 320 ft long by 30 ft wide, spanning the South Skunk River on the US-30. Fig. 5b shows the experimental setup with the two sensors installed under the deck in the longitudinal direction along with the data acquisition system secured on the first girder from the west side.

Fig. 5a shows the frequency response acquired using the patches. Results show resonant frequencies at 0.63 Hz and 4.2 Hz. Those results are consistent with a previous SHM campaign conducted in 2010 on the same bridge using fiber optic sensors,[18] where the authors identified a quasi-static frequency around 0.5 Hz and a fundamental frequency around 4.2 Hz.

4.2 Data Measurements

Currently, single data acquisition systems capable of measuring and acquiring the differential capacitance between two patches are being used. The objective is to ameliorate the measurement

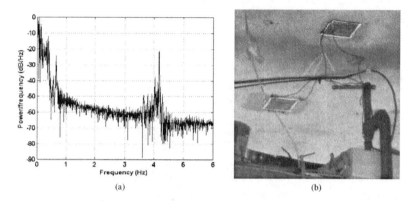

Figure 5: a) power spectra density of the capacitance signal; and b) experimental setup.

process by designing electronics capable of sequentially measuring differential capacitances between adjacent patches, throughout an entire array of patches. Using this principle, only one data acquisition system would be needed in a field application, substantially reducing the costs associated with the electronics. It follows that the performance of the capacitive strain sensor is strongly dependent upon the performance of the electronic circuits that provide a readout of the sensor array. Because strain information is being encoded in very small changes in capacitance, a readout circuit that accurately provides an output signal in function of small changes in capacitance is required.

Here, a dedicated integrated circuit solution is presented. The circuitry needs to measure the differential capacitance of a large number of small capacitors (approximately 300 pF), while minimizing intrusive electronics in the presence of large parasitics and environmental noise. Thus, a highly accurate, fast sampling circuitry is required. There exist a large number of transducers that encode the sensing information in capacitor differences, with considerable work reported on associated measurement electronics.[19-21] However, much of this work focuses on the difference of two capacitors rather than on differential capacitances between adjacent capacitors in a large array. Also, conversely to the proposed sensing method, the vast majority of applications measure only large changes in capacitance.

The schematic of the proposed circuitry is shown in Fig. 6, in simplified form of a stray-insensitive readout circuit. The circuit provides an output voltage proportional to the difference of two adjacent capacitors C_1 and C_2, where V_{IN} is a constant DC or AC voltage, ϕ_1 and ϕ_2 are complimentary non-overlapping clocks, C_F, R_1 and R_F are a capacitor and two resistors used for calibration respectively, and $A1$ and $A2$ are amplifiers, with:

$$V_{OUT} = \frac{(C_2 - C_1)}{C_F} V_{IN}(1 + \frac{R_F}{R_1})$$ (2)

Capacitors C_1 and C_2 are assumed to be equal: $C_1 = C_2 = C_{NOM}$, which can be done using appropriate manufacturization and calibration. The circuit is designed to provide a full-scale output with capacitor changes of approximately 0.1% of C_{NOM}.

The proposed data measurement method has been tested on two patches installed on a fiberglass

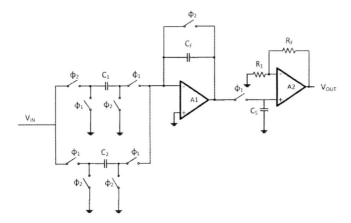

Figure 6: schematic of the differential capacitance measurement circuitry.

plate. The plate was clamped from one edge as shown in Fig. 7b. During the free vibration experiment, an initial displacement of 25.4 mm (1 in) was induced at the free edge of the plate and released. Fig. 7a shows the frequency response of the differential capacitance measurements. The measurement method captured the first five vibration modes. The spacings between modes are conform to the analytical solution given in Ref.[22] These results demonstrate the good dynamic performance of the measurement method.

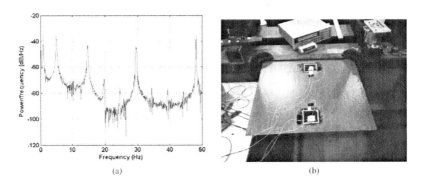

Figure 7: a) power spectra density of the differential capacitance signal; and b) experimental setup.

5 CONCLUSION

A novel sensing method has been presented for automating the SHM process of wind turbine blades. The method uses a flexible membrane constructed from an array of soft capacitors, each capacitor constituting an individual strain gauge. Such membrane allows discretization of strain over large areas, analogous to biological skin.

Results presented in this paper demonstrate the capacity of the membrane to measure strains, detect cracks, and measure frequency responses. Current advances in the development of this innovative technology include a field test to assess the performance of the sensor material in a harsh environment, and the development of new electronics for measuring and acquiring differential capacitance between adjacent patches. Preliminary laboratory testing using the new circuitry has been successful.

The proposed bio-inspired membrane is a promising solution for an automatic condition assessment of blades. Its application includes several benefits: better scheduling of maintenance and inspection programs, reduced life-cycle costs of wind turbines, reduce safety hazards associated with blade failures, and collection of data for enhanced engineering of blades.

REFERENCES

[1] M. Khan, M. Iqbal, and F. Khan, "Reliability and condition monitoring of a wind turbine," in *Electrical and Computer Engineering, 2005. Canadian Conference on*, pp. 1978–1981, IEEE, 2005.

[2] C. Ciang, J. Lee, and H. Bang, "Structural health monitoring for a wind turbine system: a review of damage detection methods," *Measurement Science and Technology*, vol. 19, p. 122001, 2008.

[3] B. Sørensen, L. Lading, P. Sendrup, M. McGugan, C. Debel, O. Kristensen, G. Larsen, A. Hansen, J. Rheinländer, and J. Rusborg, *Fundamentals for remote structural health monitoring of wind turbine blades-a preproject*. Risø National Laboratory, 2002.

[4] R. Hyers, J. McGowan, K. Sullivan, J. Manwell, and B. Syrett, "Condition monitoring and prognosis of utility scale wind turbines," *Energy Materials: Materials Science and Engineering for Energy Systems*, vol. 1, no. 3, pp. 187–203, 2006.

[5] G. Harvey and J. Jones, "Small turbine blade inspection using laser strain techniques," *Insight*, vol. 51, no. 3, p. 137, 2009.

[6] B. Sørensen, K. Branner, H. Stang, E. Jensen, E. Lund, T. Jacobsen, and K. Halling, "Improved design for large wind turbine blades of fibre composites (phase 2)-summary report," tech. rep., Danmarks Tekniske Universitet, Risø Nationallaboratoriet for Bæredygtig Energi Roskilde, 2005.

[7] L. Lading, M. McGugan, P. Sendrup, J. Rheinländer, and J. Rusborg, "Fundamentals for remote structural health monitoring of wind turbine blades - a preproject, annex b," *Risø-R-1341 (EN) Report*, 2002.

[8] D. Adams, J. White, M. Rumsey, and C. Farrar, "Structural health monitoring of wind turbines: method and application to a hawt," *Wind Energy*, 2011.

[9] P. Harrey, B. Ramsey, P. Evans, and D. Harrison, "Capacitive-type humidity sensors fabricated using the offset lithographic printing process," *Sensors and Actuators B: Chemical*, vol. 87, no. 2, pp. 226–232, 2002.

[10] Y. Zhang, "Piezoelectric paint sensor for real-time structural health monitoring," in *Proc. of SPIE*, vol. 5765, p. 1095, 2005.

[11] B. Zhang, Z. Zhou, K. Zhang, G. Yan, and Z. Xu, "Sensitive skin and the relative sensing system for real-time surface monitoring of crack in civil infrastructure," *Journal of intelligent material systems and structures*, vol. 17, no. 10, pp. 907–917, 2006.

[12] S. Sarles and D. Leo, "Membrane-based biomolecular smart materials," *Smart Materials and Structures*, vol. 20, p. 094018, 2011.

[13] S. Laflamme, K. Kollosche, V. Kollipara, H. Saleem, and G. Kofod, "Large-Scale Surface Strain Gauge for Health Monitoring of Civil Structures," in *Proc. of SPIE - NDE*, p. to be presented, 2012.

[14] S. Laflamme, M. Kollosche, J. J. Connor, and G. Kofod, "Large-scale capacitance sensor for health monitoring of civil structures," in *Proc. of 5WCSCM*, 2010.

[15] H. Stoyanov, M. Kollosche, D. N. McCarthy, and G. Kofod, "Molecular composites with enhanced energy density for electroactive polymers," *Journal of Material Chemistry*, vol. 20, pp. 7558–7564, 2010.

[16] S. Laflamme, M. Kollosche, J. J. Connor, and G. Kofod, "Soft capacitive sensor for structural health monitoring of large-scale systems," *Structural Control and Health Monitoring*, 2010.

[17] M. Kollosche, H. Stoyanov, S. Laflamme, and G. Kofod, "Strongly enhanced sensitivity in elastic capacitive strain sensors," *Journal of Material Chemistry*, vol. 21, no. 23, pp. 8292–8294, 2011.

[18] P. Lu, B. Phares, L. Greimann, and T. Wipf, "Bridge structural health-monitoring system using statistical control chart analysis," *Transportation Research Record: Journal of the Transportation Research Board*, vol. 2172, no. 1, pp. 123–131, 2010.

[19] H. Hu, T. Xu, and S. Hui, "A high-accuracy, high-speed interface circuit for differential-capacitance transducer," *Sensors and Actuators A: Physical*, vol. 125, no. 2, pp. 329–334, 2006.

[20] T. Constandinou, J. Georgiou, and C. Toumazou, "A micropower front-end interface for differential-capacitive sensor systems," in *Circuits and Systems, 2008. ISCAS 2008. IEEE International Symposium on*, pp. 2474–2477, IEEE, 2008.

[21] B. George and V. Kumar, "Analysis of the switched-capacitor dual-slope capacitance-to-digital converter," *Instrumentation and Measurement, IEEE Transactions on*, vol. 59, no. 5, pp. 997–1006, 2010.

[22] J. Dugundji, "Simple expressions for higher vibration modes of uniform Euler beams," *AIAA Journal*, vol. 26, pp. 1013–1014, Aug. 1988.

ROBOTIC AND MULTIAXIAL TESTING FOR THE DETERMINATION OF THE CONSTITUTIVE CHARACTERIZATION OF COMPOSITES

John Michopoulos, Naval Research Laboratory, Code 6394, Washington, DC, USA

Athanasios Iliopoulos, George Mason University, resident at NRL, Washington, DC, USA

John Hermanson, Forest Products Laboratory, Madison, Winsconsin, USA

ABSTRACT

As wind energy production drives the manufacturing of wind turbine blades, the utilization of glass and carbon fiber composites as a material of choice continuously increases. Consequently, the needs for accurate structural design and material qualification and certification as well as the needs for aging predictions further underline the need for accurate constitutive characterization of composites. In the present paper we describe an outline of a recently developed methodology that utilizes mutliaxial robotically controlled testing combined with design optimization for the automated constitutive characterization of composite materials for both the linear and non-linear regimes. Our approach is based on the generation of experimental data originating from custom-developed mechatronic material testing systems that can expose specimens to multidimensional loading paths and can automate the acquisition of data representing the excitation and response behavior of the specimens involved. Material characterization is achieved by minimizing the difference between experimentally measured and analytically computed system responses as described by strain fields and surface strain energy densities. Small and finite strain formulations based on strain energy density decompositions are developed and utilized for determining the constitutive behavior of composite materials. Examples based on actual data demonstrate the successful application of design optimization for constitutive characterization. Validation experiments and their comparisons to theoretical predictions demonstrate the power of this approach.

INTRODUCTION

The process of designing any structure interacting with a fluid, such that of a wind turbine blade in an air-stream, requires the ability to model the interaction for the two continua in the applicable multiphysics context. The aero-structural behavior of such a system can be described by a system of partial differential equations (PDEs)[1].

In fact, if we let $\tilde{\chi}_t$ denote a continuous mapping from a reference fluid configuration $\Omega_F(t)|_{t\ 0} \subset \mathbb{R}^3$ to a current fluid configuration $\Omega_F(t) \subset \mathbb{R}^3$ such that

$$\chi_t : \Omega_F(t)|_{t\ 0} \longrightarrow \Omega_F(t), x(\xi,t) = \chi_t(\xi), \tag{1}$$

where $t \in [0,\infty]$ denotes time, $x(\xi,t)$ denotes the time-dependent position vector of a fluid point, ξ its position in reference configuration, and $J = |\partial x/\partial \xi|$ denotes the Jacobian determinant of the deformation gradient, then the relevant PDEs take the form,

$$\left.\frac{\partial(Jw)}{\partial t}\right|_\xi + J\nabla_x \cdot \left(F(w) - \left.\frac{\partial x}{\partial t}\right|_\xi w\right) = J\nabla_x \cdot R(w) + JS(w), \text{ in } \Omega_F \tag{2a}$$

$$\rho_S \frac{\partial^2 u_S}{\partial t^2} - \nabla_x \tilde{\sigma}_S(\tilde{\varepsilon}_S) = b, \text{ in } \Omega_S \tag{2b}$$

$$\rho_f \left.\frac{\partial^2 x}{\partial t^2}\right|_\xi - \nabla_\xi \tilde{\sigma}_f(x - x|_{t\ 0}) = 0, \text{ in } \Omega_F \tag{2c}$$

Equation (2a) represents the arbitraty Lagrangean Eulerian (ALE) form of the Navier-Stokes equations where w denotes the conservative fluid state vector, F and R denote the convective and diffusive fluxes, respectively. The quantity $S(w)$ denotes the source term associated with a potential turbulence model. Equa-

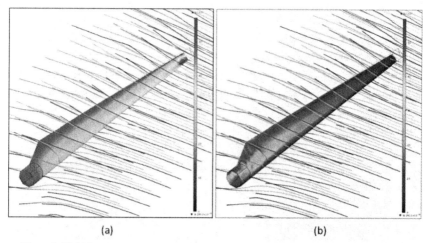

(a) (b)

Figure 1. Fluid-Structure Interaction results with velocity stream lines superimposed. Fluid pressure distribution (a) and Von-Mises stresses distributions (b) for a typical wind turbine blade.

tion (2b) is the equation governing the dynamics of the structure represented by the domain $\Omega_S(t) \subset \mathbb{R}^3$. In this equation the quantities ρ_S, $\tilde{\sigma}_S$, $\tilde{\varepsilon}_S$, u_S and b represent the density, the second order stress and strain tensors, and the displacement and body forces vectors respectively. The field governed by the Eq. (2c) does not have any direct physical origin but it is necessary as it provides algebraic and physical closure to the system and it describes the dynamics of the fluid mesh motion by casting it within the formalism of a fictitious or pseudo-structural subsystem. Tilded symbols denote second order tensor field state variables. The above equations are completed with their Dirichlet, Neumann or Cauchy boundary conditions as described elsewhere[1-3]

For the case of NACA 4-series based wind turbine blade, Fig. 1 shows a typical result produced by solving this system of equations (2) under the assumption that the material behaves elastically, for low wind velocities such as the assumptions of laminar flow can be employed.

As the ability to solve this formalism is essential for the design of wind turbine blades, the knowledge of the analytical form of the constitutive functional $\tilde{\sigma}_S(\tilde{\varepsilon}_S)$ is an essential, critical and enabling element. When this functional is linear it is known as Hooke's constitutive law. In the case where this relationship is not linear due to strain induced damage, these equations represent a more general case that allows for modeling degradation of materials.

For this reason (and in the context of aerospace naval structures), during the past decade we have embarked in an effort to automate massive multiaxial testing of composite coupons in order to determine the constitutive behavior of the bulk composite material used to make these coupons. In this paper we are reporting on an overview of a methodology and the first successful campaign of experiments for this purpose.

The approach is motivated by the data-driven requirements of employing design optimization principles for determining the constitutive behavior of composite materials as described in our recent work[4,5]

The constitutive characterization of composite materials has been traditionally achieved through conventional uniaxial tests and used for determining elastic properties. Typically, extraction of these properties, involve uniaxial tests conducted with specimens mounted on uniaxial testing machines, where the major

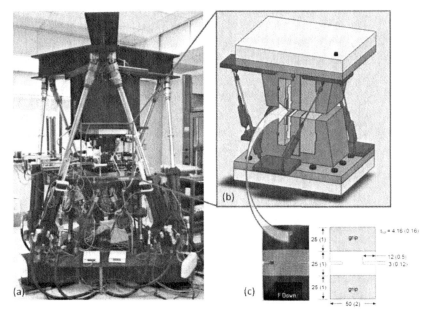

Figure 2. View of the current state of NRL66.3 6-DOF mechatronically automated systems (a); view of a CAD drawing of the grip assembly (b); typical specimen view with geometrical specifications in mm (inches) (c).

orthotropic axis of any given specimen is angled relative to the loading direction. In addition, specimens are designed such that a homogeneous state of strain is developed over a well defined area, for the purpose of measuring kinematic quantities[6,7]. Consequently, the use of uniaxial testing machines imposes requirements of using multiple specimens, griping fixtures and multiple experiments. The requirement of a homogeneous state of strain frequently imposes restrictions on the sizes and shapes of specimens to be tested. Consequently, these requirements result in increased cost and time, and consequently to inefficient characterization processes.

To address these issues and to extend characterization to non-linear regimes, multi-degree of freedom automated mechatronic testing machines, which are capable of loading specimens multiaxially in conjunction with energy-based inverse characterization methodologies, were introduced at the Naval Research Laboratory (NRL)[8–12]. This introduction was the first of its kind and has continued through the present[13–15]. The most recent prototype of these machines is shown in Fig. 2.

The energy-based approach associated with mechatronic testing, although it enables multiaxial loading and inhomogeneous states of strain, still requires multiple specimens. It is significant to state however, that these specimens are tested in a automated manner with high throughput of specimens per hour, which have reached a rate of 30 specimens per hour.

The recent development of flexible full-field displacement and strain measurements methods has afforded the opportunity of alternative characterization methodologies[16–19]. Full-field optical techniques, such as Moire and Speckle Interferometry, Digital Image Correlation (DIC), and Meshless Random Grid Method

(MRGM), which measure displacement and strain fields during mechanical tests, have been used mostly for elastic characterization of various materials[19–22]. The resulting measurements are used for identification of constitutive model constants, via the solution of an appropriately formed inverse problem, with the help of various computational techniques.

Although there are many ways to approach this problem here we are focusing on a mixed numerical/experimental method that identifies the material's elastic constants by minimizing an objective function formed by the difference between the full-field experimental measurements and the corresponding analytical model predictions via an optimization method[8,9,12–15,19,22–24].

Our approach is based on energy conservation arguments, and it can be classified according to computational cost in relation to the iterative use of FEA or not. It is important to clarify that digitally acquired images are processed by software[25] that implements the MRGM[19,26–30] and is used to extract the full-field displacement and strain field measurements as well as the boundary displacements required for material characterization. Reaction forces and redundant boundary displacement data are acquired from displacement and force sensors integrated with NRL's multiaxial loader called NRL66.3[31]. In an effort to address the computational cost of the FEA-in-the-loop approaches, the authors have initiated a dissipated and total strain energy density determination approach that has recently been extended to a framework that is derived from the total potential energy and the energy conservation, which can be applied directly with full field strain measurement for characterization[32–37].

In the section that follows we present an overview of the small strain formulation (SSF) of the general strain energy density approach followed by the finite strain formulation (FSF), that are intended to capture both the linear and non-linear constitutive behavior of composite materials with or without damage. The paper continuous with a description of the computational application of design optimization implementations based on these two formulations. A description of he experimental campaign and representative results follow. Finally, conclusions are presented.

COMPOSITE MATERIAL CONSTITUTIVE RESPONSE REPRESENTATIONS

For the general case of a composite material system we consider that a modified anisotropic hyperelastic strain energy density (SED) function can be constructed to encapsulate both the elastic and the inelastic responses of the material. However, certain classes of composite materials reach failure after small strains and some under large strains. For this reason we give two example formalisms, one involving a small (infinitesimal) strain formulation (SSF) and another involving a finite (large) strain formulation (FSF).

Small Strain Formulation

For the SSF case we have introduced a SED function that, in its most general form, can be represented as a scaled Taylor expansion of the Helmholtz free energy of a deformable body, which is expressed initially in terms of small strain invariants, and eventually of an additive decomposition in terms of a recoverable and an irrecoverable SED that can be expressed by

$$U_{SSF} = U_{SSF}^R(S;\varepsilon_{ij}) + U_{SSF}^I(D;\varepsilon_{ij}). \tag{3}$$

Clearly, all the second order monomials of strain components will be forming the recoverable part $U_{SSF}^R(S;\varepsilon_{ij})$ and the higher order monomials will be responsible for the irrecoverable part $U_{SSF}^I(D;\varepsilon_{ij})$. The resulting constitutive law is given by

$$\sigma_{ij} = \partial U_{SSF}/\partial \varepsilon_{ij} = \partial(U_{SSF}^R(S;\varepsilon_{ij}) + U_{SSF}^I(D;\varepsilon_{ij}))/\partial \varepsilon_{ij} \tag{4}$$

A general expression which provides the strain dependent version of Eq. 3 after the expansion of the strain invariants[4], is given by

$$U_{SSF} = U_{SSF}^R(S;\varepsilon_{ij}) + U_{SSF}^I(D;\varepsilon_{ij}) = \frac{1}{2}s_{ijkl}\varepsilon_{ij}\varepsilon_{kl} + d_{ijkl}(\varepsilon_{ij})\varepsilon_{ij}\varepsilon_{kl} \tag{5}$$

where s_{ijkl} are the components of the elastic stiffness tensor (Hooke's tensor) and $d_{ijkl}(\varepsilon_{ij})$ are strain-dependent damage functions, which fully define irrecoverable (or dissipated) strain energy density given by enforcing the dissipative nature of energy density. The quantity $d_{ijkl}(\varepsilon_{ij})$ can be defined in a manner analogous to that employed for the 1D system described elsewhere[4] and is given by

$$d_{ijkl}(\varepsilon_{ij}) = s_{ijkl}(1 - e^{(-\frac{\varepsilon_{ij}}{q_{ij}})^{p_{ij}}/(ep_{ij})}) \tag{6}$$

A follow-up series expansion and subsequent drop of all terms except the first is enough in capturing almost all of the characteristics of dissipative behavior, yielding,

$$d_{ijkl}(\varepsilon_{ij}) = s_{ijkl}\sum_{m=1}^{n}(-1)^m\left(\frac{1}{ep_{ij}}\right)^m\frac{\varepsilon_{ij}^{mp_{ij}}}{m!q_{ij}^{mp_{ij}}} \simeq -s_{ijkl}\left(\frac{1}{ep_{ij}}\right)\frac{\varepsilon_{ij}^{p_{ij}}}{q_{ij}^{p_{ij}}} \tag{7}$$

Thus the irrecoverable part of the energy in Eq. 3 becomes

$$U_{SSF}^I(D;\varepsilon_{ij}) = U_{SSF}^I(s_{ijkl},p_{ij},q_{ij};\varepsilon_{ij}) = -s_{ijkl}\frac{1}{e(2+p_{ij})p_{ij}q_{ij}^{p_{ij}}}\varepsilon_{ij}^{1+p_{ij}}\varepsilon_{kl} \tag{8}$$

Next, substituting Eq. 6 into Eq. 3 yields

$$U_{SSF} = U_{SSF}^R(S;\varepsilon_{ij}) + U_{SSF}^I(D;\varepsilon_{ij}) =$$
$$= \frac{1}{2}s_{ijkl}\varepsilon_{ij}\varepsilon_{kl} - s_{ijkl}\frac{1}{e(2+p_{ij})p_{ij}q_{ij}^{p_{ij}}}\varepsilon_{ij}^{1+p_{ij}}\varepsilon_{kl} \tag{9}$$

Applying Eq. 4 on Eq. 7, and employing Voight[6] notation for the case of a general orthotropic material, yields the constitutive relation

$$\begin{bmatrix} \sigma_{xx} \\ \sigma_{yy} \\ \sigma_{zz} \\ \sigma_{xz} \\ \sigma_{yz} \\ \sigma_{xy} \end{bmatrix} = \begin{bmatrix} \tilde{s}_{xx} & s_{xy} & s_{xz} & 0 & 0 & 0 \\ s_{xy} & \tilde{s}_{yy} & s_{yz} & 0 & 0 & 0 \\ s_{xz} & s_{yz} & \tilde{s}_{zz} & 0 & 0 & 0 \\ 0 & 0 & 0 & \tilde{s}_{xz} & 0 & 0 \\ 0 & 0 & 0 & 0 & \tilde{s}_{yz} & 0 \\ 0 & 0 & 0 & 0 & 0 & \tilde{s}_{xy} \end{bmatrix} \begin{bmatrix} \varepsilon_{xx} \\ \varepsilon_{yy} \\ \varepsilon_{zz} \\ \varepsilon_{xz} \\ \varepsilon_{yz} \\ \varepsilon_{xy} \end{bmatrix} \tag{10}$$

where:

$$\tilde{s}_{ij} = s_{ij}\left(1 - \bar{d}_{ij}\right) \tag{11}$$

and

$$\bar{d}_{ij} = \frac{1}{ep_{ij}q_{ij}^{p_{ij}}}\varepsilon_{ij}^{1+p_{ij}} \tag{12}$$

The orthotropic symmetry requirements involve material parameters that are the 9 elastic s_{ij} constants and $6 \times 2 = 12$ damage constants p_{ij}, q_{ij} for a total of 21 parameters. Clearly, when the quantities $\bar{d}_{ij}$ do not depend on the strains and they are constants, Eq. 8 reduces to most of the continuous damage theories given by various investigators in the past[38-41]. For a transversely isotropic material the number of material parameters drops to 5+10=15 for a 3D state of strain and to 4+8=12 for a plane stress state.

Finite Strain Formulation

The FSF can be written in a double additive decomposition manner. The first being the decomposition of the recoverable and irrecoverable SED, and the second being the decomposition between the volumetric (or dilatational) W_v and the distortional (or isochoric) W_d parts of the total SED. This decomposition is expressed by:

$$U_{SSF} = U_{SSF}^R(\alpha_i; J, \bar{C}) + U_{SSF}^I(\alpha_i, \beta_i; J, \bar{C}) =$$
$$= [W_v(J) + W_d(\bar{C}, A \otimes A, B \otimes B)] - [d_v W_v(J) + d_d W_d(\bar{C}, A \otimes A, B \otimes B)] \tag{13}$$

where α_i, β_i are the elastic and inelastic material parameters of the system, respectively. A rearrangement of these decompositions, such as the volumetric vs. distortional decomposition, which appears on the highest expression level, leads to an expression introduced in[42], i.e.,

$$U_{FSF} = (1 - d_v)W_v(J) + (1 - d_d)W_d(\bar{C}, A \otimes A, B \otimes B), \tag{14}$$

with the damage parameters $d_k \in [0,1], k \in [v,d]$ defined as

$$d_k = d_{ka}^\infty \left[1 - e^{\left(-\frac{a_k(t)}{\eta_{ka}} \right)} \right] \tag{15}$$

where $a_k(t) = \max_{s \in [0,t]} W_k^o(s)$ is the maximum energy component reached so far, and d_{ka}^∞, η_{ka} are two pairs of parameters controlling the energy dissipation characteristics of the two components of SED. In this formulation, $J = \det F$ is the Deformation Gradient, $\bar{C} = F^T F$ is the right Cauchy Green (Green deformation) tensor, A, B are constitutive material directions in the undeformed configuration, and $A \otimes A, B \otimes B$ are microstructure structural tensors expressing fiber directions. Each of the two components of SED are defined as

$$W_v(J) = \frac{1}{d}(J-1)^2 W_d(\bar{C}, A \otimes A, B \otimes B) =$$
$$= \sum_{i=1}^{3} a_i(\bar{I}_1 - 3)^i + \sum_{j=1}^{3} b_j(\bar{I}_2 - 3)^j + \sum_{k=1}^{6} c_k(\bar{I}_4 - 1)^k +$$
$$+ \sum_{l=2}^{6} d_l(\bar{I}_5 - 1)^l + \sum_{m=2}^{6} e_m(\bar{I}_6 - 1)^m + \sum_{n=2}^{6} f_n(\bar{I}_7 - 1)^n + \sum_{o=2}^{6} g_o(\bar{I}_8 - (A \cdot B)^2)^o \tag{16}$$

where the strain invariants are defined as follows:

$$\begin{aligned}
\bar{I}_1 &= tr\bar{C}, \quad \bar{I}_2 = \tfrac{1}{2}(tr^2\bar{C} - tr\bar{C}^2) \\
\bar{I}_4 &= A \cdot \bar{C}B, \quad \bar{I}_5 = A \cdot \bar{C}^2 B \\
\bar{I}_6 &= B \cdot \bar{C}B, \quad \bar{I}_7 = B \cdot \bar{C}^2 B, \quad \bar{I}_8 = (A \cdot B)A \cdot \bar{C}B
\end{aligned} \tag{17}$$

The corresponding constitutive behavior is given by the second Piola-Kirchhoff stress tensor according to[39]

$$S = 2\frac{\partial U_{FSF}}{\partial C} \tag{18}$$

or the usual Cauchy stress tensor according to

$$\sigma_{FSF} = \frac{2}{J}F \cdot \frac{\partial U_{FSF}}{\partial C} \cdot F^T. \tag{19}$$

Under the FSF formulation the material characterization problem involves determining the 36 coefficients (at most) of all monomials when the sums in the expression of distortional SED are expanded in Eq. 16, in addition to the compressibility constant d and the 4 parameters used in Eq. 15. It follows that potentially there can be a total of 41 material constants.

COMPUTATIONAL IMPLEMENTATION

In order to determine the material parameters the inverse problem at hand is solved through a design optimization approach that is described by the logic depicted in Fig. 3. The implementation of this logic involved a computational infrastructure that controlled by Matlab, where the foreword solution of the in-

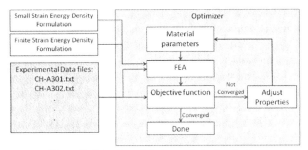

Figure 3. Overview of design optimization logic.

stantaneous FEA was accomplished by ANSYS. Due to the costly calculations we created a virtual cluster that essentially used the multiple cores available in three systems, via a spawning 102 virtual machines capable of running 102 instances of ANSYS in a parallel fashion, whereas each process was responsible for a different subdomain of the design space as described in[43].

Details if the FEM model used for the specimens constructed and tested are not presented here due to the space limitations but they can be found in[4,5].

Two objective functions were constructed. Both utilized the fact that through the REMDIS-3D software[44], developed by our group, one can obtain full field measurements of the displacement and strain fields over any deformable body as an extension of the REMDIS-2D software that is based on the MRGM[19,25–30]. Thus, our experimental measurements for the formation of the objective functions were chosen to be the strains at the nodal points of the FEM discretization. The first objective function chosen was based entirely on strains and is given by

$$J^\varepsilon = \sum_{k=1}^{N} \left(\sum_{i=1}^{2} \sum_{j=i}^{2} \left(\left[\varepsilon_{ij}^{\text{exp}} \right]_k - \left[\varepsilon_{ij}^{\text{fem}} \right]_k \right) \right)^2, \tag{20}$$

the second objective function is given in terms of surface strain energy density according to

$$J^U \approx \oint_{\partial\Omega} \left(U^{\text{exp}} - U^{\text{fem}} \right)^2 dS \approx$$
$$\approx \oint_{\partial\Omega} \left(\sum_{i=1}^{2} \sum_{j=i}^{2} \sum_{m=1}^{2} \sum_{n=m}^{2} \left(s_{ijmn} \varepsilon_{ij}^{\text{exp}} \varepsilon_{mn}^{\text{exp}} - s_{ijmn}^{\text{fem}} \varepsilon_{ij}^{\text{fem}} \varepsilon_{mn}^{\text{fem}} \right) \right)^2 dS \tag{21}$$

where $\left[\varepsilon_{ij}^{\text{exp}} \right]_k$, $\left[\varepsilon_{ij}^{\text{fem}} \right]_k$ are the experimentally determined and the FEM produced components of strain are at node k. The quantities $U^{\text{exp}}, U^{\text{fem}}$ are the surface strain energy densities formulated by using the experimental strains and the FEM produced strains respectively.

Both objective functions were implemented, and we utilized both the DIRECT global optimizer[45], which is available for Matlab[46], and a custom developed Monte-Carlo optimizer, also implemented in Matlab.

EXPERIMENTAL CAMPAIGN

Based on an analysis that falls outside the scope of the present paper[47], we identified 72 proportional loading paths to sample the 6-DoF space defined by the 3 translations and 3 rotations that can be applied by the moveable grip of the NRL66.3. We selected to use a single specimen per loading path and then repeat the process. That requires 144 specimens per material system. Each material system was a defined to be a balanced laminate with an alternating angle of fiber inclination per ply relative to the vertical axis of the

TABLE I: *Engineering properties of AS4/3501-6 laminae*

Ref.	E_{11}[GPa]	E_{22} [GPa]	ν_{12}	ν_{23}	G_{12}[GPa]
Daniels[6]	147.0	10.3	0.27	0.28	7.0
Present	125.0	10.8	0.27	0.32	7.96

specimen that is perpendicular to the axis defined by the two notches of the specimen. Four angles were used $+/-15, +/-30, +/-60, +/-75$ degrees and therefore $4 \times 144 = 576$ specimens were constructed. The actual material used to make the specimens was AS4/3506-1 epoxy resin/fiber respectively.

All specimens were tested by using NRL66.3 in May of 2011. In May 13, 2011, the system achieved a throughput of 20 test/hour for a total of 132 tests. On May 27 the system achieved a throughput of 25 tests/hour for a total of 216 tests. On June 15, 2011 the system reached its peak throughput of 28 tests/hour. All 1152 tests were completed in 12 work-days. The test yielded 13 TB of data from the sensors and the cameras of the system. Typical experimental data are shown in a more detailed reference[43].

Using the collected data and the optimization approach outlined earlier for the case of the SSF we identified the elastic constants as shown in Table 1, in comparison to those of[6]. By running FEA for the cases that correspond to the specific loading path corresponding to an experiment we can now compare the predicted distribution of any component of the strain or stress tensor. In order to demonstrate how well the FSF formulation can capture the behavior of the characterization coupons used to obtain the data utilized in the characterization process we are presenting typical examples in terms of the distributions of ε_{yy} as measured by the MRGM (left column) and as predicted by the FSF theory (right column), for both the front (top row) and the back (bottom row) of Figs. 4, 5.

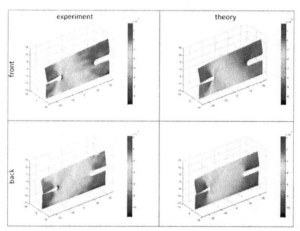

Figure 4. Comparison of (ε_{yy}) field between measured (left) and the identified model by using the FSF for the case of in plane rotation and torsion of a $+/-30$ degrees laminate

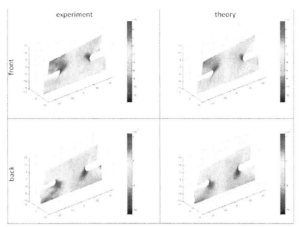

Figure 5. Comparison of (ε_{yy}) field between measured (left) and the identified model by using the FSF for the case of in-plane rotation and torsion of a +/- 60 degrees laminate

VALIDATION

The characterization methodology described here and in our previous publications, utilized the data described herein along with the data form other validation tests performed on structures of different shape and layups than that of the characterization coupons, to perform validation comparisons. A discussion of this activity is presented elsewhere[48] in more detail.

Here we present validation based on the typical double notched characterization specimen described in[43], but for loading paths that were not used for determining the constitutive model itself.

As a representative validation prediction we are presenting the results for two loading paths. It is important to emphasize that the experimental data obtained from the associated tests were not used in determining the constitutive constants (material parameters) that are fixing the associated constitutive model.

Figure 6 shows the three strain field distributions at the same loading step of a load path that involves tension, bending and torsion about the vertical axis.

Figure 7 shows the three strain field distributions at the same loading step of a load path that involves in plane rotation and torsion. All predicted strain field distributions for both of these cases, show that the predicted values are within 4% error of the experimental ones. This is within the error of the RemDiS-3D experimental method for that level of strain, except very few areas of high strain gradients.

CONCLUSIONS AND PLANS

We have described and outline of a multiaxial automated composite material characterization methodology that utilizes multiaxial experiments performed with the NRL66.3 mechatronic system.

Design optimization methodologies for the determination of the constitutive response of composite materials with or without damage has been employed. Strain energy density and full field strain based approaches have been utilized to incorporate massive full field strain measurements from specimens loaded by a custom-made multiaxial loading machine. We have formulated objective functions expressing the difference between the experimentally observed behavior of composite materials under various loading conditions, and the simulated behavior via FEA, which are formulated in terms of strain energy density functions

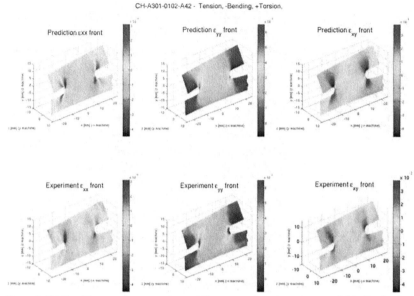

Figure 6. Predicted (top row) vs. experimental (bottom row) strain component fields distributions for a characterization coupon under combined tension, bending and torsion loading

of a particular structure under identical loading conditions.

Two formalisms involving small strains and finite strains have been utilized in a manner that involves both additive decomposition of recoverable and irrecoverable strain energy density. This was done in order to address both the elastic and inelastic response of composite materials due to damage. The finite strain formulation further involves a volumetric and distortional energy decomposition.

Representative results of the characterization have been compared with the associated experimental ones to demonstrate how well they agree.

Finally, for the validation purposes, we have presented a comparison of behavior predictions from FEA models based on the characterized constitutive behavior for the SSF vs. experimental results of characterization coupons not used for the characterization.

The coincidence between experiment and prediction for all tests conducted, clearly suggests our characterization methodology is successful in identifying the proper constitutive behavior and the associated material parameters.

This success provides confidence into pursuing our plans for utilizing the energy-based constitutive characterization to formulate failure criteria and address the very important issue of predicting failure for design and maintainability purposes of wind turbine blade applications.

ACKNOWLEDGEMENT

The authors acknowledge support for this work by the Office of Naval Research and the Naval Research Laboratory's core funding.

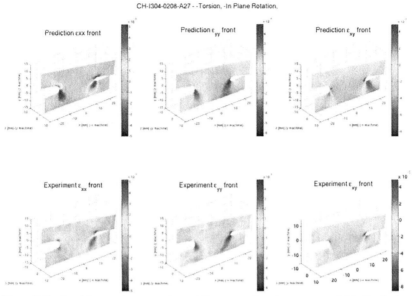

Figure 7. Predicted (top row) vs. experimental (bottom row) strain component fields distributions for a characterization coupon under combined in-plane rotation and torsion loading

REFERENCES

[1] J. Michopoulos, C. Farhat, and J. Fish, "Modeling and simulation of multiphysics systems," Journal of Computing and Information Science in Engineering **5**, 198–213 (2005).

[2] M. Lesoinne and C. Farhat, in *11th AIAA Computational Fluid Dynamics Conference* (AIAA, Orlando, Florida, 1993) pp. AIAA Paper No. 93–3325 1993.

[3] P. Geuzaine, G. Brown, and C. Farhat, "Three-field based nonlinear aeroelastic simulation technology: Status and application to the flutter analysis of an f-16 configuration," in *40th Aerospace Sciences Meeting and Exhibit*, AIAA Paper 2002-0870 (Reno, Nevada, USA, 2002) pp. 14–17.

[4] J. Michopoulos, J. Hermanson, A. Iliopoulos, S. Lambrakos, and T. Furukawa, "Data-driven design optimization for composite material characterization," Journal of Computing and Information Science in Engineering **11** (2011).

[5] J. Michopoulos, J. C. Hermanson, A. Iliopoulos, S. Lambrakos, and T. Furukawa, "Overview of constitutive response characterization for composite materials via data-driven design optimization," in *Proceedings of the ASME 2011 International Design Engineering Technical Conferences & Computers and Information in Engineering Conference IDETC/CIE, August 29-31, 2011, Washington, DC, USA* (2011).

[6] I. M. Daniel and O. Ishai, *Engineering Mechanics of Composite Materials*, 2nd ed. (Oxford University Press, USA, 2005).

[7] L. Carlsson, D. F. Adams, and R. B. Pipes, *Experimental Characterization of Advanced Composite Materials, Third Edition*, 3rd ed. (CRC Press, 1986).

[8] P. Mast, L. Beaubien, M. Clifford, D. Mulville, S. Sutton, R. Thomas, J. Tirosh, and I. Wolock, "A semi-automated in-plane loader for materials testing," Experimental Mechanics **23**, 236–241 (1983), 10.1007/BF02320415.

[9]P. W. Mast, G. E. Nash, J. G. Michopoulos, R. W. Thomas, R. Badaliance, and I. Wolock, "Experimental determination of dissipated energy density as a measure of strain-induced damage in composites," Tech. Rep. Tec. Rpt. NRL/FR/6383–92-9369 (Naval Research Laboratory, Washington, DC, USA, 1992).

[10]P. Mast, G. Nash, J. Michopoulos, R. Thomas, I. Wolock, and R. Badaliance, "Characterization of strain-induced damage in composites based on the dissipated energy density part iii. general material constitutive relation," Theoretical and Applied Fracture Mechanics 22, 115–125 (1995).

[11]P. Mast, G. Nash, J. Michopoulos, R. Thomas, I. Wolock, and R. Badaliance, "Characterization of strain-induced damage in composites based on the dissipated energy density part II. Composite specimens and naval structures," Theoretical and Applied Fracture Mechanics 22, 97–114 (1995).

[12]P. Mast, G. Nash, J. Michopoulos, R. Thomas, I. Wolock, and R. Badaliance, "Characterization of strain-induced damage in composites based on the dissipated energy density Part I. Basic scheme and formulation," Theoretical and Applied Fracture Mechanics 22, 71–96 (1995).

[13]J. Michopoulos, "Recent advances in composite materials: In honor of s.a. paipetis," (Kluwer Academic Press, 2003) Chap. Computational and Mechatronic Automation of Multiphysics Research for Structural and Material Systems, pp. 9–21.

[14]J. Michopoulos, "Mechatronically automated characterization of material constitutive respone," in Proceedings of the 6th World Congress on Computational Mechanics (WCCM-VI) (Tsinghua University Press and Springer, Beijing China, 2004) pp. 486–491.

[15]J. Michopoulos, J. Hermanson, and T. Furukawa, "Towards the robotic characterization of the constitutive response of composite materials," Composite Structures 86, 154–164 (2008).

[16]F. Pierron, G. Vert, R. Burguete, S. Avril, R. Rotinat, and M. R. Wisnom, "Identification of the orthotropic elastic stiffnesses of composites with the virtual fields method: Sensitivity study and experimental validation," Strain 43, 250–259 (2007).

[17]S. Cooreman, D. Lecompte, H. Sol, J. Vantomme, and D. Debruyne, "Elasto-plastic material parameter identification by inverse methods: Calculation of the sensitivity matrix," International Journal of Solids and Structures 44, 4329 – 4341 (2007).

[18]T. Furukawa and J. Michopoulos, "Computational design of multiaxial tests for anisotropic material characterization," International Journal for Numerical Methods in Engineering 74, 1872–1895 (2008).

[19]J. G. Michopoulos, A. P. Iliopoulos, and T. Furukawa, "Accuracy of inverse composite laminate characterization via the mesh free random grid method," ASME Conference Proceedings 2009, 367–374 (2009).

[20]L. Bruno, F. M. Furgiuele, L. Pagnotta, and A. Poggialini, "A full-field approach for the elastic characterization of anisotropic materials," Optics and Lasers in Engineering 37, 417 – 431 (2002).

[21]T. Schmidt, J. Tyson, and K. Galanulis, "Full-field dynamic displacement and strain measurement using advanced 3d image correlation photogrammetry: Part 1," Experimental Techniques 27, 47–50 (2003).

[22]J. Kajberg and G. Lindkvist, "Characterisation of materials subjected to large strains by inverse modelling based on in-plane displacement fields," International Journal of Solids and Structures 41, 3439 – 3459 (2004).

[23]M. H. H. Meuwissen, C. W. J. Oomens, F. P. T. Baaijens, R. Petterson, and J. D. Janssen, "Determination of the elasto-plastic properties of aluminium using a mixed numerical-experimental method," Journal of Materials Processing Technology 75, 204 – 211 (1998).

[24]J. Molimard, R. Le Riche, A. Vautrin, and J. Lee, "Identification of the four orthotropic plate stiffnesses using a single open-hole tensile test," Experimental Mechanics 45, 404–411 (2005), 10.1007/BF02427987.

[25]J. G. Michopoulos and A. P. Iliopoulos, "A computational workbench for remote full field 2d displacement and strain measurements," ASME Conference Proceedings 2009, 55–63 (2009).

[26]N. P. Andrianopoulos and A. P. Iliopoulos, "Displacements measurement in irregularly bounded plates using mesh free methods," in Proc. 16th European Conference of Fracture, Alexandroupolis, Greece, July 3–7 (2006).

[27] A. P. Iliopoulos, J. G. Michopoulos, and N. P. Andrianopoulos, "Performance sensitivity analysis of the Mesh-Free Random Grid method for whole field strain measurements," ASME Conference Proceedings **2008**, 545–555 (2008).

[28] A. P. Iliopoulos, J. G. Michopoulos, and N. P. Andrianopoulos, "Performance analysis of the Mesh-Free Random Grid Method for full-field synthetic strain measurements," Strain , In Print (2010).

[29] A. Iliopoulos and J. Michopoulos, "Sensitivity analysis of the mesh-free random grid method for measuring deformation fields on composites," in *Proceedings of the 17th International Conference on Composite Materials, ICCM-17, Edinburgh* (The British Composites Society: Edinburgh International Convention Centre (EICC), Edinburgh, UK, 2009, 27–31 Jul 2009).

[30] A. P. Iliopoulos and J. G. Michopoulos, "Effects of anisotropy on the performance sensitivity of the Mesh-Free random grid method for whole field strain measurement," ASME Conference Proceedings **2009**, 65–74 (2009).

[31] J. G. Michopoulos, J. C. Hermanson, and A. Iliopoulos, "Towards a recursive hexapod for the multidimensional mechanical testing of composites," in *Proceedings of the ASME 2010 International Design Engineering Technical Conferences & Computers and Information in Engineering Conference IDETC/CIE 2010*, DETC2010/CIE-28699 (2010).

[32] J. Michopoulos and T. Furukawa, "Multi-level coupling of dynamic data-driven experimentation with material identification," Lecture Notes in Computer Science (including subseries Lecture Notes in Artificial Intelligence and Lecture Notes in Bioinformatics) **4487 LNCS**, 1180–1188 (2007).

[33] T. Furukawa, J. Michopoulos, and D. Kelly, "Elastic characterization of laminated composites based on multiaxial tests," Composite Structures **86**, 269–278 (2008).

[34] H. Man, T. Furukawa, and D. Kellermann, "Implicit constitutive modeling based on the energy principles," in *Proceedings of XXII ICTAM, 25-59 August 2008, Adelaide 2008* (2008).

[35] T. Furukawa and J. Michopoulos, "Online planning of multiaxial loading paths for elastic material identification," Computer Methods in Applied Mechanics and Engineering **197**, 885–901 (2008).

[36] J. Michopoulos, T. Furukawa, and S. Lambrakos, "Data-driven characterization of composites based on virtual deterministic and noisy multiaxial data," in *2008 ASME International Design Engineering Technical Conferences and Computers and Information in Engineering Conference, DETC 2008*, Vol. 3 (New York City, NY, 2009) pp. 1095–1106.

[37] T. Furukawa and J. W. Pan, "Stochastic identification of elastic constants for anisotropic materials," International Journal for Numerical Methods in Engineering **81**, 429–452 (2010).

[38] R. Haj-Ali and H.-K. Kim, "Nonlinear constitutive models for FRP composites using artificial neural networks," Mechanics of Materials **39**, 1035 – 1042 (2007).

[39] P. Haupt, *Continuum Mechanics and Theory of Materials*, 2nd ed. (Springer, 2002).

[40] F. Van Der Meer and L. Sluys, "Continuum models for the analysis of progressive failure in composite laminates," Journal of Composite Materials **43**, 2131–2156 (2009), http://jcm.sagepub.com/content/43/20/2131.full.pdf+html.

[41] J. W. Ju, J. Chaboche, and G. Z. Voyiadjis, *Damage Mechanics in Engineering Materials*, 1st ed. (Elsevier Science, 1998).

[42] M. Kaliske, "A formulation of elasticity and viscoelasticiy for fibre reinforced material at small and finite strains," Comput. Meth. Appl. Mech. Eng. **185**, 225–243 (2000).

[43] J. Michopoulos, J. C. Hermanson, and A. Iliopoulos, "First industrial strength multi-axial robotic testing campaign for composite material characterization," in *Proceedings of the ASME 2012 International Design Engineering Technical Conferences & Computers and Information in Engineering Conference IDETC/CIE, August 12-15, 2012, Chicago, IL, USA* (2012) submitted.

[44] J. Michopoulos and A. Iliopoulos, "A computational workbench for remote full field 3d displacement and strain measurements," in *Proceedings of the ASME 2011 International Design Engineering Technical*

Conferences & Computers and Information in Engineering Conference IDETC/CIE, August 29-31, 2011, Washington, DC, USA (2011).

[45]D. R. Jones, C. D. Perttunen, and B. E. Stuckman, "Lipschitzian optimization without the Lipschitz constant," J. Optim. Theory Appl. **79**, 157–181 (1993).

[46]The Mathworks, "Matlab," http://www.mathworks.com.

[47]A. P. Iliopoulos and J. G. Michopoulos, "Loading subspace selection for multidimensional characterization tests via computational experiments," ASME Conference Proceedings **2010**, 101–110 (2010).

[48]J. Michopoulos, H. J. C., and A. Iliopoulos, "Robotic and multiaxial testing for the constitutive characterization of composites," in *Materials Challenges in Alternative & Renewable Energy, February 26-March 1, 2012, Clearwater, FL, USA* (2012).

Author Index